Green Energy and Technology

Climate change, environmental impact and the limited natural resources urge scientific research and novel technical solutions. The monograph series Green Energy and Technology serves as a publishing platform for scientific and technological approaches to "green"—i.e. environmentally friendly and sustainable—technologies. While a focus lies on energy and power supply, it also covers "green" solutions in industrial engineering and engineering design. Green Energy and Technology addresses researchers, advanced students, technical consultants as well as decision makers in industries and politics. Hence, the level of presentation spans from instructional to highly technical.

Indexed in Scopus.

Indexed in Ei Compendex.

Abdel-Mohsen O. Mohamed ·
Brendan C. O'Kelly · Amin Soltani

Sustainability in Ground Improvement: The Case of Xanthan Gum Biopolymer

Springer

Abdel-Mohsen O. Mohamed
Uberbinder Limited
Oxford, UK

EX Scientific Consultants
Abu Dhabi, United Arab Emirates

Brendan C. O'Kelly
Department of Civil, Structural
and Environmental Engineering
Trinity College Dublin
Dublin, Ireland

Amin Soltani
Institute of Innovation, Science
and Sustainability
Federation University
Churchill, VIC, Australia

ISSN 1865-3529 ISSN 1865-3537 (electronic)
Green Energy and Technology
ISBN 978-3-031-75312-1 ISBN 978-3-031-75313-8 (eBook)
https://doi.org/10.1007/978-3-031-75313-8

This Springer imprint is published by the registered company Springer Nature Switzerland AG
The registered company address is: Gewerbestrasse 11, 6330 Cham, Switzerland

I dedicate this work to my divine wife, who has provided me with endless support over the last 40 years, and to our two beloved daughters.

Abdel-Mohsen O. Mohamed

In memory of my father, Chris Ambrose O'Kelly

Brendan C. O'Kelly

To my beloved wife, Mahdieh, and my dear parents for their unwavering support and encouragement.

Amin Soltani

Preface

Soils are the most common of all materials encountered in construction operations, yet their inherent engineering attributes often fall short of the requirements for reliable infrastructure development. Soil stabilization using chemical admixtures is a ground-improvement technique routinely employed in geoengineering practice. It involves altering the characteristics of the natural soil through physicochemical processes to meet specific engineering-design requirements. Conventionally, calcium-based binders, namely cement and lime, are employed to enhance the mechanical properties of soils. Although generally effective, from a geoengineering perspective, such binders are not environmentally sustainable materials, as their production and application are associated with high energy and CO_2 emissions footprints. The cement industry accounts for 5–8% of global CO_2 emissions, with the production of one-ton cement generating approximately one-ton of CO_2. The global cement production has risen from 1.39 billion tons in 1995 to 4.1 billion tons in 2023, indicating the steady growth in the construction industry. Within the field of geotechnical engineering, the cement content used to stabilize soils varies depending on the desired strength and type of soil, although 5–10% by weight of dry soil is typically used, resulting for about 0.2% of the global CO_2 emissions. As such, there has been a longstanding motivation to reduce our dependence on cementitious materials. Additionally, cement and lime are not always effective soil stabilization solutions, as their efficacy can be limited by leaching, and they are often unsuitable for treating soils containing organic matter, salts, and sulfates. The interplay of these factors, along with the need to sustainably expand civil infrastructure to meet ever-growing societal needs, underscores the importance of exploring and developing more sustainable soil stabilization binders.

Sustainable soil stabilization solutions aim to maintain a perfect balance between infrastructure performance and the social, economic, and ecological processes required to sustain human equity, diversity, and the functionality of natural systems. To this end, biopolymers are exhibiting great promise as a financially competitive and green alternative for calcium-based binders. Biopolymers are either chemically synthesized from biological matter or biosynthesized by living organisms, and like their synthetic counterparts, they consist of monomeric units interconnected by strong covalent bonds. Recent experimental studies have shown that soils stabilized by polysaccharide-type biopolymers, such as xanthan gum (XG), exhibit a variety of promising physical and mechanical treatments, such as improved water-retention capacity, reduced compressibility and hydraulic conductivity, enhanced shear strength, and improved resistance against wind/water erosion. Despite these advancements, the existing literature reveals numerous inconsistencies, and a thorough understanding of the behavior and properties of XG-treated soils under diverse loading and environmental conditions remains somewhat elusive. More importantly, there remains a notable gap in understanding how different factors affect the interactions between XG and various soil types throughout the processes of mixing, curing, and later environmental exposure. This book represents the first of its kind, offering a comprehensive, fundamental overview of the current state of XG usage for sustainable ground improvement, while also identifying future research directions toward addressing existing gaps in knowledge and application.

The book is structured into seven chapters, each exploring a distinct dimension of XG as a sustainable soil stabilization solution, from its fundamental properties and interactions with various soil types, to its practical implications and future research opportunities. This structured approach aims to provide a holistic understanding of XG's true potential and limitations for ground improvement applications. Chapter 1 sets the stage by examining the key challenges associated with the application of XG for soil stabilization. It delves into the fundamental issues—such as environmental conditions, soil variability, and the scaling-up of laboratory findings to field applications—that must be addressed to harness the full potential of XG in sustainable ground improvement practices. Chapter 2 provides the essential background on XG's origin and chemical composition. It details the biosynthesis of XG, highlighting the factors influencing its production, and describes the unique structural features that contribute to XG's efficacy as a soil stabilizer. Chapter 3 focuses on the rheological properties of XG in aqueous solutions. It discusses the viscosity behavior and flow characteristics critical to XG's application for soil stabilization and offers an in-depth analysis of how these attributes can be manipulated to optimize XG's performance in various geoengineering contexts. Chapter 4 investigates the physicochemical interactions between XG and different soil types. It explores the mechanisms through which XG interacts with the soil particles, including adsorption, bonding, and the formation of cohesive networks. The chapter also highlights how soil attributes, such as mineralogy, texture/gradation and pore-fluid chemistry, among others, affect these interactions. Building on the understanding of soil–XG interactions, Chap. 5 examines the stability of the cohesive networks formed by XG within soil matrices. It covers the factors affecting network stability, such as environmental conditions and soil properties, and discusses strategies to enhance the longevity and effectiveness of XG-treated soils. Chapter 6 provides a detailed analysis of the geomechanical properties of XG-treated soils. It focuses on the improvement of soil strength and the modification of soil compressibility characteristics achieved through XG stabilization. The chapter illustrates how various factors, such as XG concentration and its method of application, among others, affect the mechanical performance of soil–XG systems. The final chapter (Chap. 7) addresses the sorption behavior of XG within soil matrices and its impact on the hydraulic conductivity of XG-treated soils. It discusses the implications of XG sorption on the soil's hydraulic conductivity and its potential to reduce water-induced erosion. The chapter also considers the environmental aspects of using XG in ground improvement, particularly regarding water flow and contaminant migration.

We hope this book serves as both a roadmap and a source of inspiration for using XG to foster sustainable ground improvement practices. We envision biopolymers, like XG, will not only address traditional geotechnical engineering problems but will also drive innovation across various sectors. By highlighting the interdisciplinary research needed to enhance XG-based soil stabilization methods, we aim to encourage further research and development in this promising area.

Oxford, UK/Abu Dhabi, United Arab Emirates — Abdel-Mohsen O. Mohamed
Dublin, Ireland — Brendan C. O'Kelly
Churchill, Australia — Amin Soltani

Contents

1 Unraveling the Challenges of Xanthan Gum Soil Stabilization 1
1.1 Introduction 1
1.2 Traditional Soil Stabilization Techniques: An Overview 1
1.2.1 Calcium-Based Binders 2
1.2.2 Enzyme-Based Solutions 2
1.2.3 Microbial-Based Solutions 2
1.2.4 Polymer-Based Additives 3
1.3 XG Soil Stabilization 4
1.3.1 XG Benefits 6
1.3.2 XG Applications 6
1.3.3 Technical Issues 6
1.3.4 Economical Issues 8
1.4 Challenges and Considerations 9
1.4.1 Effectiveness 9
1.4.2 Environmental Implications 10
1.4.3 Compatibility with Other Soil Additives or Treatments 12
1.4.4 Cost Considerations 14
1.4.5 Long-Term Durability Considerations 16
1.5 Summary and Concluding Remarks 18
1.6 The Way Forward 19
References 19

2 Xanthan Gum Production and Structure 25
2.1 Introduction 25
2.2 Xanthan Gum Production 26
2.3 XG Structure 28
2.3.1 Types of XG Structural Conformations 32
2.3.2 Effect of Treatment Methods 34
2.3.3 Effect of Manufacturing Method 39
2.3.4 Effect of Electrolyte Concentration 40
2.3.5 Effect of pH 41

2.3.6 Effect of pH and Temperature 44
2.3.7 Effect of Carbon Source 44
2.4 Surface Charges 44
2.5 Activation Energy 45
2.6 Biodegradation Potential 47
2.7 Summary and Concluding Remarks 48
References 49

3 Rheology of Xanthan Gum Aqueous Solution 57
3.1 Introduction 57
3.2 XG Structural Characteristics 58
3.3 Rheological Behavior in a Steady Shear Flow 58
3.3.1 Viscosity 58
3.3.2 Controlling Parameters 62
3.4 Rheological Behavior in a Dynamic Shear Flow 70
3.4.1 Oscillatory Shear Testing and Theory 70
3.4.2 Controlling Properties 72
3.5 XG Surface Charge and Interaction 84
3.6 XG Hydrogels Swelling and Strength 86
3.7 Summary and Concluding Remarks 90
3.8 Recommendations 92
References 93

4 Xanthan Gum–Soil Interactions 99
4.1 Introduction 99
4.2 Nature of the Interactions Between Clay Soils and Xanthan 100
4.2.1 Physical Interactions 100
4.2.2 Chemical Interactions 101
4.3 Interaction Mechanisms and Bonding 102
4.4 Mode of the Interaction Between Xanthan and Montmorillonite Clay 105
4.5 Mode of the Interaction Between Xanthan and Kaolinite Clay 107
4.6 Mode of Interaction Between Xanthan and Clay Mineral Oxides in Suspension 110
4.6.1 Interaction with Silica 110
4.6.2 Interaction with Alumina 112
4.6.3 Interaction with Calcium Carbonate 114
4.6.4 Interaction with Nanometal Oxides 116
4.7 XG Interaction with Heavy Metals in Suspension 118
4.8 Summary and Concluding Remarks 119
References 120

5 Stability of Xanthan Gum–Soil Cohesive Networks 125
5.1 Introduction 125
5.2 Pore Fluid Viscosity 126
5.2.1 XG–Deionized Water (DW) 126
5.2.2 XG–Soil Natural Pore Fluid 127
5.2.3 Solubility of XG in Deionized Water (DW) 128
5.2.4 Pore Fluid Concentration and Clay Mineralogy 129
5.2.5 Temperature 130
5.2.6 Salt Concentration 131
5.3 Emulsion Stability 131
5.3.1 Fluid Consistency and Flow Behavior 132
5.3.2 Aggregate Surface Area 134
5.3.3 Dispersivity 135
5.3.4 Plasticity Indices 139
5.3.5 Soil Water Potential 144
5.3.6 Compressive Strength 146
5.4 Chemical Bonding 150
5.4.1 Types of Inter-particle Forces 150
5.4.2 Energy of Interaction and Bonding Elasticity 150
5.5 Summary and Concluding Remarks 159
References 160

6 Strength and Compressibility of Xanthan Gum-Treated Soils 163
6.1 Introduction 163
6.2 Nature of the XG Gel 164
6.3 Controlling Parameters 166
6.3.1 XG Mixing Methods with Soils 166
6.3.2 Soil Type 168
6.3.3 Dry Density 177
6.3.4 Moisture Content 178
6.3.5 Curing Environment 181
6.3.6 Soil pH 184
6.4 Compressibility and Swelling 184
6.4.1 Effect of XG Content on Saturated Dispersive Soils 184
6.4.2 Effect of Moisture State on XG–Kaolinite (Non-dispersive) Clay 185
6.4.3 Control of XG Swelling 186
6.5 Strength Dependence on XG Concentration and Soil's Physical Molding State 186
6.6 Recommended Testing Protocol for XG-Treated Soils 192
6.6.1 XG Mixing Methods 192
6.6.2 XG Concentration 193
6.6.3 Physicochemical Testing Evaluation 193
6.7 Summary and Concluding Remarks 194
References 196

7 Xanthan Gum (XG) Sorption and Hydraulic Conductivity of XG-Treated Soils . . . 199
7.1 Introduction . . . 199
7.2 XG Sorption by Clay Minerals . . . 200
7.2.1 Physicochemical Characteristics of Selected Clay Minerals . . . 200
7.2.2 XG Sorption Equilibrium Time . . . 201
7.2.3 XG Concentration Effect on Clay Minerals Sorption . . . 201
7.2.4 Electrolyte Concentration Effect on XG Sorption by Clay Minerals . . . 202
7.2.5 Electrolyte Type Effect on XG Sorption by Clay Minerals . . . 203
7.2.6 pH Effects on XG Sorption by Clay Minerals . . . 205
7.3 Influence of Polysaccharides on Flocculation/Aggregation of Clay Minerals . . . 205
7.3.1 Montmorillonite . . . 205
7.3.2 Kaolinite . . . 207
7.3.3 Natural Sediments . . . 212
7.4 Effects of XG on Basic Geotechnical Properties . . . 221
7.5 Effect of XG on Soil HC . . . 223
7.5.1 Questionable Reported Results . . . 223
7.5.2 Controlling Parameters . . . 225
7.6 Summary and Concluding Remarks . . . 252
References . . . 253

About the Authors

Prof. Dr. Eng. Abdel-Mohsen O. Mohamed earned his Ph.D. from McGill University, Canada, where he was later employed as the Associate Director of the Geotechnical Research Centre and Lecturer at the Department of Civil Engineering and Applied Mechanics. He has held many senior positions in the United Arab Emirates (UAE), including Associate Provost and Chief Academic Officer at Zayed University, Dean of Research and Graduate Studies at Abu Dhabi University, and Research Director at the UAE University. He has 14 patents in areas of sustainable use of elemental sulfur and alkaline solid wastes, production of sulfur cement and concrete, carbon sequestration and utilization for treatment of solid wastes, and stabilization of sand dunes. He has authored and coauthored 21 scientific books and has published over 385 papers in refereed journals and international conference proceedings. He has been the recipient of several university and nationwide research accolades. He is the General Managing Director of EX Scientific Consultants, Abu Dhabi, UAE, Senior Advisor, Uberbinder, Inc., Seattle, Washington, USA, and an Editorial Board Advisor for number of Scientific Journals.

Research gate profile: https://www.researchgate.net/profile/Abdel_Mohsen_Mohamed/scores.

Dr. Brendan C. O'Kelly began his academic career in 1994 as an Assistant Lecturer in Civil Engineering at University College Dublin, Ireland, going on to earn his Ph.D., specializing in geotechnical engineering, while employed as the Pierce-Newman Fellow in Civil Engineering and a Research Fellow in Soil Mechanics (1996–2000) at the same institution. He then worked as a geotechnical engineer with Scott Wilson Consulting Engineers, UK, before joining the Department of Civil, Structural and Environmental Engineering, Trinity College Dublin, as a Lecturer in Geotechnical Engineering in 2001. He was promoted to Associate Professor in Civil Engineering, and awarded the title of a Fellow of Trinity College Dublin, in 2010. He is an Associate Editor for the journal *Environmental Geotechnics* (ICE) and serves as an Editorial Board Member for a number of international journals. His main research interests are soil stabilization/ground improvement, the Atterberg's consistency limits and soil classification, and the geomechanical behavior/properties of various problematic soils, including peat and other highly organic soils, on which he has authored or co-authored over 180 peer-reviewed journal articles.

http://people.tcd.ie/bokelly

Dr. Amin Soltani is a Senior Lecturer and Undergraduate Program Coordinator of Civil Engineering in the Institute of Innovation, Science and Sustainability at Federation University, Australia. He also holds an Honorary/Adjunct Fellow affiliation with the Department of Infrastructure Engineering at the University of Melbourne, Australia. Amin completed his Ph.D. in Civil and Environmental Engineering (specializing in Geotechnical Engineering) at the University of Adelaide, Australia. Before his current appointment, Amin was a Research Fellow in Geotechnical Engineering at the Universities of Melbourne and Adelaide, where he played a significant leading role in the advancement of numerous high-profile research projects in collaboration with reputable industry partners, city councils, road authorities, and state government agencies in South Australia and Victoria. Amin is a Chartered Professional Engineer with Engineers Australia. His principal research interests lie in the fields of Experimental (and Physics-Informed, Data-Driven) Soil Mechanics and Sustainable Geomaterials; the latter involving the development of novel high-performance geomaterials for a variety of civil engineering applications, including, for instance, road infrastructure, thereby fostering sustainable and resilient infrastructure systems for future generations. Amin has been the recipient of several university and nationwide research awards, including the prestigious VESKI Victoria Fellowship in 2022, which initiated his journey in biopolymers. He has co-authored over 70 peer-reviewed journal articles and serves as Editorial Board Member for a number of international journals.

https://federation.edu.au/institutes-and-schools/iiss/staff-profiles/staff-profiles/amin-soltani

https://federation.edu.au/research/find-an-expert/amin-soltani

Abbreviations

AFM	Atomic force microscope
BFWM	Bentonite-free water-based mud
CEC	Cation exchange capacity
CMC	Carboxymethyl cellulose
DDL	Diffused double layer
DI	Distilled (water)
DW	Deionized water
ESA	External surface area (of aggregate)
FTIR	Fourier transfer infrared
HC	Hydraulic conductivity
LAOS	Large-amplitude oscillatory shear (test/flow)
LBG	Locust bean gum
LCA	Life cycle assessment
LL	Liquid limit
NP	Nanoparticle
OMC	Optimum moisture content (for compaction)
PAM	Polyacrylamide
PG	Performance grade
PI	Plasticity index
PL	Plastic limit
QNM	Quantitative nanomechanics (testing)
RTFO	Rolling thin film oven (binder)
SAOS	Small-amplitude oscillatory shear (test/flow)
SDS	Sodium dodecyl sulfate
SEM	Scanning electron microscopy
SL	Shrinkage limit
SSA	Specific surface area
TOC	Total organic carbon
UV	Ultraviolet
XG	Xanthan gum
XG_{Denat}	Denatured xanthan gum

XG_{Nat}	Native xanthan gum
XG_{Renat}	Renatured xanthan gum
ZPC	Zero point of charge

Chapter 1
Unraveling the Challenges of Xanthan Gum Soil Stabilization

1.1 Introduction

Soil stabilization is a crucial aspect of civil engineering and construction projects, aimed at improving the engineering properties of soil materials. Xanthan gum (XG), a biopolymer derived from fermentation processes, has gained widespread attention as a potential soil-stabilizing agent due to its unique rheological properties. This chapter provides an overview of the traditional soil stabilization techniques, followed by outlining the potential benefits and main technical issues of employing XG as a soil-stabilizing agent. In addition, the chapter discusses the overall challenges and considerations associated with the utilization of XG for soil stabilization purposes, including its effectiveness, environmental impact, cost implications, and compatibility with different soil types. By examining these factors, engineers and researchers can make informed decisions regarding the application of XG in soil stabilization projects. The chapter concludes with an outline of the topics that are discussed in the later chapters of this book.

1.2 Traditional Soil Stabilization Techniques: An Overview

Soil stabilization is a prevalent ground-improvement technique routinely employed in geotechnical engineering practice. Nowadays, numerous soil stabilization technologies, such as calcium-based agents (cement, lime, cement kiln dust, fly ash, and granulated blast-furnace slag), geopolymers, sodium silicates, organic polymers (polyurethane, acrylamides and natural bitumen), resins, sulfonated oils (or surfactants), geosynthetics, and biopolymers, are being used in geotechnical engineering practice.

A. M. O. Mohamed et al., *Sustainability in Ground Improvement: The Case of Xanthan Gum Biopolymer*, Green Energy and Technology,
https://doi.org/10.1007/978-3-031-75313-8_1

1.2.1 Calcium-Based Binders

Calcium-based additives, particularly cement and lime, are considered among the most popular options (Soltani et al. 2020); however, they have led to significant production of greenhouse gasses (e.g., about 8–10% of the global CO_2 emissions), which are directly related to global warming. For example, in pavement construction, a single percentile reduction in cement content usage could reduce embedded energy per kilometer of road by 300 MJ and the carbon footprint by 44,000 kg of CO_2 (Williamson and Cortes 2014). To reduce cement usage, new sustainable technologies, e.g., employing bio-enzymes (Chang et al. 2016), CO_2 absorption (Ngu et al. 2019; Mohamed et al. 2022), microbial-induced calcite precipitation (Shahrokhi-Shahraki et al. 2015; Omoregie et al. 2019; Liu et al. 2020), and biopolymers (Khatami and O'Kelly 2013; Ghasemzadeh et al. 2021; Soltani et al. 2021; Azimi et al. 2024), have been investigated.

1.2.2 Enzyme-Based Solutions

Enzyme-induced calcium carbonate precipitation (Chu et al. 2013; Zomorodian et al. 2023), as well as microbial biopolymer accumulation methods, has been used to stimulate microbes to generate biopolymer within the soil matrix (Kwon and Ajo-Franklin 2013; Noh et al. 2016; Ham et al. 2018). In such methods, biofilm formation (i) enhances both the stability (i.e., increasing shear strength) and flow properties (i.e., decreasing hydraulic conductivity (HC)) of the stabilized soils (Baveye et al. 1998; Dunsmore et al. 2004; Abdullah et al. 2020; Matsubara 2021; Sadeghian et al. 2021) and (ii) helps mitigate the liquefaction hazard, due to the degree of soil saturation reduction arising from the production of insoluble gases in the soil pore voids (Rebata-Landa and Santamarina 2012; He et al. 2013).

1.2.3 Microbial-Based Solutions

Similarly, in the microbial-induced calcite precipitation process, bacteria are used to precipitate calcite ($CaCO_3$) in the soil matrix to cause improvements in the mechanical attributes (DeJong et al. 2013; Omoregie et al. 2019, 2020; Bahmani et al. 2019) and reduce the HC of the treated soils, thereby alleviating the soil liquefaction potential by either bioclogging, or allowing the induced excess pore-water pressures to dissipate (Mujah et al. 2017; Choi et al. 2020). However, the field application of microbial-induced calcite precipitation techniques requires highly specialized sets of environmental and microbial growth conditions—that is, invariably coarse-grained soil treatment medium (Dhami et al. 2016; Tang et al. 2020; Sharma et al. 2021), in-place mitigation processes for the by-product ammonia (Achal and Mukherjee 2015), and techniques to ensure the uniform distribution of precipitated calcite throughout the treated bulk soil mass (Mujah et al. 2017).

1.2.4 Polymer-Based Additives

Polymers are macromolecular materials that are composed of repeating sub-units (or monomers), and which are classified into: (i) synthetic, i.e., produced from petroleum products and (ii) natural, i.e., produced naturally in the environment by plants and living organisms (Stal 2011). They have been successfully used in different civil engineering applications related to the construction industry (Chang et al. 2020). Biopolymers are classified based on the source of their raw materials, as (Niaounakis 2015; Sujatha and O'Kelly 2023): (i) plant-derived products, i.e., mostly polysaccharides, including guar, lignin, agar, beta-glucan, alginate and carrageenan; (ii) microorganism-based products, i.e., derived through bacterial fermentation processes, such as XG, gellan gum, and dextran; and (iii) animal-derived biopolymers, i.e., produced from animal sources (e.g., chitin, chitosan, and casein).

Table 1.1 presents a summary of these biopolymers in terms of their sources, electrical charge type and water solubility. Notably, most plant-derived polysaccharides suffer from availability issues, such as price fluctuation and harvesting; however, their production is relatively cheap and depends on the growing season. In contrast, microorganism-based biopolymer products are biosynthesized under controlled environmental conditions, do not require large production space, and can be produced in a reasonable time period, resulting in stable chemical characteristics, high purity, and widespread availability in the marketplace.

It is known that XG plays an important role in various industrial applications, as a thickener and emulsion stabilizer. For example, in the petroleum industry, XG is an important component of water-based drilling fluid formulations due to its excellent rheological properties and thermal stability (Navarrete et al. 2000; Phatak 2011). In addition, XG has been used together with modified starch to control the fluid loss of drilling fluids (Jones and Hughes 1996). Table 1.2 lists some current industrial uses of XG.

In general, polysaccharide stabilizers (i.e., XG) offer increased confidence in reproducing manufactured products due to their chemical versatility, which provides control over preparation methodologies (Chang et al. 2020). Polysaccharide stabilizers are used to: (i) provide different biological functions (energy storage, structural support, gelling agent); hence they offer a variety of chemical functions (Torres et al. 2019); (ii) tolerate saline conditions across a broad pH range; and (iii) enhance durability when stored for long time periods due to their high stability (Qureshi et al. 2017).

Table 1.1 Classification of biopolymers and their related physicochemical properties

Biopolymer	Source	Charge type	Water solubility
Plant-based biopolymers			
Guar gum	Guar plant	Cationic/anionic	Soluble
Lignin and its derivatives	Cell walls of trees and barks, waste by-product of pulp and paper industry	Mainly anionic	Insoluble
Agar gum	Marine algae	Neutral	Soluble at ~85 °C
Beta glucan	Cell walls of yeast and cellulose in plants	Cationic	Soluble
Sodium alginate	Brown algae	Anionic	Soluble
Carrageenan	Red edible seaweeds	Anionic	Soluble
Microorganism-based biopolymers			
Xanthan gum	Glucose or sucrose fermentation by the *X. campestris* bacterium	Anionic	Soluble
Gellan gum	Bacterial fermentation of *Sphingomonas* elodea	Anionic	Soluble at > 90 °C
Polylysine	Bacterial fermentation	Cationic	Soluble
Dextran	Synthetized from sucrose by certain lactic-acid bacteria	Anionic/cationic	Insoluble
Animal-derived biopolymers			
Chitosan	Waste crustacean shells	Cationic	Soluble
Casein	Dairy product wastes	Anionic	Insoluble

1.3 XG Soil Stabilization

XG is a high-molecular-weight polysaccharide composed of glucose, mannose and glucuronic acid units. The special properties of XG—namely, high viscosity that increases with increasing pyruvate content (Callet et al. 1987), distinct pseudoplasticity, high water-binding capacity, high temperature stability (Seright and Henrici 1990), insensitive solution properties to monovalent and polyvalent salts (i.e., no precipitation) (Tam and Tiu 1993), and high shear stability—have contributed to increasing interest among the geotechnical and geo-environmental engineering communities regarding its use as a soil-stabilizing agent.

The unique properties of XG—namely, (i) pseudoplastic behavior, meaning its viscosity reduces under shear stress, making it easy to mix with soil, and (ii) hydrophilicity, enabling it to absorb and retain water—make it very beneficial for stabilization of arid and sandy soils. The significant potential use of XG and other various biopolymers in soil stabilization has been reported (Ramdas et al. 2021; Losini et al. 2021; Fatehi et al. 2021). Examples include the studies on XG (Ayeldeen

Table 1.2 Typical industrial uses of XG biopolymer (adopted from Garcia-Ochoa et al. (2000))

Application	Concentration (% w/w)	Functionality
Salad dressings	0.1–0.6	Emulsion stabilizer, suspending agent, dispersant
Dry mixes	0.05–0.25	Eases dispersion in hot or cold water
Syrups, toppings, relishes, sauces	0.05–0.25	Thickener, heat stability and uniform viscosity
Beverages (fruit and non-fat dry milk)	0.05–0.25	Stabilizer
Dairy products	0.5–0.7	Stabilizer, viscosity control of mix
Baked goods	0.1–0.5	Stabilizer, facilitates pumping
Frozen foods	0.05–0.25	Improves freeze/thaw stability
Pharmaceuticals (creams and suspensions)	0.1–1.1	Emulsion stabilizer, uniformity in dosage formulations
Cosmetic (denture cleaners, shampoos and lotions)	0.2–1.2	Thickener and stabilizer
Agriculture (additive in animal feed and pesticide formulations)	0.03–0.43	Suspension stabilizer, improved sprayability, reduced drift, increased cling and permanence
Textile printing and dyeing	0.2–0.7	Control of rheological properties of paste, preventing dye migration
Ceramic glazes	0.3–0.8	Prevents agglomeration during grinding
Slurry explosives[1]	0.3–1.3	Thickens formulations, improves heat stability (in combination with gellan gum)
Petroleum production	0.1–0.5	Lubricant or friction reduction in drill-hole
Enhanced oil recovery	0.05–2.05	Reduces water mobility by increasing viscosity and decreasing permeability

[1] Slurry explosives generally comprise inorganic oxidizing salt, a liquid solvent, disperser or carrier for said salt and fuel

et al. 2017), and gellan gum (Chang et al. 2015), which have shown remarkable improvements in the mechanical properties of both sandy and clayey soils.

1.3.1 XG Benefits

The use of XG in soil stabilization has several benefits, such as: (i) *improved cohesion*: XG enhances inter-particle cohesion of the soil, thereby increasing its stability and reducing erosion; (ii) *water retention*: XG's hydrophilic nature aids the soil in retaining moisture, preventing desiccation and also promoting plant growth; (iii) *environmental friendliness*: Compared to traditional soil stabilizers like cement and lime, XG is biodegradable and non-toxic, minimizing environmental impact; and (iv) *cost-effectiveness*: XG can be cost-effective compared to traditional stabilizers, in that considerably reduced amounts of XG (i.e., lower material transportation costs in comparison with the calcium-based stabilization techniques) are needed to produce equivalent soil stabilization performance.

1.3.2 XG Applications

XG can be used in a variety of civil engineering applications, such as: (i) *road construction*: XG can be used to stabilize subgrade soils, improving the load-bearing capacity of roads and reducing maintenance costs; (ii) *agriculture*: XG can be applied in agriculture to stabilize soil for crop cultivation, prevent soil erosion, and enhance water retention in soils; (iii) *erosion control*: XG can be sprayed on slopes and embankments to stabilize the surface soil layer and thereby reduce or prevent erosion, especially in areas prone to landslides; and (iv) *land reclamation*: In mining and construction sites, XG can aid in reclaiming and stabilizing disturbed soils, facilitating vegetation establishment.

1.3.3 Technical Issues

There are gaps in understanding the behavior of biopolymer materials themselves, as well as biopolymer-treated soils under diverse loading and environmental conditions. Further research is urgently required, since the degree of soil improvement, achieved in terms of ***soil consistency***, ***strength*** and ***water flow*** properties, as examples, is highly dependent on the biopolymer type and its applied concentration and method of mixing with the soil, soil type and its mineralogy and chemical composition, and curing environmental conditions, in terms of pH, temperature, humidity, and externally applied pressure. Below are some instances regarding the reported inconsistencies in the properties associated with XG-treated soils.

In relation to the ***consistency (plasticity) properties*** of the XG-stabilized soils, reported results have indicated that XG has contrary effects on plasticity, whereby the liquid limit (LL) can decrease through XG-induced particle aggregation, or alternatively the LL may increase due to XG-hydrogel formation depending on the soil type.

As such, because of the mutual interactions between XG molecules, soil minerals and the soil pore fluid, the effects of the soil mineral type, soil moisture content and pore fluid chemistry on XG behavior in treated soil must be carefully examined to fully understand the true potentials and/or limitations of these materials for soil stabilization applications.

Regarding ***strength improvement*** in XG-stabilized soils, reported studies have shown that XG can be used successfully to improve the compressive strength of sands (Chang et al. 2015; Cabalar et al. 2017; Qureshi et al. 2017), silty sands (Lee et al. 2019), clays (Chang et al. 2015; Joga and Varaprasad 2019), residual soils (Chang et al. 2015), artificial bentonite and kaolin materials (Chang et al. 2015, 2018; Fatehi et al. 2018), soft marine clays (Kwon et al. 2019), and mine tailings (Chen et al. 2013). Moreover, previous investigations have indicated that the optimum XG content for soil stabilization is in the range of 1.0–1.5% (by wt of dry soil) for curing in both ambient- and oven-drying conditions (Latifi et al. 2016), and 2.5% for curing in submerged conditions (Joga and Varaprasad 2019). However, Soldo et al. (2020) highlighted that the optimum XG content highly depends on the soil type and the soil water content. The reduction in compressive strength occurring during drying was highlighted by Chen et al. (2019). They indicated that during drying, XG promotes a reduced effect on shear strength when the water content is high, irrespective of the XG content. However, it was not possible to compare the results because of the lack of standard procedures for analyzing the XG, the untreated soils, and the XG-stabilized soils.

Regarding the ***water flow properties***, the reported HCs for XG-stabilized soils, are inconclusive. For example, sand and silty soils stabilized with 2.0 wt% XG have shown a reduction of about 10% in the HC compared to the untreated soils (Ayeldeen et al. 2016). However, the results of a mixture of XG (0.5–3.0 wt%) with sand-bentonite have shown a slight increase in the HC with increasing XG content (Biju and Arnepalli 2020), highlighting the role of the fines content (i.e., silt and clay particles) in the soil composition, as well as the XG content, on the effectiveness of the XG treatment. Notably, the increase in XG content promotes flocculation and aggregation, and an increase in the soil void and pore channel sizes, leading to an increase in the soil HC by up to one order of magnitude (Biju and Arnepalli 2020).

Moreover, in a study by Khachatoorian et al. (2003), a decrease in the HC of XG-treated Ottawa sand (0.43–0.6 mm particle-size range) over a curing period of 11 days was reported; however, results reported by Cabalar et al. (2017) for XG-treated Narli sand (0.075–1.0 mm particle-size range) did not show a clear tendency over the same curing duration. In addition, Mendonça et al. (2021) reported a slight decrease in the HC of 0.33 wt% XG-treated silty sand during the first 14 days of curing, followed by an increase for longer curing durations. The initial decrease was attributed to the hydration of the XG hydrogels; while, the increase for longer curing periods was ascribed to XG biodegradation, as well as dehydration and shrinkage of the hydrogels. Indeed, most of the reported experimental results for XG-stabilized sandy and silty soils showed an increase in the HC with an increase in the curing period, which was explained because of the expected dehydration of XG over time, and the resulting shrinkage and increasing pore sizes (Cabalar et al. 2017; Ayeldeen et al. 2016). Moreover, the gas permeability of a 3.0 wt.% XG-stabilized kaolin clay decreased by up to two orders of magnitude compared to that of the untreated clay (Ng et al. 2020), which was attributed to the higher moisture content of the clay and the increased hydration of XG that promotes pore-clogging (Ng et al. 2020).

Therefore, to overcome these questionable results, there is a need for developing standard material characterization and testing procedures.

1.3.4 Economical Issues

From an economical viewpoint, while XG has shown to be effective as a soil-stabilizing agent, the issue of the XG market price needs to be considered, irrespective of some positive reported studies (Chang et al. 2016) that advocated the price issue is no longer valid. Based on data from 2015 to 2020, due to the production in large quantities and increased use of XG, the unit cost of XG has reduced significantly over the last thirty years, with, for instance, the cost of adding 0.5 wt% XG for soil stabilization (i.e., 5 kg of XG per ton of dry soil) decreased from approximately 70–10 USD per ton of dry soil (Chang et al. 2016).

Considering the current price of XG as 2.0–3.0 USD/kg (Chang and Cho 2019), a mixture of 2 wt% XG would typically cost on average 80 USD to treat 1.0 m^3 of sand, whereas stabilization by 5 wt% ordinary Portland cement would typically cost approx. 90 USD/m^3 of sand. Hence, stabilization using XG would be less costly than that of the calcium-based treatment. Notably, since XG-based stabilization of cohesive soils requires less quantities of XG (i.e., less than 1 wt%), the anticipated cost would be decreased further.

Nevertheless, it is also worth noting that the current market prices of biopolymers are projected for use in the food and pharmaceutical industries that require very high purity and, accordingly, results in a significantly higher biopolymer production cost. Therefore, the specific purification method used, which is costly, is determined by the end user of the XG product and the economics. For example, the use of XG in enhanced oil recovery requires removal of particles such as cells that could clog up the porous oil-bearing rock, whereas the XG used as a food additive should be free of the biomass and reagents used in the recovery process. Biopolymer product specifications are less restrictive for uses in textile processing. As such, appropriate specifications should be developed for XG utilized in the geotechnical and geo-environmental engineering sector (as a soil-stabilizing agent), where the same high levels of XG purity are most likely not required; hence, the price of XG can practically be decreased (Smitha and Sachan 2016; Chang et al. 2018).

1.4 Challenges and Considerations

Overall, XG presents a promising green alternative for soil stabilization. However, its use comes with certain challenges and considerations that need to be addressed.

1.4.1 Effectiveness

One of the primary challenges associated with using XG for soil stabilization is assessing its effectiveness in different soil types and conditions. The efficacy of XG can vary depending on various factors, such as soil composition, moisture content and environmental conditions. Therefore, extensive laboratory testing and field trials are necessary to determine the optimal XG dosage and application methods for achieving desired soil stabilization outcomes.

1.4.1.1 Soil Composition

Different soil types exhibit distinct properties that can affect how XG interacts with them. In other words, cohesive (fine-grained) soils respond differently to XG compared to cohesionless (coarse grained) soils. For instance, clay soils typically have finer particles and higher plasticity, which may require different XG dosages or application techniques to achieve effective stabilization compared to sandy soils. Moreover, soils with different mineralogical compositions may exhibit different levels of compatibility with XG, further complicating the assessment of XG effectiveness for soil stabilization purposes.

1.4.1.2 Moisture Content

The moisture content of the soil plays a crucial role in determining the effectiveness of XG stabilization. Optimal moisture levels are necessary to facilitate the dispersion and hydration of XG molecules within the soil matrix. However, excessive moisture may lead to over-hydration or leaching of the XG stabilizing agent, reducing its effectiveness. Conversely, insufficient moisture may hinder the dispersion and activation of XG, resulting in inadequate soil stabilization. As such, it is essential to carefully control and monitor soil moisture levels during the stabilization process to achieve optimal results.

1.4.1.3 Environmental Conditions

Environmental factors, such as temperature, humidity and exposure to ultraviolet (UV) radiation, can influence the stabilization performance of XG. High temperatures may accelerate the degradation of XG molecules, reducing their effectiveness over time. Similarly, prolonged exposure to UV radiation can cause photodegradation of XG, compromising its stabilizing properties. Moreover, variations in humidity levels can affect the moisture content of the soil, thereby influencing the hydration and dispersion of XG within the soil matrix. Therefore, it is essential to consider the prevailing environmental conditions during the planning, design, and implementation of XG-based soil stabilization projects.

1.4.1.4 Laboratory Testing and Field Trials

Given the complex interactions occurring between XG and soil materials under different conditions, extensive laboratory testing and subsequent field trials are necessary to assess the XG effectiveness and to optimize its performance in various soil stabilization applications. Laboratory experiments allow researchers to investigate the rheological properties of XG solutions, as well as their interaction with various soil types under controlled conditions. These tests may include rheological analysis, soil consolidation testing, and shear strength measurements to evaluate the mechanical properties of the XG-stabilized soils.

Field trials provide valuable insights into the performance of XG in real-world applications, where factors such as site-specific ground conditions, construction methods, and long-term durability can be evaluated. By conducting field trials across diverse soil types and environmental conditions, engineers can validate laboratory findings and refine stabilization techniques to achieve desired outcomes. Moreover, long-term monitoring of XG-stabilized soil sites enables engineers to assess the durability and effectiveness of the XG treatment over time, identifying its true potential and limitations, as well as areas for future improvement.

In summary, assessing the effectiveness of XG for soil stabilization requires a comprehensive understanding of its interactions with various soil types under different environmental conditions. Through a combination of laboratory testing and field trials, engineers and researchers can optimize XG stabilization techniques and develop sustainable solutions for enhancing the stability and performance of soil materials in various engineering applications.

1.4.2 Environmental Implications

While XG is considered environmentally friendly compared to traditional stabilizing agents such as cement and lime, its production process and disposal of excess or unused XG can still have environmental implications. The fermentation process used to produce XG requires energy and resources, and the disposal of excess/unused XG may pose challenges in terms of biodegradability and environmental impact. Accordingly, it is essential to consider the overall environmental footprint of XG-based soil stabilization methods.

1.4.2.1 Fermentation Process

XG is produced through a fermentation process, using carbohydrates as substrates, by the bacterium *Xanthomonas campestris*. While fermentation is a natural process, it requires specific environmental conditions, including temperature, pH, and nutrient availability. Large-scale production of XG necessitates significant energy inputs for maintaining optimal fermentation conditions, which may contribute to energy consumption and greenhouse gas emissions.

1.4.2.2 Resource Utilization

The production of XG requires the cultivation of *X. campestris* bacteria on carbohydrate-rich substrates, such as corn, wheat, or soybean. The cultivation of these crops involves agricultural practices that may have environmental implications, including land use change, water consumption, pesticide use, and fertilizer runoff. Moreover, the extraction and purification of XG from fermentation broths involve chemical processes that consume additional resources and may generate waste byproducts.

1.4.2.3 Disposal and Biodegradability

(a) *Excess or Unused XG*: In soil stabilization applications, excess or unused XG may be disposed of through various methods, including landfilling, incineration, or composting. However, the biodegradability of XG can vary depending on various factors, such as environmental conditions, microbial activity, and the presence of enzymes capable of breaking down polysaccharides. While XG is considered biodegradable under suitable conditions, its degradation rate may be relatively slow compared to other organic materials, leading to concerns about its persistence in the environment.

(b) *Environmental Impact:* Improper disposal of XG or its byproducts may have adverse environmental impacts, including soil and water contamination, depletion of natural resources, and habitat degradation. Landfilling of XG waste may contribute to methane emissions and leachate contamination, posing risks to human health and the ecosystem. Incineration of XG waste may release pollutants into the atmosphere, including greenhouse gases, particulate matter, and toxic compounds. Therefore, careful management of XG waste is necessary to minimize its environmental footprint and ensure sustainable soil stabilization practices.

1.4.2.4 Overall Environmental Footprint

(a) *Life Cycle Assessment* (LCA): Conducting a comprehensive LCA of XG-based soil stabilization methods is essential for evaluating their overall environmental footprint. LCA considers the environmental impacts associated with each stage of a product's life cycle, including raw material extraction, production, transportation, use, and disposal. By quantifying resource consumption, energy use, emissions, and waste generation throughout the life cycle, LCA provides insights into the environmental sustainability of XG-based soil stabilization compared to conventional stabilizing agents, such as cement or lime.

(b) *Sustainable Practices*: Implementing sustainable practices in XG production, such as using renewable energy sources, optimizing resource utilization, and minimizing waste generation, can help mitigate its environmental impacts. Additionally, exploring alternative sources of XG, such as microbial fermentation of agricultural residues or wastewater streams, may offer opportunities to reduce the environmental footprint of XG production while valorizing waste materials.

In conclusion, while XG is considered environmentally friendly compared to traditional stabilizing agents, its production process and disposal of excess/unused XG can still have environmental implications. By considering the overall environmental footprint of XG-based soil stabilization methods and implementing sustainable practices throughout the production and disposal chain, engineers and researchers can minimize environmental impacts and promote the adoption of environmentally sustainable soil stabilization techniques.

1.4.3 Compatibility with Other Soil Additives or Treatments

Another consideration when using XG for soil stabilization is its compatibility with other soil additives or treatments. Certain chemicals or compounds present in the soil may interact with XG, affecting its performance and stability. Compatibility testing should be conducted to assess any potential adverse reactions between XG and other agents present in the soil.

1.4.3.1 Chemical Interactions

When using XG for soil stabilization, it is crucial to consider potential chemical interactions between XG and other additives or treatments applied to the soil. Certain chemicals or compounds present in the soil, such as salts, minerals, organic matter, and contaminants, may interact with XG molecules, affecting their performance and stability. These interactions can lead to changes in the rheological properties, hydration behavior, and bonding mechanisms of XG in the soil matrix.

1.4.3.2 Effect on Stabilization Performance

Incompatibility between XG and other soil additives or treatments can compromise the effectiveness of soil stabilization efforts. For example, interactions with certain chemicals may inhibit the hydration or dispersion of XG molecules, resulting in the development of inadequate bonding and cohesion within the soil matrix. Conversely, some additives may enhance the performance of XG by promoting better dispersion, hydration or cross-linking of polymer chains. Understanding these interactions is essential for optimizing stabilization techniques and achieving desired engineering properties in stabilized soils.

(a) *Compatibility Testing*: To assess the compatibility of XG with other soil additives or treatments, compatibility testing should be conducted under controlled laboratory conditions. Compatibility tests involve mixing XG with various additives or treatments at different concentrations and observing any changes in rheological properties, chemical stability, or performance characteristics. Common compatibility tests include rheological analysis, viscosity measurements, gelation studies, and microscopy imaging to evaluate the interactions between XG and other additives.
(b) *Optimizing Formulations*: Based on the compatibility test results, engineers can optimize soil stabilization formulations by adjusting the dosage, composition, or application sequence of XG and other additives. For example, modifying the pH, ionic strength, or soil electrolyte concentration can mitigate adverse chemical interactions and improve the compatibility of XG with soil treatments. Additionally, selecting compatible additives or synergistic combinations can enhance the overall effectiveness of soil stabilization techniques, while minimizing potential risks of incompatibility.
(c) *Field Validation*: After conducting compatibility testing in the laboratory, it is essential to validate the performance of optimized soil stabilization formulations under field conditions. Field trials allow engineers to assess the practical feasibility and effectiveness of using XG with other additives or treatments for soil stabilization purposes in real-world applications. By monitoring stabilized soil sites over time, and evaluating key engineering properties, including strength, permeability, and durability, engineers can verify the compatibility of XG-based stabilization methods and identify any unforeseen challenges or opportunities for future improvement.

In summary, compatibility with other soil additives or treatments is a critical consideration when using XG for soil stabilization. By conducting compatibility testing and optimizing stabilization formulations, engineers can ensure the effective integration of XG with other additives or treatments and achieve sustainable solutions for enhancing soil stability in various engineering applications.

1.4.4 Cost Considerations

The cost of XG can vary depending on factors such as purity, production methods and market demand. While XG may offer advantages for soil stabilization applications in terms of better engineering performance and reduced environmental impact, its higher upfront cost compared to conventional soil-stabilizing agents, such as cement and lime, may deter some project stakeholders. Therefore, a cost–benefit analysis should be conducted to evaluate the economic feasibility of using XG for soil stabilization projects.

1.4.4.1 Production Methods and Purity

The cost of XG can vary significantly depending on the production methods employed and the purity of the final product. XG can be produced through both conventional fermentation processes and advanced biotechnological methods. Conventional fermentation processes typically require lower capital investment but may yield lower purity XG, which can affect its performance and price. On the other hand, advanced biotechnological methods, such as recombinant DNA technology, may lead to higher purity but can incur higher production costs due to technological complexities and regulatory requirements.

1.4.4.2 Market Demand and Supply

Like any commodity, the cost of XG is influenced by market demand and supply dynamics. Fluctuations in global demand for XG, driven by factors such as shifts in consumer preferences, changes in industrial applications and geopolitical developments, can affect its market price. Additionally, the availability of raw materials for XG production, such as carbohydrate substrates and microbial cultures, can affect production costs and ultimately the price of XG.

1.4.4.3 Quality and Performance

Higher purity XG products may command a premium price due to their superior quality and performance characteristics. Purer XG typically exhibits better rheological properties, stability, and compatibility with soil materials, leading to more reliable and consistent results in soil stabilization applications. While higher purity XG may entail higher upfront costs, its performance benefits may justify the investment by reducing the need for additional additives or remediation measures.

1.4.4.4 Comparison with Conventional Stabilizing Agents

Despite its potential advantages in terms of better engineering performance and reduced environmental impact, XG may have a higher upfront cost compared to conventional soil-stabilizing agents, such as cement and lime. Cement and lime are widely used for soil stabilization due to their availability, affordability and proven track record in engineering applications. However, these traditional soil-stabilizing agents may have drawbacks, such as higher carbon footprint, longer curing times, and limited applicability in certain soil conditions.

1.4.4.5 Cost–Benefit Analysis

(a) *Economic Feasibility Assessment*: Conducting a cost–benefit analysis is essential to evaluate the economic feasibility of using XG for soil stabilization projects. A cost–benefit analysis involves comparing the total costs associated with using XG, including procurement, application, and maintenance costs, against the anticipated benefits, such as improved soil stability, reduced construction time, and lower long-term maintenance expenses.
(b) *Consideration of Life Cycle Costs*: In addition to upfront costs, it is important to consider the life cycle costs associated with XG-based soil stabilization methods. Life cycle cost analysis considers the total costs incurred over the entire lifespan of a project, including initial investment, operational expenses, and ongoing maintenance costs. By quantifying the long-term economic implications of using XG, engineers can make informed decisions regarding its economic viability compared to conventional soil-stabilizing agents.
(c) *Risk Assessment*: Assessing the risks associated with using XG for soil stabilization purposes is crucial for mitigating potential cost overruns and project delays. Factors such as market volatility, regulatory compliance, technological uncertainties, and performance variability should be carefully evaluated to identify potential risks and develop contingency plans. Incorporating risk management strategies into the cost–benefit analysis helps ensure that project stakeholders can make informed decisions and effectively manage financial risks throughout the project lifecycle.

In summary, while XG may offer advantages in terms of better performance and reduced environmental impact for soil stabilization applications, its higher upfront cost compared to conventional soil-stabilizing agents warrants a thorough cost–benefit analysis. By considering factors such as XG production methods, market dynamics, quality and performance, engineers can evaluate the economic feasibility of using XG and hence make informed decisions to optimize soil stabilization project outcomes, while maximizing value for stakeholders.

1.4.5 Long-Term Durability Considerations

Ensuring the long-term durability of XG-stabilized soil systems is essential for the success of ground improvements engineering projects. Factors such as aging, weathering, and mechanical loading can affect the stability and performance of XG-treated soils over time. Long-term monitoring and maintenance strategies should be implemented to assess the ongoing effectiveness of XG stabilization and address any deterioration or degradation issues.

1.4.5.1 Aging Effects

Over time, XG-stabilized soils may undergo aging processes that can affect their mechanical properties and stability. Aging effects can result from factors such as exposure to environmental conditions (e.g., temperature variations, moisture fluctuations), biological activity (e.g., microbial degradation), and chemical reactions (e.g., oxidation, hydrolysis). These processes may lead to changes in the microstructure, moisture retention, and strength characteristics of XG-treated soil, potentially affecting its long-term performance.

1.4.5.2 Weathering

Weathering processes, including freeze–thaw cycles, wetting and drying cycles, and exposure to UV radiation, can contribute to the degradation of XG and its interactions with soil particles. Freeze–thaw cycles, common in cold climates, can induce frost heave and formation of ice lens within stabilized soil, leading to mechanical damage and loss of cohesion. Wetting and drying cycles can cause undesirable shrink–swell behavior in clayey soils, affecting the integrity of XG bonds and increasing susceptibility to erosion. UV radiation exposure can accelerate the photodegradation of XG molecules, reducing their effectiveness in stabilizing XG-treated surficial soil deposits.

1.4.5.3 Mechanical Loading

The application of external mechanical loads, such as vehicle traffic, building foundations, or earthworks, imposes stresses on stabilized soil structures and deposits, which may influence the performance and durability of XG-treated soil. Prolonged or repetitive loading can lead to consolidation, settlement, or shear deformation in stabilized soil layers, affecting their load-bearing capacity and stability over time. Additionally, dynamic loading from heavy machinery or seismic events may trigger vibration-induced liquefaction or cyclic degradation in XG-treated soil, necessitating reinforcement or other mitigation measures.

1.4.5.4 Long-Term Monitoring and Maintenance Strategies

(a) *Periodic Inspection*: Implementing a program of regular inspection and monitoring is recommended for assessing the ongoing effectiveness and durability of XG stabilization in soil. Periodic visual inspections, supplemented by geotechnical testing and instrumentation, can help identify signs of distress, deterioration, or performance degradation in XG-stabilized soil structures and deposits. Monitoring parameters, such as settlement, deformation, moisture content, shear strength, and permeability, provide valuable insights into the behavior and condition of XG-treated soil over time.

(b) *Performance Evaluation*: Conducting periodic performance evaluations allows engineers to assess the structural integrity, stability and functionality of XG-treated soil deposits in real-world conditions. Performance assessments may include field testing, laboratory analysis, and numerical modeling to quantify key engineering properties and evaluate compliance with design criteria and performance standards. By comparing measured performance indicators against project requirements, engineers can identify any deviations or deficiencies, and subsequently implement corrective actions, as necessary.

(c) *Maintenance and Remediation*: Implementing proactive maintenance and remediation strategies is essential for preserving the long-term durability and performance of XG-stabilized soil structures and deposits. Maintenance activities may include routine repairs, surface treatments, vegetation management and drainage improvements to mitigate erosion, stabilize slopes and maintain drainage conditions. In cases where deterioration or degradation is observed, remediation measures such as reapplication of XG, soil reinforcement, or soil replacement may be necessary to restore stability and functionality.

(d) *Adaptive Management*: Adopting an adaptive management approach allows project stakeholders to respond to changing environmental conditions, evolving performance requirements and emerging challenges over the lifespan of XG-stabilized soil structures and deposits. By continuously monitoring performance, collecting feedback and adjusting management practices accordingly, engineers can optimize the long-term durability, resilience, and sustainability of XG-based soil stabilization solutions.

In summary, ensuring the long-term durability of soil stabilized with XG requires proactive planning, ongoing monitoring, and adaptive management strategies. By addressing factors such as aging, weathering, mechanical loading, and implementing comprehensive monitoring and maintenance programs, engineers can maximize the effectiveness and longevity of XG-treated soil structures and deposits, ultimately enhancing the success and sustainability of engineering projects.

1.5 Summary and Concluding Remarks

Soil stabilization using XG presents various challenges and considerations that engineers and researchers must address to ensure successful project outcomes. The effectiveness of XG can vary depending on a variety of factors, such as soil composition, moisture content and environmental conditions. Extensive laboratory testing and field trials are necessary to determine the optimal XG dosage and application methods for achieving desired stabilization outcomes across different soil types and environmental conditions.

Additionally, the environmental impacts of XG production and disposal of excess/unused XG should be carefully evaluated. While XG is considered environmentally friendly compared to traditional soil-stabilizing agents, its production process requires energy and resources, with the disposal of excess/unused XG posing possible challenges in terms of biodegradability and environmental impact. Therefore, it is essential to consider the overall environmental footprint of XG-based soil stabilization methods and implement sustainable practices to minimize environmental impacts.

Compatibility with other soil additives or treatments is another critical consideration when using XG for soil stabilization. Certain chemicals or compounds present in the soil may interact with XG, affecting its performance and stability. Compatibility testing should be conducted to assess any potential adverse reactions between XG, and other agents present within the soil, and optimize XG-based stabilization formulations accordingly.

Furthermore, the cost implications of using XG for soil stabilization purposes should be evaluated through a comprehensive cost–benefit analysis. While XG may offer advantages in terms of better engineering performance and reduced environmental impact, its higher upfront cost compared to conventional stabilizing agents (e.g., cement and lime) may deter some project stakeholders. Hence, it is essential to assess the economic feasibility using XG and consider factors such as XG production methods, market dynamics, quality, and performance.

Ensuring the long-term durability of soil stabilized with XG is crucial for the success of engineering projects. Factors such as aging, weathering, and mechanical loading can affect the stability and performance of XG-treated soil over time. Long-term monitoring and maintenance strategies should be implemented to assess the ongoing effectiveness of XG stabilization and address any deterioration or degradation issues.

In conclusion, soil stabilization using XG offers promising benefits in terms of improved engineering properties, environmental sustainability, and versatility. However, it is essential to address the challenges and considerations associated with

XG-based soil stabilization to ensure successful project outcomes. By carefully evaluating various factors, such as effectiveness, environmental impact, compatibility, cost implications, and long-term durability, engineers and researchers can make informed decisions regarding the application of XG in soil stabilization projects. Through proactive planning, ongoing monitoring and adaptive management strategies, the potential of XG to enhance soil stability and engineering performance can be realized, contributing to the advancement of sustainable and resilient infrastructure development.

1.6 The Way Forward

The preceding discussion highlights that there are a number of parameters that significantly impact the overall engineering performance (e.g., consistency, mechanical and hydraulic-flow properties) of XG-stabilized soils, including XG concentration/content, the soil type, chemical and mineralogical compositions and moisture content (i.e., hydration level), and the soil-XG-water mixing method (i.e., wet or dry mixing). In addition, other important factors, such as the nature and composition of the soil pore fluid, should be considered because they affect: (i) the XG structural conformation (i.e., random, disordered or ordered); (ii) the interactions (i.e., hydrogen bonding and cation exchange) between the XG and charged surfaces of clay particles; and (iii) the soil-XG structural formation (i.e., dispersed/aggregated).

Therefore, this book provides a detailed analysis of:

(a) the physicochemical and structural properties of the XG biopolymer;
(b) the conformational structures of XG and their rheological properties;
(c) the consistency properties of XG-stabilized soils and energy of interaction between XG and soils;
(d) the governing parameters that affect the strength properties of XG-stabilized soils;
(e) the durability of XG-stabilized soils under variable environmental conditions; and
(f) the water flow properties of XG-stabilized soils and their controlling parameters.

References

Abdullah HH, Shahin MA, Walske ML (2020) Review of fly-ash-based geopolymers for soil stabilization with special reference to clay. Geosciences 10(7):249. https://doi.org/10.3390/geosciences10070249

Achal V, Mukherjee A (2015) A review of microbial precipitation for sustainable construction. Constr Build Mater 93:1224–1235. https://doi.org/10.1016/j.conbuildmat.2015.04.05

Ayeldeen KM, Abdelazim MN, Mostafa AES (2016) Evaluating the physical characteristics of biopolymer/soil mixtures. Arab J Geosci 9:371. https://doi.org/10.1007/s12517-016-2366-1d

Ayeldeen M, Negm A, El-Sawwaf M, Kitazume M (2017) Enhancing mechanical behaviors of collapsible soil using two biopolymers. JRMGE 9(2):329–339. https://doi.org/10.1016/j.jrmge.2016.11.007

Azimi M, Soltani A, Mirzababaei M, Jaksa, MB, Ashwath N (2024) Biopolymer stabilization of clayey soil. J Rock Mech Geotech Eng 16(7):2801–2812. https://doi.org/10.1016/j.jrmge.2023.12.020

Bahmani M, Fatehi H, Noorzad A, Hamedi J (2019) A study on biological soil improvement using new environmental bacteria isolated from northern Iran. Environ Geotech 9(8):1–16. https://doi.org/10.1680/jenge.18.00176

Baveye P, Vandevivere P, Hoyle BL, DeLeo PC, de Lozada DS (1998) Environmental impact and mechanisms of the biological clogging of saturated soils and aquifer materials. Crit Rev Environ Sci Technol 28:123–191. https://doi.org/10.1080/10643389891254197

Biju MS, Arnepalli DN (2020) Effect of biopolymers on permeability of sand-bentonite mixtures. J Rock Mech Geotech Eng 12(5):1093–1102. https://doi.org/10.1016/j.jrmge.2020.02.004

Cabalar AF, Wiszniewski M, Skutnik Z (2017) Effects of xanthan gum biopolymer on the permeability, odometer, unconfined compressive and triaxial shear behavior of a sand. Soil Mech Found Eng 54:356–361. https://doi.org/10.1007/s11204-017-9481-1

Callet F, Milas M, Rinaudo M (1987) Influence of acetyl and pyruvate contents on rheological properties of xanthan in dilute solution. J Biol Macromol 9(5):291–293. https://doi.org/10.1016/0141-8130(87)90068-7

Chang I, Im J, Cho GC (2016) Introduction of microbial biopolymers in soil treatment for future environmentally friendly and sustainable geotechnical engineering. Sustainability 8(3):251. https://doi.org/10.3390/su8030251

Chang I, Im J, Prasidhi AK, Cho GC (2015) Effects of Xanthan gum biopolymer on soil strengthening. Constr Build Mater 74(9):65–72. https://doi.org/10.1016/j.conbuildmat.2014.10.026

Chang I, Kwon YM, Im J, Cho GC (2018) Soil consistency and interparticle characteristics of xanthan gum biopolymer–containing soils with pore-fluid variation. Can Geotech J 56:1206–1213. https://doi.org/10.1139/cgj-2018-0254

Chang I, Lee M, Tran ATP, Lee S, Kwon YM, Im J, Cho GC (2020) Review on biopolymer-based soil treatment (BPST) technology in geotechnical engineering practices. Transp Geotech 24:100385. https://doi.org/10.1016/j.trgeo.2020.100385

Chang I, Cho GC (2019) Shear strength behavior and parameters of microbial gellan gum-treated soils: from sand to clay. Acta Geotech 14:361–375. https://doi.org/10.1007/s11440-018-0641-x

Chen R, Zhang L, Budhu M (2013) Biopolymer stabilization of mine tailings. J Geotech Geoenviron Eng 139(10):1802–1807. https://doi.org/10.1061/(ASCE)GT.1943-5606.000090

Chen C, Wu L, Perdjon M, Huang X, Peng Y (2019) The drying effect on xanthan gum biopolymer treated sandy soil shear strength. Constr Build Mater 197:271–279. https://doi.org/10.1016/j.conbuildmat.2018.11.120

Choi SG, Chang I, Lee M, Lee JH, Han JT, Kwon TH (2020) Review on geotechnical engineering properties of sands treated by microbially induced calcium carbonate precipitation (MICP) and biopolymers. Constr Build Mater 246:118415. https://doi.org/10.1016/j.conbuildmat.2020.118415

Chu J, Ivanov V, Stabnikov V, Li B (2013) Microbial method for construction of an aquaculture pond in sand. Géotechnique 63(10):871–875. https://doi.org/10.1680/geot.SIP13.P.007

DeJong J, Soga K, Kavazanjian E, Burns S, Van Paassen L, Al Qabany A, Aydilek A, Bang S, Burbank M, Caslake LF, Chen CY, Cheng X, Chu J, Ciurli S, Esnault-Filet A, Fauriel S, Hamdan N, Hata T, Inagaki Y, Jefferis S, Kuo M, Laloui L, Larrahondo J, Manning DAC, Martinez B, Montoya BM, Nelson DC, Palomino A, Renforth P, Santamarina JC, Seagren EA, Tanyu B, Tsesarsky M, Weaver T (2013) Biogeochemical processes and geotechnical applications: progress, opportunities and challenges. Geotechnique 63(4):287–301. https://doi.org/10.1680/geot.SIP13.P.017

Dhami NK, Reddy MS, Mukherjee A (2016) Significant indicators for biomineralisation in sand of varying grain sizes. Constr Build Mater 104:198–207. https://doi.org/10.1016/j.conbuildmat.2015.12.023

Dunsmore BC, Bass CJ, Lappin-Scott HM (2004) A novel approach to investigate biofilm accumulation and bacterial transport in porous matrices. Environ Microbiol 6(2):183–187. https://doi.org/10.1046/j.1462-2920.2003.00546.x

Fatehi H, Abtahi SM, Hashemolhosseini H, Hejazi SM (2018) A novel study on using protein-based biopolymers in soil strengthening. Constr Build Mater 167:813–821. https://doi.org/10.1016/j.conbuildmat.2018.02.028

Fatehi H, Ong DEL, Yu J, Chang I (2021) Biopolymers as green binders for soil improvement in geotechnical applications: a review. Geosciences 11(7):291. https://doi.org/10.3390/geosciences11070291

Garcia-Ochoa F, Santos VE, Casas JA, Gomez E (2000) Research review paper: xanthan gum: production, recovery, and properties. Biotechnol Adv 18(7):549–579. https://doi.org/10.1016/S0734-9750(00)00050-1

Ghasemzadeh H, Mehrpajouh A, Pishvaei M (2021) Laboratory analyses of kaolinite stabilized by vinyl polymers with different monomer types. Eng Geol 280:105938. https://doi.org/10.1016/j.enggeo.2020.105938

Ham SM, Chang I, Noh DH, Kwon TH, Muhunthan B (2018) Improvement of surface erosion resistance of sand by microbial biopolymer formation. J Geotech Geoenviron Eng 144(7):06018004

He J, Chu J, Ivanov V (2013) Mitigation of liquefaction of saturated sand using biogas. Géotechnique 63(4):267–275. https://hdl.handle.net/10356/98137

Joga JR, Varaprasad BJS (2019) Sustainable improvement of expansive clays using xanthan gum as a biopolymer. Civ Eng J 5(9):1863–1903

Jones TGJ, Hughes TL (1996) In: Schramm LL (ed) Suspensions: fundamentals and applications in the petroleum industry. Advances in Chemistry Series, American Chemical Society, Chapter 10, pp 463–562. ISBN:0841231362, 9780841231368

Khachatoorian R, Petrisor IG, Kwan CC, Yen TF (2003) Biopolymer plugging effect: laboratory-pressurized pumping flow studies. J Pet Sci Eng 38(1–2):13–21. https://doi.org/10.1016/S0920-4105(03)00019-6

Khatami HR, O'Kelly BC (2013) Improving mechanical properties of sand using biopolymers. J Geotech Geoenviron Eng 139:1402–1406. https://doi.org/10.1061/(ASCE)GT.1943-5606.0000861

Kwon TH, Ajo-Franklin JB (2013) High-frequency seismic response during permeability reduction due to biopolymer clogging in unconsolidated porous media. Geophysics 78(6):117–127. https://doi.org/10.1190/geo2012-0392.1

Kwon YM, Chang I, Lee M, Cho GC (2019) Geotechnical engineering behavior of biopolymer-treated soft marine soil. Geomech Eng 17(5):453–464. https://doi.org/10.12989/gae.2019.17.5.453

Latifi N, Horpibulsuk S, Meehan CL, Abd Majid MZ, Tahir MM, Mohamad ET (2016) Improvement of problematic soils with biopolymer-an environmentally friendly soil stabilizer. J Mater Civ Eng 29(2). https://doi.org/10.1061/(ASCE)MT.1943-5533.0001706

Lee S, Chung M, Park HM, Song KI, Chang I (2019) Xanthan gum biopolymer as soil-stabilization binder for road construction using local soil in Sri Lanka. J Mater Civ Eng 31(11):06019012–06019021. https://doi.org/10.1061/(ASCE)MT.1943-5533.0002909

Liu S, Wang R, Yu J, Peng X, Cai Y, Tu B (2020) Effectiveness of the anti-erosion of an MICP coating on the surfaces of ancient clay roof tiles. Constr Build Mater 243:118202

Losini AE, Grillet AC, Bellotto M, Woloszyn M, Dotelli G (2021) Natural additives and biopolymers for raw earth construction stabilization–a review. Constr Build Mater 304(2):1123. https://doi.org/10.1016/j.conbuildmat.2021.124507

Matsubara H (2021) Stabilization of weathered limestone surfaces using microbially enhanced calcium carbonate deposition. Eng Geol 284:106044. https://doi.org/10.1016/j.enggeo.2021.106044

Mendonça A, Morais PV, Pires AC, Chung AP, Oliveira PJV (2021) Reducing soil permeability using bacteria-produced biopolymer. Appl Sci 11:7278. https://doi.org/10.3390/app11167278
Mohamed AMO, El Gamal MM, Hameedi S (2022) Sustainable utilization of carbon dioxide in waste management. Elsevier, p 606
Mujah D, Shahin MA, Cheng L (2017) State-of-the-art review of bio-cementation by microbially induced calcite precipitation (MICP) for soil stabilization. Geomicrobiol J 34:524–537. https://doi.org/10.1080/01490451.2016.1225866
Navarrete RC, Dearing HL, Constien VG, Marsaglia KM, Seheult JM, Rodgers PE (2000) Experiments in fluid loss and formation damage with xanthan-based fluid while drilling. IADC/SPE 62732. Asia Pacific Drilling Technology Conference, Kuala Lumpur, Malaysia. https://doi.org/10.2118/62732-MS
Ng CWW, So PS, Lau SY, Zhou C, Coo JL, Ni JJ (2020) Influence of biopolymer on gas permeability in compacted clay at different densities and water contents. Eng Geol 272(11):105631. https://doi.org/10.1016/j.enggeo.2020.105631
Ngu LH, Song JW, Hashim SS, Ong DEL (2019) Lab-scale atmospheric CO_2 absorption for calcium carbonate precipitation in sand. Greenh Gases Sci Technol 9(3):519–528. https://doi.org/10.1002/ghg.1869
Niaounakis M (2015) Biopolymers: applications and trends; William Andrew, Norwich, NY, USA
Noh DH, Ajo-Franklin JB, Kwon TH, Muhunthan B (2016) P and S wave responses of bacterial biopolymer formation in un- consolidated porous media. J Geophys Res Biogeosci 121(4):1158–1177. https://doi.org/10.1002/2015JG003118
Omoregie AI, Palombo EA, Ong DEL, Nissom PM (2019) Biocementation of sand by *Sporosarcina pasteurii* strain and technical-grade cementation reagents through surface percolation treatment method. Constr Build Mater 228:116828. https://doi.org/10.1016/j.conbuildmat.2019.116828
Omoregie AI, Palombo EA, Ong DEL, Nissom PM (2020) A feasible scale-up production of Sporosarcina pasteurii using cus- tom-built stirred tank reactor for in-situ soil biocementation. Biocatal Agric Biotechnol 24:101544. https://doi.org/10.1016/j.bcab.2020.101544
Phatak A (2011) Patent WO2011123249-A2
Qureshi MU, Chang I, Al-Sadarani K (2017) Strength and durability characteristics of biopolymer-treated desert sand. Geomech Eng 12(5):785–801. https://doi.org/10.12989/gae.2017.12.5.785
Ramdas VM, Mandree P, Ramchuran S, Lalloo R, Mgangira M, Mukaratirwa S (2021) Review of current and future bio-based stabilization products (enzymatic and polymeric) for road construction materials. Transp Geotech 27:7158. https://doi.org/10.1016/j.trgeo.2020.100458
Rebata-Landa V, Santamarina JC (2012) Mechanical effects of biogenic nitrogen gas bubbles in soils. J Geotech Geoenviron Eng 138(2):128–137. https://doi.org/10.1061/(ASCE)GT.1943-5606.000057
Sadeghian F, Haddad A, Jahandari S, Rasekh H, Ozbakkaloglu T (2021) Effects of electrokinetic phenomena on the load-bearing capacity of different steel and concrete piles: a small-scale experimental study. Can Geotech J 58(5):741–746. https://doi.org/10.1139/cgj-2019-0650
Seright RS, Henrici BJ (1990) Xanthan stability at elevated temperatures. SPE Reservoir Eng 5:52–60
Shahrokhi-Shahraki R, Zomorodian SMA, Niazi A, O'Kelly BC (2015) Improving sand with microbial-induced carbonate precipitation. Proc Inst Civ Eng Ground Improv 168:217–230. https://doi.org/10.1680/grim.14.00001
Sharma M, Satyam N, Reddy KR (2021) Effect of freeze-thaw cycles on engineering properties of bio-cemented sand under different treatment conditions. Eng Geol 284:106022. https://doi.org/10.1016/j.enggeo.2021.106022
Smitha S, Sachan A (2016) Use of agar biopolymer to improve the shear strength behavior of sabarmati sand. Int J Geotech Eng 10:387–400. https://doi.org/10.1080/19386362.2016.1152674
Soldo A, Miletić M, Auad ML (2020) Biopolymers as a sustainable solution for the enhancement of soil mechanical properties. Sci Rep 10:267. https://doi.org/10.1038/s41598-019-57135-x

Soltani A, Deng A, Taheri A, Mirzababaei M, Jaksa MB (2020) A dimensional description of the unconfined compressive strength of artificially cemented fine-grained soils. J Adhes Sci Technol 34(15):1679–1703. https://doi.org/10.1080/01694243.2020.1717804
Soltani A, Raeesi R, Taheri A, Deng A, Mirzababaei M (2021) Improved shear strength performance of compacted rubberized clays treated with sodium alginate biopolymer. Polymers 13(5):764. https://doi.org/10.3390/polym13050764
Stal LJ (2011) Biopolymer. In: Gargaud M, Amils R, Quintanilla JC, Cleaves HJ, Irvine WM, Pinti DL, Viso M (eds) Encyclopedia of astrobiology. Springer, Berlin/Heidelberg, Germany, pp 199–200. https://doi.org/10.1007/978-3-642-11274-4
Sujatha ER, O'Kelly BC (2023) Biopolymer based soil treatment for geotechnical engineering applications. In: Thomas S, AR A, Jose Chirayil C, Thomas B (eds) Handbook of biopolymers. Springer, Singapore, pp 609–626. https://doi.org/10.1007/978-981-16-6603-2_22-1
Tam KC, Tiu C (1993) Rheology of water-soluble polymers: a comparative study on the effect of monovalent salt. Polym Plast Technol Eng 32:123. https://doi.org/10.1080/03602559308020157
Tang CS, Yin LY, Jiang NJ, Zhu C, Zeng H, Li H, Shi B (2020) Factors affecting the performance of microbial-induced carbonate precipitation (MICP) treated soil: a review. Environ Earth Sci 79:1–23. https://doi.org/10.1007/s12665-020-8840-9
Torres FG, Troncoso OP, Pisani A, Gatto F, Bardi G (2019) Natural polysaccharide nanomaterials: an overview of their immunological properties. Int J Mol Sci 20:102. https://doi.org/10.3390/ijms20205092
Williamson S, Cortes DD (2014) Dimensional analysis of soil–cement mixture performance. Géotechnique Lett 4(1):33–38. https://doi.org/10.1680/geolett.13.00082
Zomorodian SMA, Nikbakht S, Ghaffari H, O'Kelly BC (2023) Enzymatic-induced calcite precipitation (EICP) method for improving hydraulic erosion resistance of surface sand layer: a laboratory investigation. Sustainability 15(6):5567. https://doi.org/10.3390/su15065567

Chapter 2
Xanthan Gum Production and Structure

2.1 Introduction

In nature, microbial polysaccharides are essential components of biofilms, which are the main support form of life on earth (Flemming et al. 2016), and microorganisms' lifestyle (Penesyan et al. 2019). These biofilms contain microbial cells that are fixed within a self-produced matrix of extracellular polymeric substances, which are composed of polysaccharides, proteins, lipids, and nucleic acids (Powell et al. 2018). The relative composition and molecular structure of these polymeric substances vary depending on the producing microorganism and the environmental conditions (e.g., fluid chemistry, pressure, and temperature). Therefore, depending on the potential industrial application, different industrial fermentation processes can be adopted.

Microorganisms produce polysaccharides for different applications, such as enhancement of structural properties and environmental protection. Concerning the former application, microbial polysaccharides can provide a variety of functions: (i) increase the viscosity of aqueous liquids, the water retention capacity, and the adhesion capacity of microorganisms to each other, as well as to interacting surfaces (Alaa 2018); (ii) facilitate cell-to-cell interactions (Dos Santos et al. 2018); and (iii) provide a structure, with viscoelastic properties, that increases biofilm resistance to external loads (Peterson et al. 2015).

The preceding characteristics are pertinent to the application of microbial polysaccharides for soil stabilization. However, for the case of protecting the environment, the biofilm acts as a protective matrix in which both the polysaccharide-producing microorganisms and co-living cells are encapsulated; hence increasing the stability of the biopolymer. Such an enhanced functional property aids microorganisms to protect themselves, in extreme environments (Yin et al. 2019), against ultraviolet radiation (de Carvalho 2017), arid land extreme temperatures (Kent et al. 2018), pH variations (Narayanan et al. 2016), and salinity (Wang et al. 2019).

This chapter delves into details related to the production of xanthan gum (XG) and the factors that impact its structural conformations and rheological properties,

A. M. O. Mohamed et al., *Sustainability in Ground Improvement: The Case of Xanthan Gum Biopolymer*, Green Energy and Technology,
https://doi.org/10.1007/978-3-031-75313-8_2

which are pertinent to the understanding of the soil-XG interactions and stabilization mechanisms.

2.2 Xanthan Gum Production

The production of XG involves the fermentation of glucose by *Xanthomonas campestris*, followed by downstream processing to isolate and purify the gum. These processes involve the following steps:

(a) *Inoculum preparation*: A seed culture of *X. campestris* is grown in a suitable medium under controlled conditions to obtain a high cell density inoculum.
(b) *Fermentation*: The inoculum is transferred to a fermentation medium containing glucose as the carbon source and other nutrients. There are many parameters that impact the growth of the microorganism and hence XG production, including the type of bioreactor used, mode of operation (batch or continuous), medium composition, and culture conditions (temperature, pH, dissolved oxygen concentration) (Lopes De Monaco et al. 2015; Habibi and Khosravi-Darani 2017). Both the culture environment and operational conditions influence the molecular structure of the manufactured XG. In addition, since *X. campestris* needs micro- and macro-nutrients for their growth, Na^+, K^+, Ca^{2+}, and Fe^{3+} salts are used as micronutrients; while carbon (glucose or fructose) and nitrogen are used as the macronutrients. Finally, high pressure is used to inactivate the microorganisms.
(c) *Recovery and purification*: After the fermentation is complete, the broth is subjected to a series of processing steps to recover and purify the XG. The steps typically include clarification, precipitation, washing, and drying, as follows:
    (i) *Clarification*: The fermentation broth is clarified to remove cell debris and other insoluble impurities by centrifugation or filtration.
    (ii) *Precipitation*: The XG is precipitated from the clarified broth by the addition of an alcohol, such as isopropyl alcohol or ethanol. The precipitation is performed under controlled temperature and pH conditions to obtain a high yield and purity of XG.
    (iii) *Washing*: The precipitated XG is washed with alcohol and then water to remove residual impurities and alcohol.
    (iv) *Drying*: For ease of handling, storage, and application, the washed XG is dried to produce a dry powder.

Notably, these recovery processes vary depending on the end use (e.g., oil industry, food processing, medical, pharmaceutical, or engineering). After precipitation, the XG is dried under vacuum or with forced circulation of an inert gas in batch or continuous dryers, with a final moisture content (the quantity of water contained in the XG) ranging from 10 to 15% by weight. Then, the XG is milled to specific particle sizes as well as in an agglomerated form to control both dispersion and dissolution rates. In addition, the XG production is further optimized in view of its molecular properties, such as the substitution pattern with acetyl and pyruvate groups, and rheological behavior on a genetic level (Gansbiller et al. 2019). As an example of strain genetic engineering, the XG wettability can be increased by using lower amounts of ethanol in the downstream processing (Dai et al. 2019).

Production of XG with different rheological properties is of great importance; e.g., for applications in coastal areas where high amounts of salts are present in the soil pore fluid. Commercially, there are different possibilities including, for instance, particularly salt-tolerant XG, or XG that exhibits a reduced pseudoplastic flow behavior (Fig. 2.1). The differences in the apparent viscosity can be observed especially with low shear rates, with implications for the XG grade selection, which depends on the end use. Commercial salt-tolerant XG has been developed, while maintaining an excellent viscosity-forming potential, for practical use with high concentrations of NaCl up to 20,000 ppm.

Overall, the production of XG involves a complex process that requires careful control of the fermentation conditions and downstream processing steps. However, the resulting product has numerous industrial applications, making it a valuable biopolymer.

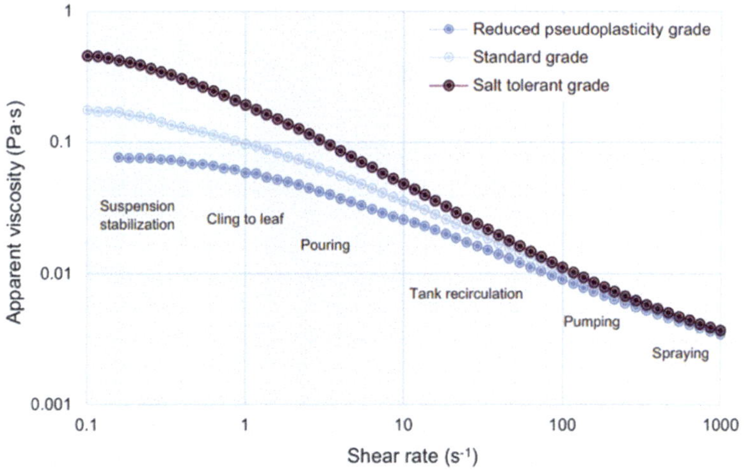

Fig. 2.1 Model flow curves for solutions of 0.1% (w/v) of different XG commercial grades in standard tap water (1 g/l NaCl, 0.15 g/l $CaCl_2.2H_2O$) (after Berninger et al. 2021). [This is an open access article under the terms of the Creative Commons Attribution-Non-Commercial-No Derives License, which permits use and distribution in any medium, provided the original work is properly cited, the use is non-commercial, and no modifications or adaptations are made]

2.3 XG Structure

XG is mainly produced from the Gram-negative bacterium *X. campestris*, which is used to form an extracellular polysaccharide of high molecular weight, referred to as xanthan. These cells are motile (move independently using metabolic energy) and aerobic (Garcia-Ochoa and Santos 2000). XG has a natural anionic structure composed of a repeating unit from pentasaccharide having D-glucose, D-mannose, and D-glucuronic acid with the molar ratio 2:2:1. The XG backbone is mainly a linear β-(1,4)-D-glucose, which is formed from the connection of terminal β-D-mannose with D-glucuronic acid via β-(1,4) in a structure that is linked to D-mannose through α-(1,2) (Jansson et al. 1975; Faria et al. 2011; Abbaszadeh et al. 2015; Wang et al. 2017; Kumar et al. 2018) (Fig. 2.2). Acetate and pyruvate groups are attached to two D-mannose units with non-stoichiometric amounts according to the production of the bacterial strain, the fermentation conditions and/or subsequent chemical modification (Cadmus et al. 1976; Sutherland 1981; Tait et al. 1986; Abbaszadeh et al. 2015).

XG also possesses a secondary structure that appears as a five-fold helical molecule, which is mainly stabilized by the non-covalent bond, measuring 4.7 nm long and 1.9 nm in diameter. The reversible ordered (helix)–disordered (coil) conformational structural transition of the XG secondary structure can be obtained by changing the temperature and the solution ionic strength (Fig. 2.3). The disorder form is favored by high temperatures and low ionic concentrations (Holzwarth 1976; Moorhouse et al. 1977; Cheetham and Mashimba 1992; Katzbauer 1998; Abbaszadeh et al. 2015). In addition, the coil-helix transition temperature mainly depends on the

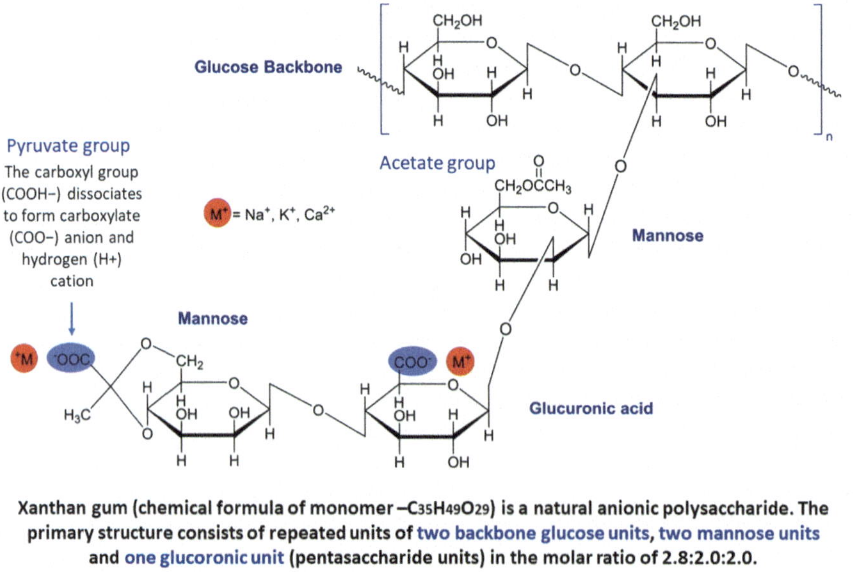

Fig. 2.2 Structure of XG bioploymer

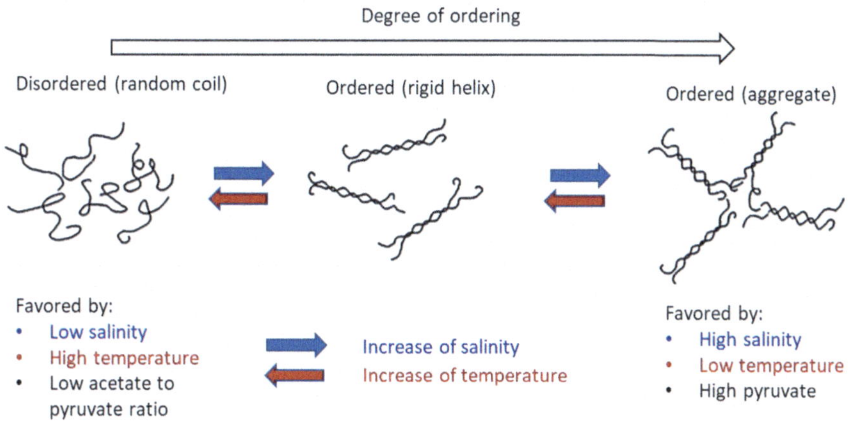

Fig. 2.3 XG conformational structures for a variety of conditions

content of acetate and pyruvate groups, ionic strength, pH level, and polymer concentration (Abbaszadeh et al. 2015). In the ordered state, the main chain and the side chain form a single-/double-helix structure through hydrogen bonding. These spiral structures also maintain their stability by electrostatic forces and steric effects. Furthermore, when the single-/double-helix structure is combined by a covalent bond, it can form a regular network of spiral copolymers that provide the molecules a strong rigidity, hence making the XG molecular chain in aqueous solution stretched and controlling the fluidity of the XG aqueous solution.

There is also a repeated disaccharide-making backbone with a single helix to form fiber (Fig. 2.4). This conformation gives stability to the structure as the trisaccharide side chains are lined up with the backbone, strengthened by non-covalent and hydrogen bonds. In solution, these side chains cover the backbone all over, such that the (1,4) bonds are protected from attacks. It is understood that this shield is the key to the excellent stability of XG.

It is worth noting that different strains or fermentation conditions give rise to differing degrees of acetylation and pyruvylation, which moderate the functionality of XG (Williams and Phillips 2004). The mannose group can contribute to producing pyruvate and energy for glycolysis (Berg et al. 2002). The pyruvate group can join the citric acid cycle pathway to produce energy (Meléndez-Hevia et al. 1996), whereas glucuronic acid produces the L. ascorbic acid (a form of vitamin C) (Valpuesta and Botella 2004) that is used in the photosynthesis process (Gallie 2013). XG behaves as polyanions (i.e., having negative charges at several sites), because along with the structure chains there are also anionic units of pyruvate and D-glucuronic acid (Petri 2015).

Previous studies have indicated that XG is not a homogeneous polymer, and that repeating units can contain acetyl, pyruvate, both groups, or neither (Kulicke and van Eikern 1992). Acetyl groups tend to stabilize the ordered form, presumably by promoting association of the side chains with the backbone through hydrogen

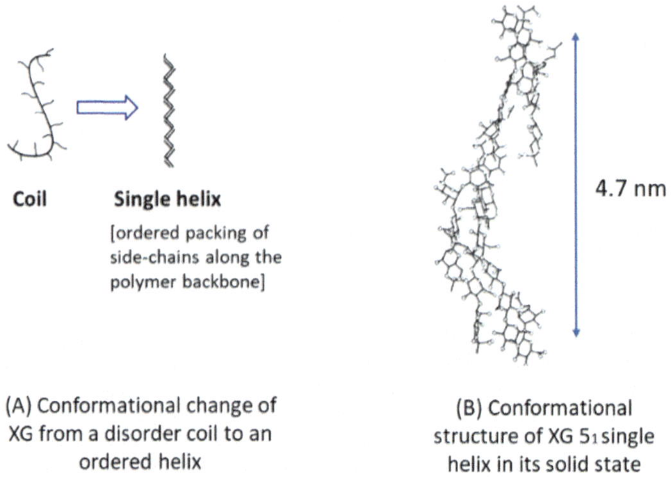

Fig. 2.4 XG conformational structure: **a** disorder coil to an ordered helix; and **b** structural arrangement in a single helix (adopted from Morris (2019))

bonding. Pyruvate groups, in contrast, tend to favor the disordered form because electrostatic repulsion between these groups is minimized when the side chains extend away from the backbone (Tako 1991). The degree of acetyl and pyruvic acetal substitution depends on the fermentation conditions and bacterial strain (Milas and Rinaudo 1984; Rochefort and Middleman 1987). The repeat unit has a mass of about 1000 Dalton (Da) and a length of about 5 nm. The ranking of thermal stability based on experimental data of the ordered conformation is as follows: pyruvic-free > native xanthan > acetyl and pyruvic-free > acetyl-free.

Due to the presence of glucuronic and pyruvic acids on the side chain, XG represents a highly charged polysaccharide (anionic) with a very rigid polymer backbone. It has a high molecular weight of approximately 2.9×10^6 to 2.9×10^7 g/mol with a narrow distribution. Based on X-ray diffraction analysis, the secondary structure of XG could be described as a five-fold helical conformation, where the backbone is stabilized by the side chains (Moorhouse et al. 1977).

XG has unique pseudoplastic rheological properties (flow rate increases, and viscosity decreases faster than normally in relation to applied shear stress) because of the natural formation of rigid-rod chains (helices) from XG molecules in aqueous solution. The aggregation and alignment of the rigid chains of XG contribute to lower entropy (i.e., a measure of the molecular disorder, or randomness, of a system) and high stability. Due to the weak association of XG rods, 3D networks could be formed leading to emulsion stability and particle suspension of XG in aqueous solution. However, with the application of shear stresses, a gradual breakdown of the network structure takes place, leading to an increase of pseudoplasticity in XG (Morris et al. 1977; Lo et al. 2007). Since XG has a polyelectrolyte microstructure, its rheological behavior is dependent on the solution's ionic strength and pH level, the XG concentration, temperature, and pressure. For instance, with increasing the ionic strength of the solution, the side chains of XG can be folded down to the backbone, thereby resulting in a more ordered (helix) and stable structure (Ahmad et al. 2014; Morris 2019).

Several factors can affect the structure of XG, which, in turn, can affect its functionality and application. The following are some of the factors that influence the structure of XG:

(a) *Molecular weight*: The molecular weight of XG affects its conformational structure. High molecular weight XG has a more extended conformation than its low molecular weight counterpart due to increased chain stiffness and reduced entanglement between chains.
(b) *Degree of acetylation* (i.e., the number of acetyl groups attached to the side chains): It affects the conformational structure of XG; higher degrees of acetylation resulting in a more compact conformation due to increased interchain repulsion and reduced chain flexibility.
(c) *pH*: The pH of the solution affects the ionization of the carboxyl groups present in XG, which, in turn, affects the conformation of the molecule. At low pH, the carboxyl groups are mostly protonated, leading to a compact conformation, while at high pH levels, they are mostly deprotonated, leading to an extended conformation.
(d) *Temperature*: Temperature also affects the conformational structure of XG, with higher temperatures leading to a more extended conformation. At high temperatures, the XG molecule is more susceptible to degradation, which can affect its viscosity and other properties.
(e) *Salt concentration*: The presence of salt ions in the solution can affect the interaction between XG molecules, leading to changes in the viscosity and other properties of the solution. At high salt concentrations, the viscosity of the solution decreases due to the shielding of electrostatic interactions between the molecules.
(f) *Shear rate*: The application of shear forces to XG solution can affect the conformation of the molecule, leading to changes in the viscosity and other properties of the solution. At low shear rates, the molecule adopts a compact conformation, while at high shear rates it adopts an extended conformation.

In summary, the conformational structure of XG is highly complex and flexible, with the backbone forming an extended coil and the side chains protruding outward from the coil. The conformation is sensitive to solution properties, which can induce changes in the shape and size of the molecule. Understanding the factors that affect the conformational structure of XG is important for optimizing its performance in various applications, including food processing, pharmaceuticals, cosmetics, agriculture and engineering.

2.3.1 Types of XG Structural Conformations

XG is present in three structural conformations, namely native, denatured, and renatured. Native XG (XG_{Nat}) refers to XG dissolved in aqueous solution containing a moderate electrolyte concentration at room temperature (Milas and Rinaudo 1986). It has a molar mass per unit contour length (M_L) of about 2000 Da/nm and a persistence length (L_P) ranging from 120 to 150 nm. Denatured XG (XG_{Denat}) refers to XG in aqueous solution at high temperature, or in an aqueous solution containing very low electrolyte concentration. Heating above the melting temperature (T_m) of 50 °C produces a disorder-flexible coil structure and reductions in molecular weight, viscosity and hydrodynamic volume. XG_{Denat} has an M_L of about 1000 Da/nm and an L_P of about 5 nm. Additional heating for a long period leads to XG degradation, further reducing its molecular weight, viscosity η, and radius of gyration R_g. Renatured XG (XG_{Renat}) refers to an XG solution that has been heated above T_m for a few minutes and then cooled quickly. For instance, Callet et al. (1987), Milas et al. (1996) and Capron et al. (1997) produced the XG_{Renat} by heating the native form for a few minutes at about 90 °C followed by immediate fast cooling. These processes produce an ordered-flexible, single-/double-helical conformation with increased hydrodynamic volume. XG_{Renat} has an M_L of about 4000 Da/nm and an L_P ranging from 50 to 125 nm. Additional information regarding the three conformational structures of XG is presented in Table 2.1.

The stiffness of the XG polymer chain has been characterized by its persistence length, which is reported as $L_P = 5$, 45, and 125 nm for disordered-flexible coil, single-helix, and rigid double-helical conformations, respectively (Rinaudo 2001). However, the XG element can be characterized by the molar mass per unit contour length (M_L), which is close to 1000 and 2000 Da/nm for single- and double-helix conformations, respectively (Sato et al. 1984a, b, c). The stiffness of the XG polymer is attributed to the intra- and intermolecular ester bonds due trans-esterification reactions and dehydration of the XG chains (Bueno et al. 2013). Esterification between XG acid groups (i.e., acetyl or pyruvyl) and OH groups occurs upon heating the XG film, which was established using FTIR spectroscopy analysis (i.e., carbonyl bands of XG corresponding to the acidic and ester forms were observed at 1643 and 1728 cm^{-1}, respectively). In addition, scanning electron microscopy analysis of the cryofracture surface indicates that XG dehydrated hydrogels have a heterogeneous porous structure with low crosslinking density and nanofibrils (about 20–30 nm in diameter).

Notably, commercial XG has been preheated by the producer or during the manufacturing process; hence, its native conformation has been destroyed and would probably have helical structures. The nature of the helical structures of XG_{Nat} and XG_{Renat} is similar (Milas and Rinaudo 1979; Capron et al. 1998). The XG_{Renat} conformation has the same molar mass as XG_{Nat}, but its M_L of the XG_{Renat} form (4000 Da/nm) is double that of XG_{Nat} (2000 Da/nm) (Chazeau et al. 1995; Milas et al. 1996).

After extraction, without any heating, XG chains arranged as single helices stabilized by Ca^{2+} ions can be irreversibly denatured (i.e., to give XG_{Denat}), changing to

Table 2.1 Types and conformation of XG

	Type of xanthan gum		
	Native xanthan gum (XG_{Nat})	Denatured xanthan gum (XG_{Denat})	Renatured xanthan gum (XG_{Renat})
Processing technique	Dissolved in aqueous solution containing a moderate electrolyte concentration at room temperature (i.e., no heating)	Dissolved in aqueous solution containing a moderate electrolyte concentration at high temperature	Dissolved in aqueous solution containing a moderate electrolyte concentration at high temperature followed by cooling
Conformation	Native conformation is a modification of the ordered conformation and has rigid helical structures	Disordered-flexible coil structures	Native conformation is a modification of the ordered conformation and has flexible double-helical structures
Electrical charge	Polyanion at pH > 4.5 due to the deprotonation of O-acetyl and pyruvyl residues	Polyanion at pH > 4.5 due to the deprotonation of O-acetyl and pyruvyl residues	Polyanion at pH > 4.5 due to the deprotonation of O-acetyl and pyruvyl residues
Molar mass per unit contour length, M_L	About 2000 Da/nm	About 1000 Da/nm	About 4000 Da/nm
Solubility	Soluble in cold and hot water	Soluble in cold and hot water	Soluble in cold and hot water
Persistence length, L_P	Ranging from 120 to 150 nm	About 5 nm	Ranging from 50 to 125 nm

coils (Chazeau et al. 1995; Capron et al. 1997; Baumgartner et al. 2008). The coiled conformation can be reversibly renatured (XG_{Renat}) to the double-helical structure. The ordered or disordered state of the XG chains is controlled by temperature and ionic strength. The helix conformation is stabilized by hydrogen bonds and destabilized by electrostatic repulsion between carboxylate groups along the chains. Thus, at low ionic strength or high temperature, chains assume coil conformation with a persistence length of about 5 nm (Tinland and Rinaudo 1989; Chazeau et al. 1995). On the other hand, under high ionic strength or low temperature, XG chains are arranged in helical conformation with a persistent length of about 35 nm (Tinland and Rinaudo 1989). The enthalpy variation involved in the transition from helix to coil conformation was estimated at 13.9 J/g (Bresolin et al. 1998).

Moreover, Gulrez et al. (2012) used the atomic force microscopy to determine the XG conformation and reported that it consists of (i) double-helical ordered XG_{Renat} state (pre-heat-treated by the manufacturer) in dilute solution conditions and (ii) random coil conformation in very dilute solution. The ordered XG_{Renat} conformation was shown to have a partially molten double helix, with more flexibility than the perfectly ordered XG_{native} double helix. Furthermore, thermal treatment of XG solution in the dilute region leads to an order–disorder transition, while irradiation of

XG solution leads to an order–disorder transition as well as the production of single-strand, low-molecular-weight XG molecules, whereas homogenization treatments by autoclaving and high-pressure cause degradation of XG.

In addition, using X-ray diffraction analysis, XG was shown to adopt a helical conformation having a five-fold symmetry and a pitch of 4.7 nm in the solid state (Moorhouse et al. 1977). However, in aqueous solution, XG underwent an order to disorder (i.e., helix to coil) transition depending on the temperature, ionic strength, pH level (Milas et al. 1996; Capron et al. 1997), and acetyl/pyruvate contents (Smith et al. 1981; Morrison et al. 2004).

The ordered molecule of XG can undergo a conformational transition to a disordered-flexible coil either by increasing the temperature above its characteristic melting temperature (T_m) or by decreasing the salinity of the solution (Hacche et al. 1987; Callet et al. 1989; Milas and Tinland 1990; Kawakami and Norisuye 1991; Oviatt and Brant 1994b; Capron et al. 1996, 1997; Milas et al. 1996; Gunning et al. 1998; Terada et al. 2005; Mehyar et al. 2008). The value of T_m largely depends on the concentration of the polymer and external salt (Rinaudo 2001). The order–disorder conformational change (denaturation) of the XG molecule in 0.01-M NaCl is reported to occur at a temperature of approximately 50 °C (Capron et al. 1997). As mentioned earlier, within the ordered structure, there exist two different kinds of conformation: native and renatured (i.e., XG_{Nat} and XG_{Renat}, respectively). The disordered structure of denatured XG (i.e., XG_{Denat}) is unanimously understood to be random coil and can be obtained by raising the temperature above T_m (> 50 °C), while prolonged heating (> 2 h) at high temperature results in degradation (Nishinari et al. 1996; Capron et al. 1997).

However, the precise structural details of the ordered conformation of XG remain a subject of some debate, with two different models proposed. That is, one group of researchers suggests an ordered single-stranded structure stabilized by alignment and packing of side chains along the polymer backbone. This hypothesis was based on number of techniques, such as light scattering, viscometry, and sedimentation and concentration-independence studies (Callet et al. 1987; Milas and Rinaudo 1986; Milas and Tinland 1990; Milas et al. 1996). On the other hand, the majority of studies have suggested an ordered coaxial double-helix conformation (Brant et al. 1985; Hacche et al. 1987; Hatakenaka et al. 1987; Kawakami and Norisuye 1991; Oviatt and Brant 1994a, b; Gunning et al. 1996; Capron et al. 1997; Morrison et al. 2004; Terada et al. 2005; Mehyar et al. 2008).

2.3.2 *Effect of Treatment Methods*

Several treatments, such as heating, autoclaving, and micro fluidizing, have been employed on polysaccharides (including XG) to overcome their associative nature to promote conformational changes. Knowledge of the effects of these processes is beneficial for optimizing the performance of XG in various formulations and applications. In a study by Capron et al. (1997), it was reported that the effect of heat treatment

(110 °C for 20 min in 0.1-M NaCl) on XG conformation showed dissociation of the double helix into single strand (i.e., denaturation). However, Lagoueyte and Paquin (1998) showed that the process of microfluidization (high-pressure homogenization) on XG solution caused degradation due to mechanical stress. Autoclaving is another method used successfully to disperse aggregated molecules and hence reduce the associative nature of polysaccharides in diluted solutions (Wang et al. 2001).

In addition, other studies have shown that after heating XG solution at 95 °C for 2.5 h, only 40% of the double-helical fraction is destroyed (Kawakami and Norisuye 1991; Mehyar et al. 2008), indicating that XG does not lose its double strand suddenly at high temperature, but rather partial melting of the double helix occurs. For example, Gulrez et al. (2012) heated XG of 4 mg/ml concentration at 60 °C and reported on the changes of the M_w, viscosity η, radius of gyration R_g, and polydispersity, as shown in Fig. 2.5.

Notably, the influence of M_w on the melt viscosity η depends on whether the polymer is above or below the onset of entanglements: (i) when η is proportional

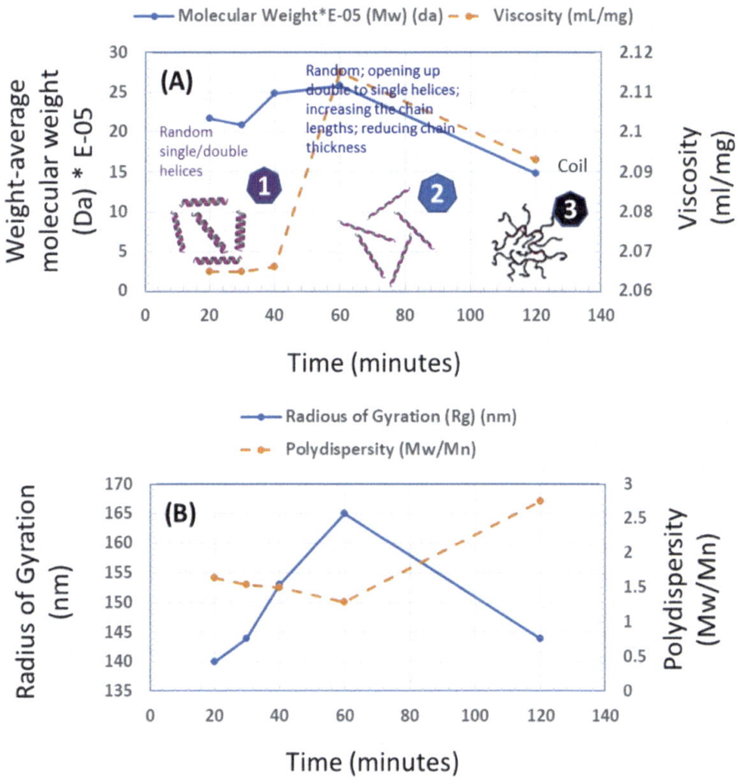

Fig. 2.5 Changes of the weight-average molecular weight M_w, intrinsic viscosity, radius of gyration and polydispersity of 4 mg/ml of XG in water heated at 60 °C (data from Gulrez et al. (2012)). *Note* M_n denotes the number-average molecular weight

to $(M_w)^1$, the polymer conformation is below entanglement molecular weight; and (ii) when η is proportional to $(M_w)^{3.4}$, the polymer conformation is above entanglement molecular weight. The increase in R_g tends to decrease chain mobility, increase strength and toughness, and increase the glass-transition temperature (T_g), attributed to the increase in chain interactions via van der Waals' attractions, as well as entanglements that resulted from the increase in R_g. These interactions tend to hold strongly the individual chains in their positions and resist deformations and polymer-matrix breakup at higher applied pressures and temperatures.

Referring to Fig. 2.5, as time increases up to 60 min, the weight-average molecular weight, viscosity and R_g increased, while the width of the molecular chain distribution (polydispersity) decreased. The increased length of the polymer chain arises from elevated interactions thorough van der Waals' attractions and entanglements, while the reduction in polydispersity value suggests disassociation of the high molecular weight aggregates. However, as time increases from 60 to 120 min, the M_w, η, and R_g decreased, but the polydispersity increased due to the enhanced flexibility of the polymer chain and its transition toward coil conformation. Notably, the M_w decreased by approximately one-third of the initial value of the 120-min heated sample, which could be attributed to dissociation of the double helix into single strand.

In addition, the role of heating methods (heat treatment, autoclaving, and irradiation) on the conformation of XG was evaluated by Gulrez et al. (2012) for food-grade commercial XG (Jungbunzlauer, Austria). The effect of heating methods on the M_L of XG at 4 mg/ml concentration in water is shown in Fig. 2.6a. The M_L for thermally treated XG is about 904 Da/nm, close to the 1000 Da/nm value reported by Capron et al. (1997) and Terada et al. (2005) for single strand. Hence, heating transformed the XG conformation from random double helices to single strands with shorter lengths and thicknesses due to dissociation of the double helices in solution. The resultant flexibility of the single strands could lead to the coil conformation. The calculated Mark–Houwink coefficient α_2 was about 0.7 (Gulrez et al. 2012), which is less 1.0 for the perfectly rigid-rod double-helical ordered native XG (Capron et al. 1997; Terada et al. 2005). Therefore, the tested XG material is in a double-helical ordered form (i.e., XG_{Renat}), having greater flexibility compared to native XG due to heat treatment and subsequent cooling.

In summary, treatment methods have great influence of the XG conformation that can be described as follows (Fig. 2.6b). For heat treatment, the XG polymer chains dissociated and M_w reduced to half of its original value; however, R_g and η remain unchanged and M_L was measured at about 100 Da/nm (i.e., opening of double helix to single strand). For high-pressure homogenization, the XG polymer chains degraded, with the values of M_w, $R_{g,}$ and η reduced significantly; however, M_L remained around 1800 Da/nm. For autoclaving at 120 °C, simultaneous dissociation of the XG polymer chain (i.e., opening of helices) and degradation occurred, producing significant reductions in M_w, $R_{g,}$ and η; however, M_L remained around 1360 Da/nm. For irradiation in dilute aqueous solution, dissociation followed by degradation took place, causing significant reductions in M_w, R_g and η, as well as M_L, which reduced to about 1180 Da/nm.

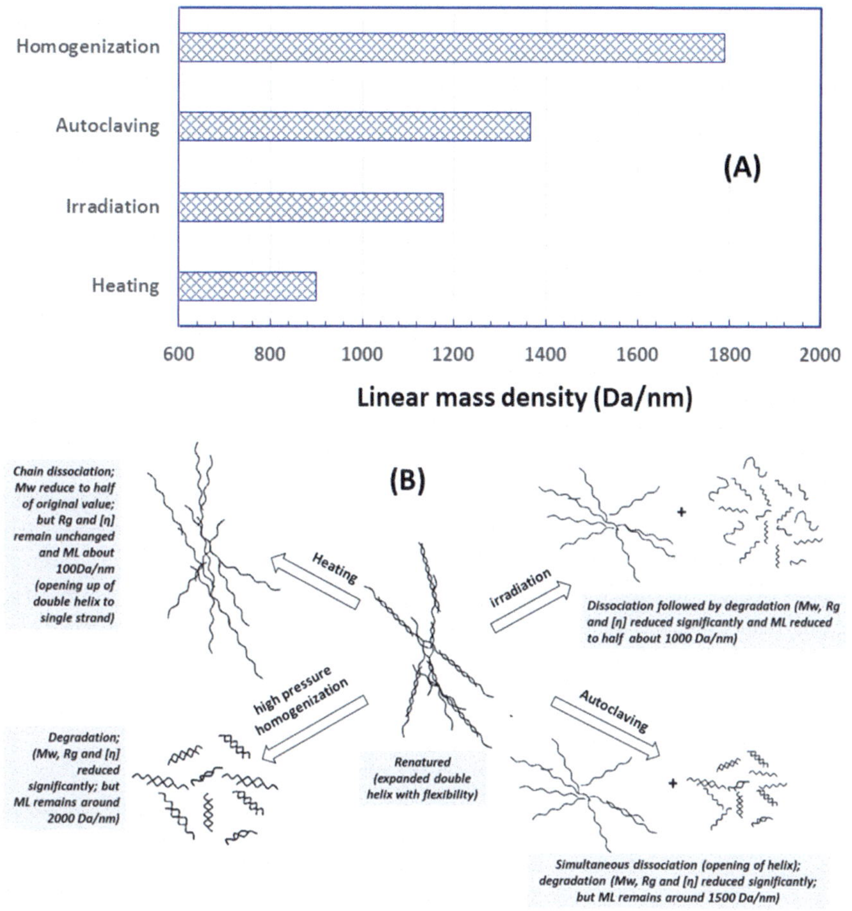

Fig. 2.6 Effects of heating methods on the M_L of XG at 4 mg/ml concentration in water and possible conformational changes (data from Gulrez et al. (2012))

It is worth noting that polymer-chain entanglement plays an important role in polymer rheological and mechanical properties (Sorbie 2013; Fried 2014). The molecular entanglement of the polymer chains, which depends on polymer concentration and molecular weight, causes the polymer molecules to form a transient network structure of different types of bonding. The molecular entanglement of the polymer is classified into different regimes: (i) dilute; (ii) semi-dilute untangled; (iii) semi-dilute entangled; (iv) concentrated untangled; and (e) concentrated entangled (Heo and Larson 2008). These regimes are differentiated by three concentration points, C^* (overlap concentration), C_e (entanglement concentration), and C^{**} (fully entanglement concentration), [$C^* < C_e <$ C**] (Fried 2014). In dilute solutions of concentration C_{ds} (i.e., less than C^*), the polymer chains are far apart and do not touch. For the semi-dilute unentangled regime of concentration C_{sdunt}, ranging between C^* and C_e, the polymer chains are closer to each other but still do not touch. Whereas for the semi-dilute entangled regime of concentration C_{sdent}, falling in the range of C_e to C^{**}, the hydrodynamic volumes begin to touch each other. While for the concentrated regime, the polymer concentration is greater than C^{**}, such that the chains are fully entangled.

The entanglement concentration C_e can be easily defined for polymers with high molecular weight (typically $M_w > 10^6$ g/mol) in comparison with that of lower molecular weight polymers (Graessley 1980). The overlap concentration C^* can be estimated using the following expression (Rubinstein and Colby 2003):

$$C^* = \frac{M_w}{\frac{4}{3}\pi N_A R_g^2} \tag{2.1}$$

where M_w = molecular weight; N_A = Avogadro's number; and R_g = radius of gyration.

It is worth noting that for a good solvent, the C_e that separates the semi-dilute unentangled and semi-dilute entangled regimes typically has a value of about ten times C^* (Graessley 1980).

The effects of heating on XG conformation can be evaluated using the R_g, the M_L, the polydispersity, and the Mark–Houwink–Kuhn–Sakurada formulae (see Fig. 2.7), as follows:

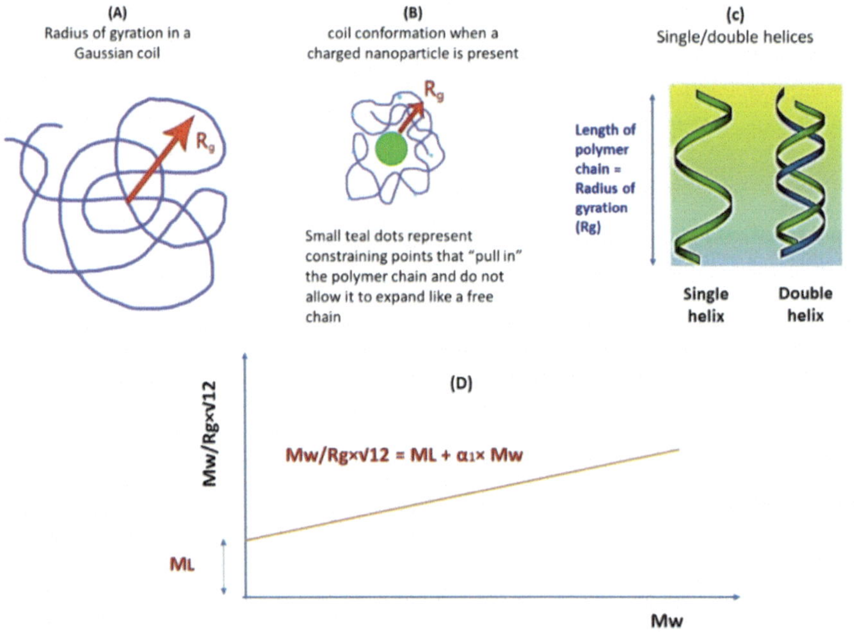

Fig. 2.7 Schematic representations of (**a**–**c**) the radius of gyration R_g, and (**d**) determining the M_L for polymers

(a) ***R_g determination***: Schematic representations of the R_g in a Gaussian coil (i.e., coil conformation when a charged ion is present), and in a single/double helix are shown in Fig. 2.7a–c. Notably, the presence of ionic species in solution tends to reduce the R_g and solution viscosity because salts have a screening effect on negative charges featuring on the polymer chain. For example, the R_g of FLOPAAM 3630 polyelectrolyte (SNF Floerger, Vaughan, Canada) reduced from about 350 nm (in distilled water) to 135 nm (in 1 mol/l of NaCl), implying a ~38% reduction (Zhang et al. 2008). The R_g can be determined experimentally with static light scattering, as well as with small angle neutron- and X-ray scattering, and is used to describe the polymer chain length.

(b) ***M_L calculation***: The M_L can be determined by knowing the weight-average molecular weight M_w and the R_g. The M_L can be determined by plotting $[M_w/R_g \times \sqrt{}12]$ on the y-axis against M_w on the x-axis and extrapolating to $M_w = 0$ (see Fig. 2.7d). The y-axis intercept gives the M_L for the polymer (Sato et al. 1984a), with the polymer approximating a perfect-rigid-rod shape as M_L approaches zero.

(c) ***Degree of polydispersity calculation***: The degree of polydispersity, which is used to describe the width of the molecular weight distribution, can be obtained from the ratio of the weight-average molecular weight M_w to the number-average molecular weight M_n.

(d) ***Determination of the XG conformation type***: The Mark–Houwink–Kuhn–Sakurada formulae define the relationship of molecular weight to intrinsic viscosity $[\eta = k'(M_w)^{\alpha 1}]$ and radius of gyration $[R_g = k''(M_w)^{\alpha 2}]$. Note that the coefficients α_1 and α_2 determine the conformation of the polymeric chain (Tombs and Hardings 1998). For compact spherical conformation, α_1 is 0, and α_2 is about 0.33. For random coil, α_1 ranges between 0.5 and 0.8, and α_2 falls between 0.5 and 0.6. For rigid-rod conformation, α_1 is about 1.8 and α_2 is approximately 1.0.

2.3.3 Effect of Manufacturing Method

Based on the atomic forced microscopy image analysis and the chain height measurements of the deposited XG onto a poly-l-lysine-coated mica surface during manufacturing, it was reported (Gunning et al. 1996) that the method by which XG is deposited on the surface (drop or dilute deposition) affects the XG structural conformation. For instance, when XG is dropped on the surfaces (i.e., drop deposition), it showed a relatively linear appearance of the molecular chains that are rigid. The height measurements indicated that in thicker regions, the polymer chain height is 1.883 nm, which is quite close to the reported XG helical diameter of about 2 nm. Whereas in the thinner regions, the chain height is approximately 0.886 nm (i.e., almost half of that in the thicker regions). In other words, these observations suggest the presence of a combination of molten single-helical structures in the thinner regions, and a double-helical structure in the thicker regions. However, when a dilute XG solution is deposited onto the surface, the molecular chain height was about 0.178 nm, which is less than that of the drop deposition case. Moreover, there were sharp curves along the chain, suggesting to its significantly higher flexibility.

The preceding highlights the impact of the deposition methods on the XG conformation that can be explained as follows. Incubation of dilute XG solution onto poly-l-lysine coated mica will result in almost immediate electrostatic binding of the negatively charged XG molecules onto the positively charged poly-l-lysine layer. This traps the solution conformation of some of the XG molecules on the coated mica surface, with any remaining unbound molecules removed by the water rinse. Whereas drop deposition causes concentration of the XG solution as the solvent evaporates. Hence, the XG molecules are likely to remain mobile, as well as hydrated until most of the water is lost, at which point the molecules finally become trapped on the mica surface by physisorption. Therefore, for the drop deposition method, the molecular mobility coupled with the increasing XG concentration induces the renatured XG (i.e., XG_{Renat}) helical conformation once dried; while, for the dilute XG solution, the electrostatically bound XG molecules appear to be trapped in the random coil conformation of denatured XG (i.e., XG_{Denat}).

2.3.4 Effect of Electrolyte Concentration

XG is highly soluble in both cold and hot water because of the polyelectrolyte nature of the XG molecule. However, intensive agitation is needed to prevent the formation of lumps. XG solutions are highly viscous, even at low polymer concentrations, which resists flow. XG solutions are pseudoplastic, or shear thinning, and the viscosity decreases with increasing shear rate. The viscosity also depends on temperature (both dissolution and measurement temperatures), the biopolymer concentration, concentration of salts and pH level. Commercial XG polymer is usually sold in powder form, which has a moisture content of about 10% by weight and an ash content of about 5–10% by weight. The difficulty of describing the molecular conformation of XG to measured solution properties is attributable to the fact that sample preparation is sensitive to both culture conditions and isolation methods (Milas et al. 1990). Other typical properties of XG are provided in Table 2.2.

Studies on the behavior of XG in distilled water indicate that it has a highly disordered (coiled) conformation because the anionic side chains repel each other (Fig. 2.8a). However, on the addition of a small amount of electrolyte (NaCl), the electrostatic repulsion among the side chains is reduced, allowing the side chains to wrap around and hydrogen bond to the backbone, such that the polymer chain straightens into a relatively rigid helical rod (Carrington et al. 1996). For dilute aqueous XG solutions with 0.1-M NaCl, formation of additional junction zones between XG chains based on the duplex motif of the native XG double strand may occur (Sato et al. 1984a, b, c). However, at high salt concentration, aggregation may take place through entanglements and the XG will acquire rod-like conformation due to increased attraction forces. In the presence of electrolyte, a colloidal network is

Table 2.2 Typical physical properties of commercial XG biopolymer (adopted from Garcia-Ochoa et al. (2000))

Property	Value/description
Physical state	Cream-colored powder
Moisture content (%)	8–15
Ash content (%)	7–12
Nitrogen (%)	0.3–1
Acetate content (%)	1.9–6.0
Pyruvate content (%)	1.0–5.7
Monovalent salts (g/l)	3.6–14.3
Divalent salts (g/l)	0.085–0.17
Viscosity (cP) (15.8 1/s, $C_P = 1$ g/l, $T_D = 25$ °C, $T_M = 25$ °C)	13–35

C_P = NaCl concentration; T_D = dissolution temperature; and T_M = measurement temperature

understood to be formed due to intermolecular hydrogen bonding among the helical rod segments and the limited polymer entanglement (Fig. 2.8b). This network of entangled stiff molecules controls the rheological properties of XG aqueous solutions.

The interactional mechanisms among XG helices in H_2O aqueous medium (Rwei and Nguyen 2014) are shown in Fig. 2.8c. The system might contain double-stranded right-handed helices of XG (ordered five-fold segments) connected by a single helix (disordered sequence segment). The double helices of XG are stiff and oriented by many intramolecular hydrogen bonds. Because of the high attractive forces between the semi-flexible disordered chains and the double helices, a mesophase (a state of matter intermediate between liquid and solid) is formed (e.g., gelatin forms a partially ordered structure in a mesophase).

2.3.5 *Effect of pH*

In a highly acidic environment, the XG chains become uncharged and partially flexible due to the protonation of the anionic carboxylic groups (Oh et al. 1999). However, in a weak acid environment, the carboxylic groups become partially protonated (pH 4) or completely deprotonated (pH 6) (Bueno et al. 2013; Bueno and Petri 2014; Shiroodi and Lo 2015). Brunchi et al. (2016) found that at pH 5, the electrostatic repulsive forces between $-COO^-$ groups located on the side chain of XG contributed to XG chain expansion and an increase of the solution viscosity at low XG concentrations; however, as XG concentration increases, the viscosity decreases.

Furthermore, the experimental results concerning the phase formation of XG/H_2O/H_3PO_3 and XG/H_2O/H_3PO_4 tertiary systems (Rwei and Nguyen 2014) have shown that there are three distinct phases and a transition region, i.e., a completely separated phase, a liquid-crystalline miscible phase, an isotropically miscible phase,

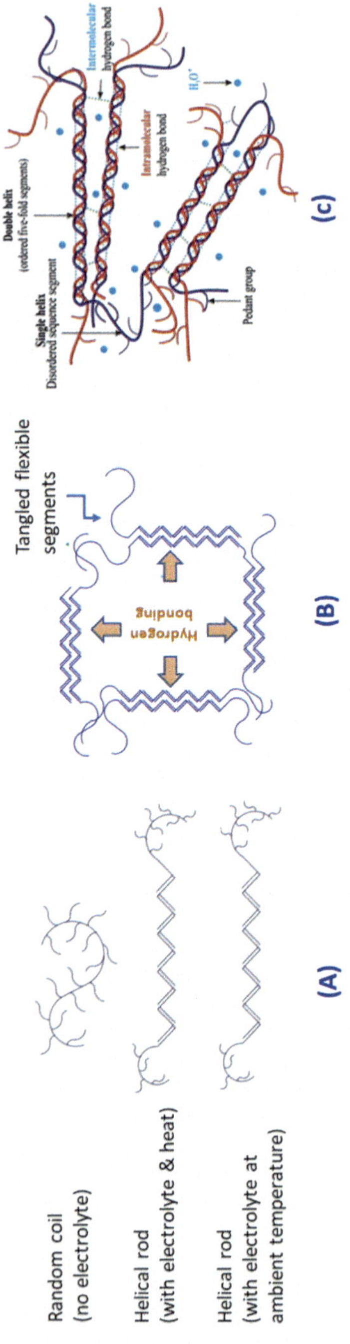

Fig. 2.8 Effects of electrolyte on XG molecular conformation: **a** highly disordered (coiled); **b** ordered with tangled-flexible segments; and **c** possible interaction mechanisms among XG helices in H_2O aqueous medium

and the separated/liquid-crystalline region. These phases are highly dependent on the XG and acid (H_3PO_3 or H_3PO_4) concentrations, and the mixture temperature. At room temperature, for XG concentrations < 4% (mass of XG/volume of the aqueous solution) and any acid concentration, the system is isotropically miscible. Furthermore, at high temperatures (25–92 °C) the system will continue to be isotropically miscible for XG concentrations < 3% with various acid concentrations. Whereas for the liquid-crystalline phase, the chain rearrangement of the rod-shaped molecules would take place at a critical XG concentration (Collings and Hird 1997; Sperling 2006), which, for instance, was determined as 4% by Rwei and Nguyen (2014).

The possible interactional mechanisms among XG helices in H_2O, and H_3PO_3 or H_3PO_4 aqueous medium (Rwei and Nguyen 2014) are shown in Fig. 2.9. Due to the presence of acids (H_3PO_3 and H_3PO_4), hydrogen bonds are broken (i.e., broken interactions) and double segments in the XG double helices are deformed (Rwei and Nguyen 2014). However, Stokke and Christensen (1996) have indicated that the XG/HCl or XG/H_2SO_4 system exhibited acid hydrolysis, which resulted in partial truncation of the pedant groups of XG helices and reduction in the molecular weight. Since the H_3PO_3 and H_3PO_4 acids are moderate, no acid hydrolysis at glycosidic linkages or esterification at the hydroxyl groups has been developed for acid concentrations of less than 60% because of the original rigid configurations of the XG helices. However, at higher concentrations close to 60%, the XG helices would be deformed into disordered phases due to the breakup of both intramolecular and intermolecular bonds (Fig. 2.9).

The effects of both esterification and acid hydrolysis on the functional groups of XG molecules induced by the acids H_3PO_3 and H_3PO_4 have been evaluated by Rwei and Nguyen (2014) using FTIR spectra. For XG in 30% acids (H_3PO_3 and H_3PO_4), the P–H stretching vibration peaks were observed at a wavelength of 2426.22 cm^{-1}. Also, the vibrations of HPO_3^{2-} and PO_4^{3-} functional groups were shown at peaks of 525.56 and 494.60 cm^{-1} for H_3PO_3 and H_3PO_4 acids, respectively. The sharp peak for PO_4^{3-} indicated the strong interaction between XG and H_3PO_4 acid.

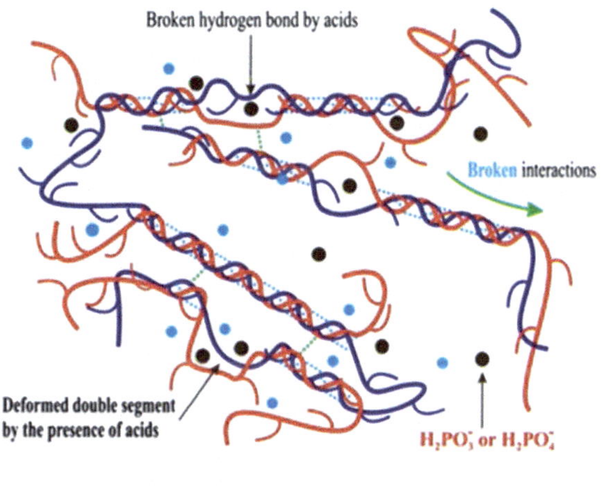

Fig. 2.9 Illustration of interactional mechanism among XG helices in H_3PO_3 or H_3PO_4 aqueous medium (adopted from Rwei and Nguyen (2014))

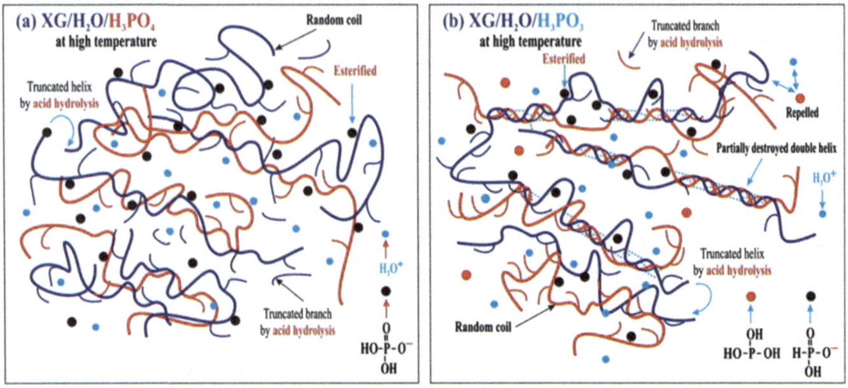

Fig. 2.10 Conformational responses of XG molecules to high temperature for **a** XG/H_3PO_4 system and **b** XG/H_3PO_3 system (adopted from Rwei and Nguyen (2014))

2.3.6 *Effect of pH and Temperature*

At high temperatures, the helices would dissociate in the network due to truncating and separating molecules leading to an isotropic behavior or to random coil formations (Fig. 2.10). Esterification and acid hydrolysis at high temperatures occur more strongly in the H_3PO_4 system (than in the H_3PO_3) due to the high degree of both oxidation and anionic concentration in the mixture.

2.3.7 *Effect of Carbon Source*

Carbon is a very important ingredient for microbial cell growth, in addition to nutrients and other growth factors. Variations in the nutrients and the substrate concentrations affect the side chain structure, molecular mass, and yield of the XG product; however, they do not affect the backbone of XG. Note that glucose and sucrose are the most common carbon sources in any fermentation medium. De-acylation affects the rheological properties by making the XG mixture behave like a gel, while it is liquid-like if the XG mixture is in its native state because destabilization of the double-helical structure increases flexibility of the side chain.

2.4 Surface Charges

Determination of zeta potential, ζ, is very useful to know the magnitude of the XG particle surface charge, as well as the stability of the XG particle in the aqueous system. A system having a zeta potential outside the limits of -30 to $+30$ mV is

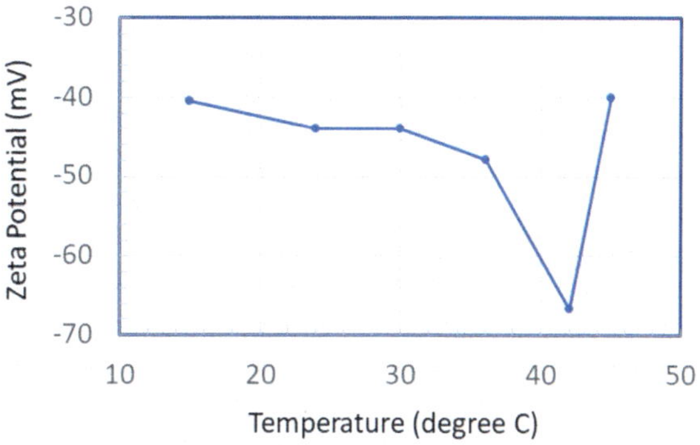

Fig. 2.11 Variation of zeta potential with temperature for 100 mg/l XG in water (data from Brunchi et al. (2016))

considered stable (i.e., the particles repel each other, and no agglomeration takes place). However, if the zeta potential varies between these two values, the system is considered unstable (i.e., the electrostatic repulsion between particles is insufficient to prevent agglomeration) (Nita et al. 2011; Morariu et al. 2012). The variation of zeta potential with temperature for XG-water aqueous solution is shown in Fig. 2.11. A small decrease of the zeta potential value (from −40.7 to −47.9 mV) was observed for increasing temperature from 15 to 36 °C. However, for temperatures above 36 °C, the zeta potential reached a minimum of − 66.7 mV at 42 °C. This behavior is attributed to the conformational changes of the XG chains that take place in the temperature range of 33–45 °C, which can be ascribed to the intensification of electrostatic repulsive interactions between the charged carboxylic groups. In other words, at higher temperature, the counterions are far away from the poly-ion and the chains become more extended (Bordi et al. 1996).

2.5 Activation Energy

The calculated flow activation energies E_t (in kJ/mol), based on viscosity variations with temperature T and XG concentration, are shown in Fig. 2.12. Since XG-water solutions contain secondary bonding, such as hydrogen bonding, the plot of the natural logarithm of viscosity [ln η] versus [1/T] is not a straight line and contains two linear dependences. Irrespective of the XG concentration, the flow activation energies below of the inflection points were almost the same as that of water (about 15 kJ/mol). However, above the inflection points, the flow activation energies were lower except for that obtained at an XG concentration of 100 mg/l. As the XG concentration in solution increases, the required flow activation energy of the polymer chain gradually decreases, since at very low XG concentration, the flow energy is controlled by the water molecules movements in solution (Brunchi et al. 2016). Notably, the flow activation energy for XG with high concentration (100 mg/l) is almost doubled.

As temperature increases, the flow activation energy decreases and then increases, because temperature increase induces a conformational transition of XG chains from

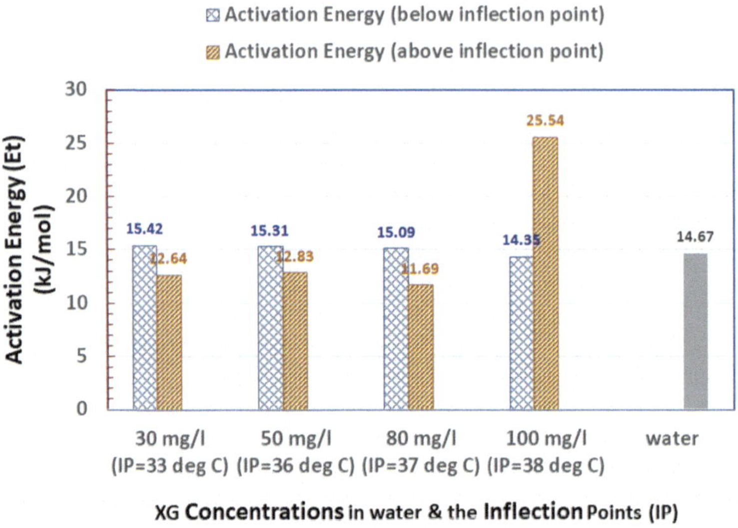

Fig. 2.12 Activation energy of xanthan gum (XG) solutions in water at various polymer concentrations based on viscosity measurements (data from Brunchi et al. (2016))

a rigid state (with lower viscosity in aqueous solution) to a more flexible conformation in a disordered state (with higher viscosity). It has also been reported (Paoletti et al. 1983; Jones et al. 1987; Liu et al. 1987; Rochefort and Middleman 1987; Bordi et al. 1996) that the chain conformation of XG in solution varies from a rigid and ordered chain at room temperature to a less ordered structure derived from a partial or complete dissociation of the strands at higher temperatures, giving rise to clusters involving different numbers of polymer chains. In addition, at lower temperatures (< 25 °C), Jones et al. (1987) reported that XG has a single-helix chain conformation structure that progresses to an extended random coil at higher temperatures (60 °C). Furthermore, at temperatures of greater than 30–35 °C, Bordi et al. (1996) reported that an aggregation process occurs with a partial separation of the double-stranded structure, while below the transition temperature, a single-helix configuration in obtained.

However, for XG concentration of 100 mg/l, the calculated activation energy associated with the electrophoretic migration of the charged particles was reported as being 15.55 and 68.77 kJ/mol for temperatures of up to 36 °C and above 36 °C, respectively (Brunchi et al. 2016). As the temperature increases, the charged particles' mobility is hindered by the aggregates formed by the XG chains; hence, a higher energy is required to mobilize the particles. Notably, the viscosity-based activation energy calculations are consistent with those calculated using electrophoretic mobility data for temperatures below 36 °C; but above this temperature, the electrophoretic calculations show almost triple that of the viscosity-based activation energy calculations. This difference could be due to the different manner by which the activation energy is calculated. In other words, the kinematic viscosity gives information about the flow of the chains under gravity, whereas electrophoretic mobility represents the velocity of a charged species' movement relative to the medium by applying an electric field (Oddy and Santiago 2004).

2.6 Biodegradation Potential

Since XG is a naturally occurring microbial polysaccharide, it can be biodegraded into oligosaccharides, monosaccharides, H_2O, and CO_2. The depolymerization of the polysaccharide is achieved by enzymatic attack (xanthanase, cellulase, and xanthan lyase) (Riaz et al. 2021). The xanthanase enzyme works on the main chain of XG without affecting the side chains. The cellulase enzyme attacks the backbone of XG and can access the backbone if the side chain was degraded; hence the XG will take a disordered form. The xanthan-lyase enzyme attacks the side chain of XG, causing degradation of linkages between XG molecules.

In addition, it has been reported that XG can be biodegraded within two days, but it can resist the attack of common cellulases (Abu Elella et al. 2020). The trisaccharide side chain of XG was found to be relatively stable against some enzymatic attacks. For example, when XG is attacked by the fungal cellulases, its cleavage backbone will be elevated in the case of the disordered helical structural form; however, its ordered form will not be affected by such an attack (Rinaudo and Milas 1980; Katzbauer 1998). Furthermore, it has been reported that XG was completely depolymerized into monosaccharides by *Bacillus* sp. strain GL1 (Nankai et al. 1999).

Therefore, the rate of biodegradation depends upon the presence of specific microorganisms that produce the required enzymes, as well as various environmental conditions (e.g., temperature, pH level and ionic concentration) (Hovland 2015). In addition, acetyl and pyruvate groups play a crucial role in the hydrolysis of XG—that is, pyruvate contributes to the viscosity, while the acetyl group contributes to the gel property. Pyruvate groups present on the terminal of XG inhibit the enzyme to attack the main chain.

Overall, however, the reported results are conflicting regarding the biodegradation potential of XG; some authors report degradation potential by bacteria present in the activated sludge within seven days (Muchova et al. 2009), while others observed a certain resistance of XG to microbial degradation up to several months (Hovland 2015). The biodegradation potential of XG (loss in its functional attributes) is an important property that mandates further in-depth research, particularly in the context that XG is being proposed/used for soil stabilization purposes, to better understand the longevity and hence durability of the XG-treated soils.

2.7 Summary and Concluding Remarks

The conformational ordering of XG is very controversial. In a study by Moorhouse et al. (1977), the X-ray diffraction patterns obtained from oriented fibers in the solid state showed that these fibers have a five-fold helix structure (i.e., pentagonal in cross section) with a pitch of 4.7 nm. However, in another study by Holzwarth and Prestridge (1977), the data was analyzed using electron microscopy, and it was shown that the disordered XG had strands of about 2 nm in width, whereas the ordered conformation had a width of about 4 nm, indicating a multistranded structure, most probably a double helix.

Furthermore, in a study by Okuyama et al. (1980), it was hypnotized that: (i) multi-stranded structures are sterically feasible (i.e., contrary to the findings by Moorhouse et al. (1977)); and (ii) a 51 double helix with the two strands running antiparallel to one another does exist, similar to the original single-helix model that shows the polymer chains are highly extended (i.e., not wound together in a tight "spiral staircase" arrangement). On the other hand, from the first-order kinetics, it was demonstrated the presence of good linearity; hence intramolecular conversion from disorder to order (i.e., coil → single helix) does occur, meaning that single-helix formation is the rate-limiting process in conformational ordering of XG.

As a result, to these conflicting results, Morris (2019) hypothesized that the conformational ordering of XG is a two-step process, described as follows:

- ***Step 1***: Conversion of disordered coils to single helices, which is stabilized by ordered packing of side chains along the polymer backbone, and the conversion process can be expressed by first-order kinetics. However, under certain conditions, the ordering process stops after step 1, because a disorder–order transition could occur without an increase in molecular weight.
- ***Step 2***: Formation of dimers, whereby both the molecular weight and the molar mass per unit contour length are doubled. The conformation of the dimers might be represented by a 5_1 antiparallel double helix.

Depending on salinity and temperature, the secondary structure (i.e., backbone conformation) is known to undergo an order to disorder (helix → coil) transition. In distilled water at 25 °C, the backbone is disordered (or partially ordered in the form of a randomly broken helix), but highly extended due to the electrostatic repulsions from the charged groups on the side chains. Due to the highly extended structure, the molecules may align and associate (due to hydrogen bonding) to form a weakly structured material. In this state, as the temperature is increased, a transition to a coil-like configuration occurs, which causes a dissociation of the molecules and a subsequent change in the rheological properties.

When salt is added to the solution at 25 °C, a disorder to order transition takes place, whereby the backbone takes on a helical conformation and the charged trisaccharide side chains collapse down onto the backbone (due to charge screening effects) and stabilize the ordered conformation. In this rod-like configuration, with the electrostatic repulsions screened, the molecules are easily aligned and strongly associate with each other (due to hydrogen bonding and ionic interactions) to form a gel-like structure. As the temperature of the solution is increased, this stabilized ordered form shifts to a disordered conformation (helix → coil) at higher temperatures (80–100 °C).

References

Abbaszadeh A, Lad M, Janin M, Morris G, MacNaughtan W, Sworn G, Foster T (2015) A novel approach to the determination of the pyruvate and acetate distribution in xanthan. Food Hydrocolloids 44:162–171. https://doi.org/10.1016/j.foodhyd.2014.08.014

Abu Elella MH, Goda ES, Gab-Allah MA, Hong SE, Pandit B, Lee S, Gamal H, Rehman A, Yoon KR (2020) Xanthan gum-derived materials for applications in environment and eco-friendly materials: a review. J Environ Chem Eng. https://doi.org/10.1016/j.jece.2020.104702

Ahmad NH, Mustafa S, Che Man YB (2014) Microbial polysaccharides and their modification approaches: a review. Int J Food Prop 18(2):332–347. https://doi.org/10.1080/10942912.2012.693561

Alaa FM (2018) Effectiveness of exopolysaccharides and biofilm forming plant growth promoting rhizobacteria on salinity tolerance of faba bean (Vicia faba L.). Afr J Microbiol Res 12(17):399–404. https://doi.org/10.5897/AJMR2018.8822

Al-Assaf S, Phillips GO, Deeble DJ, Parsons B, Starnes H, Von Sonntag C (1995) The enhanced stability of the cross-linked hylan structure to hydroxyl (OH) radicals compared with the uncross-linked hyaluronan. Radiat Phys Chem 46(2):207–217. https://doi.org/10.1016/0969-806X(95)00015-P

Baumgartner S, Pavli M, Kristl J (2008) Effect of calcium ions on the gelling and drug release characteristics of xanthan matrix tablets. Eur J Pharm Biopharm 69(2):698–707. https://doi.org/10.1016/j.ejpb.2007.11.015

Berg JM, Tymoczko JL, Stryer L, Berg J, Tymoczko J, Stryer L (2002) Biochemistry, 9th edn. WH Freeman, 1319114679, 9781319114671

Berninger T, Dietz N, Lopez OG (2021) Water-soluble polymers in agriculture: xanthan gum as eco-friendly alternative to synthetics. Microb Biotechnol 14(5):1881–1896. https://doi.org/10.1111/1751-7915.13867

Besio GJ, Leavesley IM, Prud'homme RK, Farinato R (1987) Electric birefringence measurements of native, denatured, and renatured xanthan polysaccharide. J Appl Polym Sci 33(3):825–834

Bordi F, Cametti C, Paradossi G (1996) Conformational changes of xanthan in salt-free aqueous solutions: a low-frequency electrical conductivity study. J Phys Chem 100(17):7148–7154

Brant DA, Hacche LS, Washington GE (1985) A light-scattering study of the temperature-driven conformation change in xanthan. Abs Pap Am Chem Soc 190(SEP):86-MBD

Bresolin TM, Milas M, Rinaudo M, Ganter JL (1998) Xanthan–galactomannan interactions as related to xanthan conformations. Int J Biol Macromol 23(4):263–275. https://doi.org/10.1016/s0141-8130(98)00061-0

Brunchi CE, Morariu S, Bercea M (2014) Intrinsic viscosity and conformational parameters of xanthan in aqueous solutions: salt addition effect. Colloids Surf B Biointerfaces 122:512–519. https://doi.org/10.1016/j.colsurfb.2014.07.023

Brunchi CE, Bercea M, Morariu S, Dascalu M (2016) Some properties of xanthan gum in aqueous solutions: effect of temperature and pH. J Polym Res 23:123. https://doi.org/10.1007/s10965-016-1015-4

Bueno VB, Petri DFS (2014) Xanthan hydrogel films: molecular conformation, charge density and protein carriers. Carbohyd Polym 101:897–904. https://doi.org/10.1016/j.carbpol.2013.10.039

Bueno VB, Bentini R, Catalani LH, Petri DFS (2013) Synthesis and swelling behavior of xanthan-based hydrogels. Carbohyd Polym 92(2):1091–1099. https://doi.org/10.1016/j.carbpol.2012.10.062

Cadmus M, Rogovin S, Burton K, Pittsley J, Knutson C, Jeanes A (1976) Colonial variation in *Xanthomonas campestris* NRRL B-1459 and characterization of the polysaccharide from a variant strain. Can J Microbiol 22(7):942–948. https://doi.org/10.1139/m76-136

Cadmus MC, Jackson LK, Burton KA, Plattner RD, Slodki ME (1982) Biodegradation of xanthan gum by *Bacillus sp*. Appl Environ Microbiol 44(1):5–11. https://doi.org/10.1128/aem.44.1.5-11.1982

Callet F, Milas M, Rinaudo M (1987) Influence of acetyl and pyruvate contents on rheological properties of xanthan in dilute solution. Int J Biol Macromol 9(5):291–293. https://doi.org/10.1016/0141-8130(87)90068-7

Callet F, Milas M, Rinaudo M (1989) On the role of thermal treatments on the properties of xanthan solutions. Carbohydr Polym 11(2):127–137. https://doi.org/10.1016/0144-8617(89)90020-9

Capron I, Yvon M, Muller G (1996) In vitro gastric stability of carrageenan. Food Hydrocolloids 10(2):239–244. https://doi.org/10.1016/S0268-005X(96)80040-3

Capron I, Brigand G, Muller G (1997) About the native and renatured conformation of xanthan exopolysaccharide. Polymer 38(21):5289–5295. https://doi.org/10.1016/S0032-3861(97)00079-7

Capron I, Alexandre S, Muller G (1998) An atomic force microscopy study of the molecular organisation of xanthan. Polymer 39(23):5725. https://doi.org/10.1016/S0032-3861(97)10344-5

Carrington S, Odell J, Fisher L, Mitchell J, Hartley L (1996) Polyelectrolyte behavior of dilute xanthan solutions: salt effects on extensional rheology. Polymer 37(13):2871–2875. https://doi.org/10.1016/0032-3861(96)87653-1

Chazeau L, Milas M, Rinaudo M (1995) Conformations of xanthan in solution: analysis by steric exclusion chromatography. Int J Polym Anal Charact 2:21–29

Cheetham NW, Mashimba EN (1992) Proton and carbon-13 NMR studies on xanthan derivatives. Carbohydr Polym 17(2):127–136. https://doi.org/10.1016/0144-8617(92)90106-Z

Chen X, Wang M, Yang F, Tang W, Li X (2014) Isolation and characterization of xanthan-degrading *Enterobacter* sp. nov. LB37 for reducing the viscosity of xanthan in petroleum industry. World J Microbiol Biotechnol 30(5):1549–1557. https://doi.org/10.1007/s11274-013-1578-8

Collings PJ, Hird M (1997) Introduction to liquid crystals: chemistry and physics, 1st edn. Taylor & Francis, London

Dai X, Gao G, Wu M, Wei W, Qu J, Li G, Ma T (2019) Construction and application of a *Xanthomonas campestris* CGMCC15155 strain that produces white xanthan gum. Microbiology 8(2):631. https://doi.org/10.1002/mbo3.631

de Carvalho CCCR (2017) Biofilms: microbial strategies for surviving UV exposure. Adv Exp Med Biol 996:233–239. https://doi.org/10.1007/978-3-319-56017-5_19

Djenar NS, Mulyono EWS (2018) The effect of growth medium composition on *X. campestris* metabolism in producing xanthan gum. MATEC Web Conf 156:01012. https://doi.org/10.1051/matecconf/201815601012

Dos Santos ALS, Galdino ACM, de Mello TP, Ramos LdS, Branquinha MH, Bolognese AM et al (2018) What are the advantages of living in a community? A microbial biofilm perspective! Mem Inst Oswaldo Cruz 113(9):180212. https://doi.org/10.1590/0074-02760180212

Faria S, de Oliveira Petkowicz CL, de Morais SAL, Terrones MGH, de Resende MM, de França FP, Cardoso VL (2011) Characterization of xanthan gum produced from sugar cane broth. Carbohydr Polym 86(2):469–476. https://doi.org/10.1016/j.carbpol.2011.04.063

Fei B, Wach RA, Mitomo H, Yoshii F, Kome T (2000) Hydrogel of biodegradable cellulose derivatives. I. Radiation-induced crosslinking of CMC. J Appl Polym Sci 78(2):278–283. https://doi.org/10.1002/1097-4628(20001010)78:2<278::AID-APP60>3.0.CO;2-9

Fiume MM, Heldreth B, Bergfeld WF, Belsito DV, Hill RA, Klaassen CD, Liebler DC, Marks JG Jr, Shank RC, Slaga TJ (2016) Safety assessment of microbial polysaccharide gums as used in cosmetics. Int J Toxicol 35(1 Suppl):5S-49S. https://doi.org/10.1177/1091581816651606

Flemming HC, Wingender J, Szewzyk U, Steinberg P, Rice SA, Kjelleberg S (2016) Biofilms: an emergent form of bacterial life. Nat Rev Microbiol 14:563–575. https://doi.org/10.1038/nrmicro.2016.94

Fried JR (2014) Polymer science and technology. Pearson Education Inc. ISBN 978-0-13-703955-5

Gallie DR (2013) L-ascorbic acid: a multifunctional molecule supporting plant growth and development. Scientifica (Cairo) 2013(1):795964. https://doi.org/10.1155/2013/795964

Gansbiller M, Schmid J, Sieber V (2019) In-depth rheological characterization of genetically modified xanthan-variants. Carbohydr Polym 213:236–246. https://doi.org/10.1016/j.carbpol.2019.02.055

Garcia-Ochoa F, Santos VE (2000) Oxygen transfer and uptake rates during xanthan gum production. Enzym Microb Technol 27(9):680–690. https://doi.org/10.1016/s0141-0229(00)00272-6

Garcıa-Ochoa F, Santos V, Casas J, Gomez E (2000) Xanthan gum: production, recovery, and properties. Biotechnol Adv 18(7):549–579. https://doi.org/10.1016/S0734-9750(00)00050-1

Graessley W (1980) Polymer chain dimensions and the dependence of viscoelastic properties on concentration molecular weight and solvent power. Polymer 21(3):258–262. https://doi.org/10.1016/0032-3861(80)90266-9

Gulrez SKH, Al-Assa, S, Phillips GO (2011) Hydrogels: methods of preparation, characterization and applications. In: Carpi A (ed) Progress in molecular and environmental bioengineering—from analysis and modeling to technology applications. Intech Open, London. https://doi.org/10.5772/24553

Gulrez SKH, Al-Assaf S, Fang Y, Phillips GO, Gunning AP (2012) Revisiting the conformation of xanthan and the effect of industrially relevant treatments. Carbohydr Polym 90(4):1235–1243

Gunning AP, Kirby AR, Morris VJ (1996) Imaging xanthan gum in air by ac "tapping" mode atomic force microscopy. Ultramicroscopy 63(1):1–3. https://doi.org/10.1016/0304-3991(96)00032-0

Gunning AP, Cairns P, Kirby AR, Round AN, Bixler HJ, Morris VJ (1998) Characterising semi-refined iota-carrageenan networks by atomic force microscopy. Carbohydr Polym 36(1):67–72. https://doi.org/10.1016/S0144-8617(97)00112-4

Habibi H, Khosravi-Darani K (2017) Effective variables on production and structure of xanthan gum and its food applications: a review. Biocatal Agric Biotechnol 10(4):130–140. https://doi.org/10.1016/j.bcab.2017.02.013

Hacche LS, Washington GE, Brant DA (1987) Light-scattering investigation of the temperature-driven conformation change in xanthan. Macromolecules 20(9):2179–2187

Hatakenaka K, Liu W, Norisuye T (1987) Stability of xanthan in aqueous sodium-chloride at elevated-temperature. Int J Biol Macromol 9(6):346–348. https://doi.org/10.1016/0141-8130(87)90007-9

Heo Y, Larson R (2008) Universal scaling of linear and nonlinear rheological properties of semidilute and concentrated polymer solutions. J Rheol 49(5):1117–1128. https://doi.org/10.1021/MA800521G

Holzwarth G (1976) Conformation of the extracellular polysaccharide of *Xanthomonas campestris*. Biochemistry 15(19):4333–4339. https://doi.org/10.1021/bi00664a030

Holzwarth G, Prestridge EB (1977) Multistranded helix in xanthan polysaccharide. Am Assoc Adv Sci 197(4305):757–759. https://doi.org/10.1126/science.887918

Hovland B (2015) Assessment of the biodegradability of xanthan in offshore injection water. Master's Thesis in Microbiology, Beate Hovland, University of Bergen

Jansson PE, Kenne L, Lindberg B (1975) Structure of the extracellular polysaccharide from *Xanthomonas campestris*. Carbohydr Res 45(1):275–282. https://doi.org/10.1016/S0008-6215(00)85885-1

Jones SA, Goodall DM, Cutler AN, Norton IT (1987) Application of conductivity studies and polyelectrolyte theory to the conformation and order-disorder transition of xanthan polysaccharide. Eur Biophys J 15:185–191. https://doi.org/10.1007/BF00263683

Katzbauer B (1998) Properties and applications of xanthan gum. Polym Degrad Stab 59(1–3):81–84. https://doi.org/10.1016/S0141-3910(97)00180-8
Kawakami K, Norisuye T (1991) 2nd virial-coefficient for charged rods—sodium xanthan in aqueous sodium-chloride. Macromolecules 24(17):4898–4903
Kent AG, Garcia CA, Martiny AC (2018) Increased biofilm formation due to high-temperature adaptation in marine Roseobacter. Nat Microbiol 3(9):989–995. https://doi.org/10.1038/s41564-018-0213-8
Kulicke WM, van Eikern A (1992) Determination of the microstructure of the fermentation polymer xanthan. Makromol Chem Macromol Symp 61(1):75–93
Kumar A, Rao KM, Han SS (2018) Application of xanthan gum as polysaccharide in tissue engineering: a review. Carbohydr Polym 180:128–144. https://doi.org/10.1016/j.carbpol.2017.10.009
Lagoueyte N, Paquin P (1998) Effects of microfluidization on the functional properties of xanthan gum. Food Hydrocolloids 12(3):365–371. https://doi.org/10.1016/S0268-005X(98)00004-6
Liu W, Sato T, Norisuye T, Fujita H (1987) Thermally induced conformational change of xanthan in 0.01M aqueous sodium chloride. Carbohydr Res 160:267–281. https://doi.org/10.1016/0008-6215(87)80317-8
Lo YM, Argin-Soysal S, Hsu CH (2007) Bioconversion of whey lactose into microbial exopolysaccharides. In: Yang ST (ed) Bioprocessing for value-added products from renewable resources. Elsevier, Amsterdam, ch 22, pp 559–583
Lopes De Monaco B, Lopes Lessa V, More Silva B, Da Sila Carvalho Filho MA, Schnitzler E, Gustavo Lacerda L (2015) Xanthan gum: properties, production conditions, quality and economic perspective. J Food Nutr Res 54(3):185–194
Mehyar GF, Liu Z, Han JH (2008) Dynamics of antimicrobial hydrogels in physiological saline. Carbohydr Polym 74(1):92–98
Meléndez-Hevia E, Waddell TG, Cascante M (1996) The puzzle of the Krebs citric acid cycle: assembling the pieces of chemically feasible reactions, and opportunism in the design of metabolic pathways during evolution. J Mol Evol 43(3):293–303. https://doi.org/10.1007/BF02338838
Melton LD, Mindt L, Rees DA, Sanderson GR (1976) Covalent structure of the extracellular polysaccharide from *Xanthomonas campestris*: evidence from partial hydrolysis studies. Carbohydr Res 46(2):245–257. https://doi.org/10.1016/s0008-6215(00)84296-2
Milas M, Rinaudo M (1979) Conformational investigation on the bacterial polysaccharide xanthan. Carbohydr Res 76:189–196. https://doi.org/10.1016/0008-6215(79)80017-8
Milas M, Rinaudo M (1984) On the existence of two different secondary structure for the xanthan in aqueous solutions. Polym Bull 12:507–514. https://doi.org/10.1007/BF00958656
Milas M, Rinaudo M (1986) Properties of xanthan gum in aqueous solutions—role of the conformational transition. Carbohydr Res 158:191–204. https://doi.org/10.1016/0008-6215(86)84017-4
Milas M, Tinland B (1990) Behaviour of xanthan in cadoxen. Carbohydr Polym 13(1):47–56. https://doi.org/10.1016/0144-8617(90)90050-3
Milas M, Viehweg H, Weiss A (1990) Structure and properties of xanthan versus fermentation time in emulsion system. Carbohydr Polym 13(2):119–131. https://doi.org/10.1016/0144-8617(90)90079-8
Milas M, Reed WF, Printz S (1996) Conformation and flexibility of native and re-natured xanthan in aqueous solutions. Int J Biol Macromol 18(3):211–221. https://doi.org/10.1016/0141-8130(95)01080-7
Moorhouse R, Walkinshaw MD, Arnott S (1977) Xanthan gum molecular conformation and interactions, vol 45. ACS Symposium Series, pp 90–102. In: Sandford PA, Laskin A (eds) Extracellular microbial polysaccharides. San Francisco, CA, USA: ACS Publications, ISBN 0-8412-0372-5
Morariu S, Brunchi CE, Bercea M (2012) The behavior of chitosan in solvents with different ionic strengths. Ind Eng Chem Res 51(39):12959–12966. https://doi.org/10.1021/ie301309g

Morris ER (1977) Molecular origin of xanthan solution properties. In: Sandford PA, Laskin A (eds) Extracellular microbial polysaccharides 45, 81. American Chemical Society, Washington, DC, ch 6, pp 81–89. https://doi.org/10.1021/bk-1977-0045.ch006

Morris VJ (2006) Bacterial polysaccharides. In: Stephen AM, Phillips GO, Williams PA (eds) Food polysaccharides and their applications. Taylor & Francis, London, pp 413–454

Morris ER (2019) Ordered conformation of xanthan in solutions and "weak gels": single helix, double helix—or both? Food Hydrocolloids 86:18–25. https://doi.org/10.1016/j.foodhyd.2017.11.036

Morris ER, Rees DA, Young G, Walkinshaw MD, Darke A (1977) Order-disorder transition for a bacterial polysaccharide in solution. A role for polysaccharide conformation in recognition between *Xanthomonas* pathogen and its plant host. J Mol Biol 110(1):1–16. https://doi.org/10.1016/S0022-2836(77)80095-8

Morrison NA, Clark R, Talashek T, Yuan CR (2004) New forms of xanthan gum with enhanced properties. Gums Stabilizers Food Ind 12(294):124–130. https://doi.org/10.1039/9781847551214-00124

Muchova M, Ruicka J, Julinova M, Dolezalova M, Houser J, Koutny M, Bunkova L (2009) Xanthan and gellan degradation by bacteria of activated sludge. Water Sci Technol 60(4):965–973. https://doi.org/10.2166/wst.2009.443

Nankai H, Hashimoto W, Miki H, Kawai S, Murata K (1990) Microbial system for polysaccharide depolymerization: enzymatic route for xanthan depolymerization by *Bacillus sp.* strain GL1. Appl Environ Microbiol 65(6):2520–2526. https://doi.org/10.1128/aem.65.6.2520-2526.1999

Narayanan V, Sanchezi Nogue V, van Niel EWJ, Gorwa-Grauslund MF (2016) Adaptation to low pH and lignocellulosic inhibitors resulting in ethanolic fermentation and growth of *Saccharomyces cerevisiae*. AMB Express 6(1):59. https://doi.org/10.1186/s13568-016-0234-8

Nishinari K, Watase M, Rinaudo M, Milas M (1996) Characterization and properties of gellan-kappa-carrageenan mixed gels. Food Hydrocolloids 10(3):277–283. https://doi.org/10.1016/S0268-005X(96)80002-6

Nita LE, Chiriac AP, Bercea M, Neamtu I (2011) Aspects concerning the temperature influence on the polymer/polymer interactions between poly(aspartic acid) and poly (ethylene glycol). Colloid Surf A Physicochem Eng Aspects 374:121–128. https://doi.org/10.1016/j.colsurfa.2010.11.036

Norton IT, Goodall DM, Frangou SA, Morris ER, Rees DA (1984) Mechanism and dynamics of conformational ordering in xanthan polysaccharide. J Mol Biol 175(3):371–394. https://doi.org/10.1016/0022-2836(84)90354-1

Oddy MH, Santiago JG (2004) A method for determining electrophoretic and electroosmotic mobilities using AC and DC electric field particle displacements. J Colloid Interface Sci 269(1):192–204. https://doi.org/10.1016/s0021-9797(03)00601-5

Oh MH, So JH, Yang SM (1999) Rheological evidence for the silica-mediated gelation of xanthan gum. J Colloid Interface Sci 216(2):320–328. https://doi.org/10.1006/jcis.1999.6325

Okuyama K, Arnott S, Moorhouse R, Walkinshaw MD, Atkins EDT, Wolf Ullish CH (1980) In: French AD, Gardner KCH (eds) Fiber diffraction methods: chap 26, vol 141. ACS Symposium Series, pp 411–427. https://doi.org/10.1021/bk-1980-0141.ch026, ISBN13: 9780841205895, ISBN: 9780841207424

Oviatt HW, Brant DA (1994a) The rheological behavior of semi-dilute aqueous xanthan solutions after thermal treatment. Abstr Pap Am Chem Soc 207:394–400

Oviatt HW, Brant DA (1994b) Viscoelastic behavior of thermally treated aqueous xanthan solutions in the semi-dilute concentration regime. Macromolecules 27(9):2402–2408

Penesyan A, Nagy SS, Kjelleberg S, Gillings MR, Paulsen IT (2019) Rapid microevolution of biofilm cells in response to antibiotics. NPJ Biofilms Microbiomes 5(1):34. https://doi.org/10.1038/s41522-019-0108-3

Peterson BW, He Y, Ren Y, Zerdoum A, Libera MR, Sharma PK (2015) Viscoelasticity of biofilms and their recalcitrance to mechanical and chemical challenges. FEMS Microbiol Rev 39(2):234–245. https://doi.org/10.1093/femsre/fuu008

Petri DF (2015) Xanthan gum: a versatile biopolymer for biomedical and technological applications. Appl Polym Sci 132(23):42035. https://doi.org/10.1002/app.42035

Powell LC, Pritchard MF, Ferguson EL, Powell KA, Patel SU, Rye PD (2018) Targeted disruption of the extracellular polymeric network of *Pseudomonas aeruginosa* biofilms by alginate oligosaccharides. NPJ Biofilms Microbiomes 29(4):13. https://doi.org/10.1038/s41522-018-0056-3

Riaz T, Iqbal MW, Jiang B, Chen J (2021) A review of the enzymatic, physical, and chemical modification techniques of xanthan gum. Int J Biol Macromol 186:472–489. https://doi.org/10.1016/j.ijbiomac.2021.06.196

Rinaudo M (2001) Relation between the molecular structure of some polysaccharides and original properties in sol and gel states. Food Hydrocolloids 15(4–6):433–440. https://doi.org/10.1016/S0268-005X(01)00041-8

Rinaudo M, Milas M (1980) Enzymic hydrolysis of the bacterial polysaccharide xanthan by cellulase. Int J Biol Macromol 2(1):45–48. https://doi.org/10.1016/0141-8130(80)90009-4

Rochefort WE, Middleman S (1987) Rheology of xanthan gum: salt, temperature, and strain effects in oscillatory and steady shear experiments. J Rheol 31:337–369. https://doi.org/10.1122/1.549953

Rwei SP, Nguyen TA (2014) Liquid crystalline phase in xanthan gum (XG)/H_2O/H_3PO_3 and XG/H_2O/H_3PO_4 tertiary systems: a thermal and rheological study. Cellulose 21(5):3231–3241. https://doi.org/10.1007/s10570-014-0358-4

Sato T, Kojima S, Norisuye T, Fujita H (1984a) Double-stranded helix of xanthan in dilute-solution—further evidence. Polym J 16(5):423–429

Sato T, Norisuye T, Fujita H (1984b) Double-stranded helix of xanthan: dimensional and hydrodynamic properties in 0.1 M aqueous sodium chloride. Macromolecules 17(12):2696–2700. https://doi.org/10.1021/ma00142a043

Sato T, Norisuye T, Fujita H (1984c) Double-stranded helix of xanthan in dilute solution—evidence from light-scattering. Polym J 16(4):341–350

Shiroodi SG, Lo YM (2015) The effect of pH on the rheology of mixed gels containing whey protein isolate and xanthan-curdlan hydrogel. J Dairy Res 82(4):506–512. https://doi.org/10.1017/S0022029915000370

Smith IH, Symes KC, Lawson CJ, Morris ER (1981) Influence of the pyruvate content of xanthan on macromolecular association in solution. Int J Biol Macromol 3(2):129–134. https://doi.org/10.1016/0141-8130(81)90078-7

Sorbie K (2013) Polymer-improved oil recovery. Springer Science & Business Media

Sperling LH (2006) Introduction to physical polymer science, 4th edn. Wiley, New Jersey

Stokke BT, Christensen BE (1996) Release of disordered xanthan oligomers upon partial acid hydrolysis of double-stranded xanthan. Food Hydrocoll 10(1):83–89. https://doi.org/10.1016/S0268-005X(96)80058-0

Sutherland I (1981) *Xanthomonas* polysaccharides—improved methods for their comparison. Carbohyd Polym 1(2):107–115. https://doi.org/10.1016/0144-8617(81)90003-5

Sutherland IW (1987) Xanthan lyases—novel enzymes found in various bacterial species. J Gen Microbiol 133(11):3129–3134. https://doi.org/10.1099/00221287-133-11-3129

Tait M, Sutherland I, Clarke-Sturman A (1986) Effect of growth conditions on the production, composition and viscosity of *Xanthomonas campestris* exopolysaccharide. Microbiology 132(6):1483–1492. https://doi.org/10.1099/00221287-132-6-1483

Tako MJ (1991) Synergistic interaction between deacylated xanthan and galactomannan. Carbohydr Chem 10(4):619–633. https://doi.org/10.1080/07328309108543936

Terada S, Yoshimoto H, Fuchs JR, Sato M, Pomerantseva I, Selig MK (2005) Hydrogel optimization for cultured elastic chondrocytes seeded onto a polyglycolic acid scaffold. J Biomed Mater Res Part A 75A(4):907–916. https://doi.org/10.1002/jbm.a.30505

Tinland B, Rinaudo M (1989) Dependence of the stiffness of the xanthan chain on the external salt concentration. Macromolecules 22(4):1863–1865. https://doi.org/10.1021/ma00194a058

Tombs MP, Hardings SE (1998) An introduction to polysaccharide biotechnology. Taylor & Francis, London

Valpuesta V, Botella MA (2004) Biosynthesis of L-ascorbic acid in plants: new pathways for an old antioxidant. Trends Plant Sci 9(12):573–577. https://doi.org/10.1016/j.tplants.2004.10.002

Wang Q, Wood PJ, Cui W, Ross-Murphy SB (2001) The effect of autoclaving on the dispersibility and stability of three neutral polysaccharides in dilute aqueous solutions. Carbohydr Polym 45(4):355–362. https://doi.org/10.1016/S0144-8617(00)00280-0

Wang Z, Wu J, Zhu L, Zhan X (2017) Characterization of xanthan gum produced from glycerol by a mutant strain *Xanthomonas campestris* CCTCC M2015714. Carbohydr Polym 157:521–526. https://doi.org/10.1016/j.carbpol.2016.10.033

Wang DC, Jiang CH, Zhang LN, Chen L, Zhang XY, Guo JH (2019) Biofilms positively contribute to *Bacillus amyloliquefaciens* 54-induced drought tolerance in tomato plants. Int J Mol Sci 20(24):6271. https://doi.org/10.3390/ijms20246271

Williams PA, Phillips GO (2004) Introduction to food hydrocolloid. In: Phillips GO, Williams PA (eds) Handbook of hydrocolloid. Woodhead, Cambridge, UK, pp 1–19. ISBN 978-1-84569-414-2

Yin W, Wang Y, Liu L, He J (2019) Biofilms: the microbial "protective clothing" in extreme environments. Int J Mol Sci 20(14):3423. https://doi.org/10.3390/ijms20143423

Zhang Q, Zhou JS, Zhai YA, Liu FQ, Gao G (2008) Effect of salt solutions on chain structure of partially hydrolyzed polyacrylamide. J Cent South Univ Technol 15:80–83. https://doi.org/10.1007/s11771-008-0319-x

Chapter 3
Rheology of Xanthan Gum Aqueous Solution

3.1 Introduction

Xanthan Gum (XG) is a polysaccharide that is produced by the bacterium *Xanthomonas campestris* through fermentation, as discussed in Chap. 2. The practical use of XG in various industries, including food processing, pharmaceuticals, cosmetics, agriculture, and engineering, is mainly a result of its ability to induce high viscosity at low polymer concentrations in aqueous environments. It can modify the rheological behavior of fluids (i.e., the flow and deformation of XG suspension under applied stresses) by acting as a thickener and stabilizer. For the former, XG is a highly effective thickening agent. It forms a network of long-chain polymers that entrap water and increase the viscosity of the fluid. As elaborated later, the thickening properties of XG are dependent on various factors, such as XG concentration, the solution pH and temperature, shear rate, and the presence of salts and/or other chemical additives. Moreover, the rheological behavior of XG solutions is pseudoplastic, which means that the viscosity decreases with increasing shear rate. Additionally, XG is an effective stabilizer, particularly in emulsions. It can stabilize oil-in-water emulsions by forming a protective layer around the oil droplets, preventing them from coalescing. XG can also stabilize suspensions by preventing settling and maintaining a uniform distribution of particles.

In addition, XG is a versatile polysaccharide that exhibits unique rheological properties. Its ability to modify the viscosity and stability of fluids has made it an essential ingredient for various industries. The rheological behavior of XG solutions is dependent on various factors, such as XG concentration, pH level, temperature, shear rate and salt concentration. Understanding the rheological properties of XG is important for optimizing its use in various applications, including in the fields of geotechnical and geo-environmental engineering.

A. M. O. Mohamed et al., *Sustainability in Ground Improvement: The Case of Xanthan Gum Biopolymer*, Green Energy and Technology,
https://doi.org/10.1007/978-3-031-75313-8_3

3.2 XG Structural Characteristics

As discussed in Chap. 2, XG exists in aqueous solution as a double-stranded helix conformation (Sato et al. 1984a, b, c), or as a single-stranded structure (Berth et al. 1996), depending on the bacteria origin and preparation conditions. In aqueous solution at low ionic strength, an XG molecule presents as an extended chain due to the intra-chain electrostatic repulsion. However, with the addition of a small amount of salt, shielding of the repulsion takes place, and the XG molecule undergoes contraction and forms a random-coiled structure. In addition, XG molecules are completely rigid for molecular weights of $M_w < 3 \times 10^5$ g/mol, and they are considered semi-flexible above that value (Sato et al. 1984c).

The basic XG structural characteristics, such as molecular weight, hydrodynamic dimensions, and persistence lengths are summarized in Table 3.1. The molecular weight of XG ranges from 1.8×10^6 to 8.7×10^6 g/mol. The hydrodynamic length (L) refers to the average length of the molecules in solution, which for XG varies from 700 to 1500 nm in distilled (DI) water, reducing to 120 nm in 0.1-M NaCl. The persistence length (L_p) is defined as the average projection of the end-to-end vector on a single bond direction, with the L_p of XG varying between 31 and 780 nm depending on the XG source. For cases where $L < L_p$, the molecule behaves as a rigid rod, whereas the behavior approaches that of a random coil for $L > L_p$. The hydrodynamic diameter varies from 1.6 to 4.0 nm in DI water, and between 2.0 and 2.5 nm in 0.1-M NaCl. Notably, the hydrodynamic dimensions and the persistence length vary due to the different sources of the XG biopolymer, the solvent used during its preparation, and the environmental conditions (e.g., pressure, temperature and pH level). In general, XG preserves a certain degree of toughness, which is intermediate between the double-stranded DNA and the triple-stranded collagen (Zirnsak et al. 1999).

3.3 Rheological Behavior in a Steady Shear Flow

3.3.1 Viscosity

The presence of salts in solution influences the viscosity of XG solutions. At low XG concentrations (approximately < 0.3 wt%), the viscosity decreases when a small amount of salt (approximately < 0.1 wt%) is added; while the viscosity increases at higher XG concentrations or when a larger amount of salt is included (Zatz and Knapp 1984). As discussed in Chap. 2, in the low-salt concentration regime (dilute), the addition of salt to a solution of XG in distilled water causes a disorder → order (random → helix) transition in the backbone conformation, which is always accompanied by a substantial decrease in molecular size and a corresponding decrease in the solution viscosity (Fig. 3.1).

Table 3.1 Reported hydrodynamic dimensions and persistence lengths for XG (reproduced from Zirnsak et al. 1999)

Reference	Molecular weight, M_w	Solvent type	Persistence length, L_p (nm)	Hydrodynamic dimensions	
				Length, L (nm)	Diameter, d (nm)
Whitcomb et al. (1977)	–	Distilled water	–	700–1000	1.6–4.0
Whitcomb and Macosko (1978)	7.6×10^6	Distilled water	–	1500	1.9
Rinaudo and Milas (1978)	2.0×10^6	0–1.0 M NaCl	–	600	2.0
Tinland and Rinaudo (1989)	8.7×10^6	0.01–0.3 M NaCl	31	–	–
Berth et al. (1996)	2.9×10^6	0.01 M NaCl	150	–	–
Sato et al. (1984a, b, c)	7.4×10^6	0.10 M NaCl	–	120	2.0–2.5
Jamieson et al. (1982)	5.4×10^6	1.0 M NaCl	–	1500	2.0
Muller and Lecourtier (1988)	1.8×10^6	5 g/l NaCl	50	–	–
Carriere et al. (1993)	1.8×10^6	5 g/l NaCl	460–780	–	–

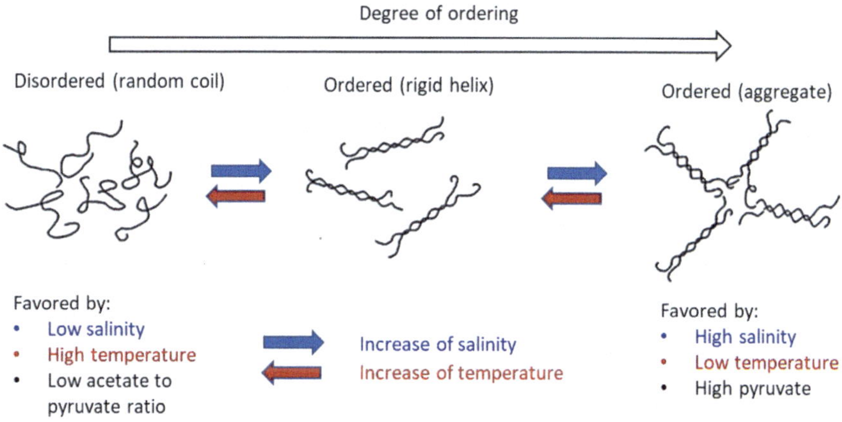

Fig. 3.1 Degree of ordering of the XG backbone conformation as function of salinity, temperature, and acetate and pyruvate contents

The addition of NaCl to XG with low concentration (< 0.2 wt%) reduces the hydrodynamic diameter of the XG molecules, thereby producing a decrease in shear viscosity of the XG gel (Wyatt and Liberatore 2010; Wyatt et al. 2011). In this case, the XG chains do not significantly interact with each other despite the presence of hydrogen bonding. However, addition of NaCl to XG with high concentrations (> 0.2 wt%) reduces the hydrodynamic diameter of the XG molecules, and at the same time increases the shear viscosity of the XG gel (Wyatt and Liberatore 2010). As such, the XG chains are contacting each other via hydrogen bonds, and forming denser XG gel solutions.

The increase of shear viscosity could be explained, as follows. In the absence of NaCl, the electrostatic repulsion between the XG chains tends to keep the XG chains at a certain distance from one another, despite the denser XG solution. Screening this repulsion by the addition of NaCl makes it possible for XG chains to come closer in contact with each other and form hydrogen bonds at high XG concentrations. The establishment of the hydrogen bonds make it difficult for the XG chains to move past one another in solution and leads to an increase in the shear viscosity (Wyatt et al. 2011). Therefore, the establishment of hydrogen bonding and increased interactions have a larger effect on the shear viscosity of the XG gel solutions than the decrease in the hydrodynamic diameter (Zhong et al. 2013).

As viscosity changes, the molecular weight changes, as illustrated in Fig. 3.2 for XG mixed with 0.1-M NaCl. The results indicate that as reaction time increases, the viscosity changed from about 3×10^1 to 1×10^4 cm^3/g (three orders of magnitude higher) and the molecular weight changed from about 7×10^4 to 7×10^6 g/mol (two orders of magnitude higher).

The differential weight fraction as a function of the molecular mass for XG at room temperature is shown in Fig. 3.3 (data from Viebke (1998)). The variations in the molecular mass are attributed to changes in processing and production, as well as to the formation of aggregates due to intermolecular attraction (Southwick et al.

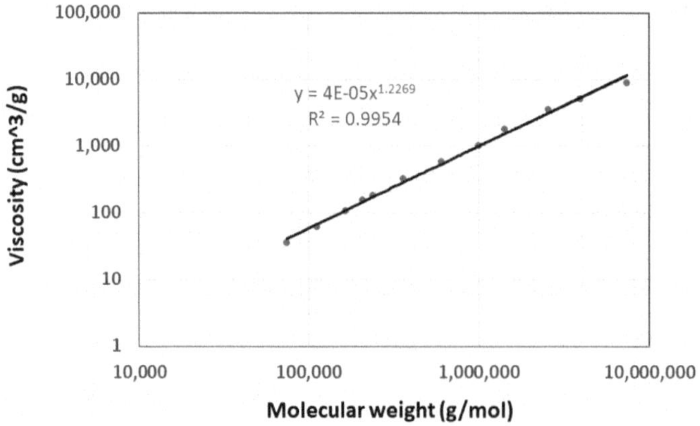

Fig. 3.2 Variations of viscosity against molecular weight for XG mixed with 0.1-M NaCl (data from Sato et al. (1984c))

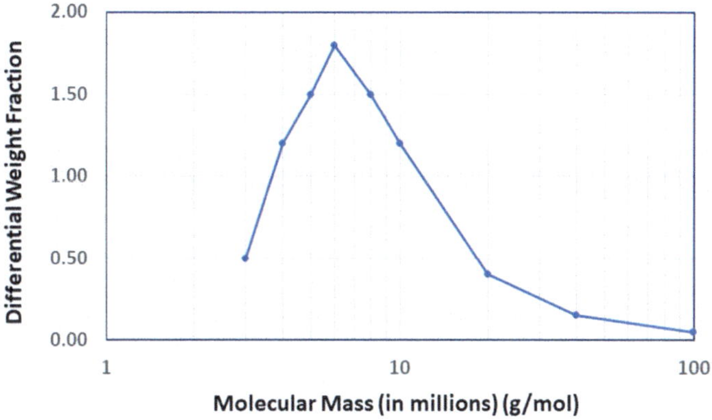

Fig. 3.3 Differential weight fraction as a function of the molecular mass for XG at room temperature (data from Viebke (1998))

1980; Cuvelier and Launay 1986). The broad molecular mass distribution may be attributed to the transition from order (helix) to disorder (coiled) conformation.

As discussed in Chap. 2, since XG is an anionic polysaccharide, it might exhibit a certain sensitivity to pH changes due to changes in its charge density; as such, the molecular associations between the single molecules are influenced (Pastor et al. 1994). However, other studies have indicated that the viscosity of XG solutions does not change significantly with changes in pH level, irrespective of the chemical compositional changes. At pH ≥ 9, XG gradually deacetylated (Tako and Nakamur 1984), while at pH < 3, XG loses the pyruvic acid acetyl groups (Bradshaw et al. 1983). However, for both cases (deacetylated or de-pyruvylated), XG demonstrated similar rheological properties as native XG, with no notable changes in the viscosity of XG solutions for pH changes between pH 1 and pH 13.

It is worth noting that the structural conformation of XG can be interpreted from viscosity measurements by adopting the Mark–Houwink theory, as discussed in Chap. 2, whereby the intrinsic viscosities (η) are plotted as functions of the weight-average molar mass (M_w) to obtain the Mark–Houwink exponent (α_1) from the ordinary regression trendline of the results (i.e., $\eta = k'(M_w)^{\alpha 1}$). For instance, referring to the η versus M_w data scatter plot shown in Fig. 3.2 for XG mixed with 0.1-M NaCl, η was found to be proportional to ${M_w}^{1.2269}$; hence, the Mark–Houwink exponent α_1 is equal to 1.2269, meaning that the solution structure is somewhere between a completely expanded (random) coil ($\alpha_1 = 0.5$–0.8) and a rigid rod ($\alpha_1 = 1.8$).

However, Arendt and Kulicke (1998) showed that the simple Mark–Houwink exponent theory could not explain this high exponent value (of $\alpha_1 = 1.2269$) for a high-molecular-weight XG biopolymer, and the chain stiffness of worm-like chains should be accounted for in the modelling exercise. Also, for XG in a 0.1-M NaCl solution, the authors reported that the viscosity was barely below that of XG in aqueous solution and decreased over the whole temperature range from 20 to 80 °C by a mere 25%. They proposed the following structural-property relationship for

XG:

$$\eta = C_{XG} \times \eta + 0.57(C_{XG} \times \eta)^2 + 0.15(C_{XG} \times \eta)^{3.5} \tag{3.1}$$

where η = intrinsic viscosity; and C_{XG} = XG concentration.

Moreover, for moderately NaCl-concentrated solutions, the relationship between the zero-shear viscosity (i.e., the viscosity of a fluid/solution at very low shear rates (approaching zero)) and molecular mass was $\eta \propto {M_w}^{5.5}$, giving $\alpha_1 = 5.5$, which is higher than that of the random-coil type of polymers where entirely physical entanglement occurs for $\alpha_1 = 3.4$ (Graessley 1982). Therefore, it is important to be aware of self-association between XG molecules, even at very low concentrations.

In the study by Arendt and Kulicke (1998), it was shown that the viscosity is high even at a low C_{XG}, and increases with increasing C_{XG}. However, it starts twisting when C_{XG} exceeds a critical concentration, because of the development of a loosely bound network arising from the nature of the hydrogen bridging between the lateral chains of adjacent XG molecules, and the molecular entanglements (Norton et al. 1984).

In addition, the XG structural confirmation depends on temperature, as discussed in Chap. 2. At low or ambient temperatures, XG exists in ordered conformation in solutions (Fig. 3.1), where the charged trisaccharide sidechain folds back around a cellulose backbone to produce rigid rod-like structure stabilized by intramolecular non-covalent bonding (Norton et al. 1984). Notably, the ordered form structure of XG still remains controversial (i.e., whether the network assemblage comprises side-by-side single helices or coaxial double helices) (Norton et al. 1984). In the ordered state, XG exhibits a high viscosity at low shear rates; however, the viscosity drops exponentially to a minimum as the shear rate increases. As temperature increases, however, the secondary structure undergoes an ordered (helix)—disordered (coil) conformation transition (Norton et al. 1984).

3.3.2 *Controlling Parameters*

The rheological behavior of XG in aqueous solution depends on the XG type, particle-size range (gradation) and concentration, shear rate, dissolution and measurement temperatures, surface tension, pH level, and the monovalent and divalent metal ion concentration. These parameters are discussed below.

3.3.2.1 Shear Rate

XG solutions have a non-Newtonian rheology (pseudoplastic behavior), with a decrease in the apparent viscosity as shear rate increases. For example, for the research-grade XG (CAS: 11138-66-2; Sigma Aldrich), the viscosity of 0.5% XG solution varies from 10×10^3 to 12×10^3 cP (1 cP = 1 mPa s), depending on

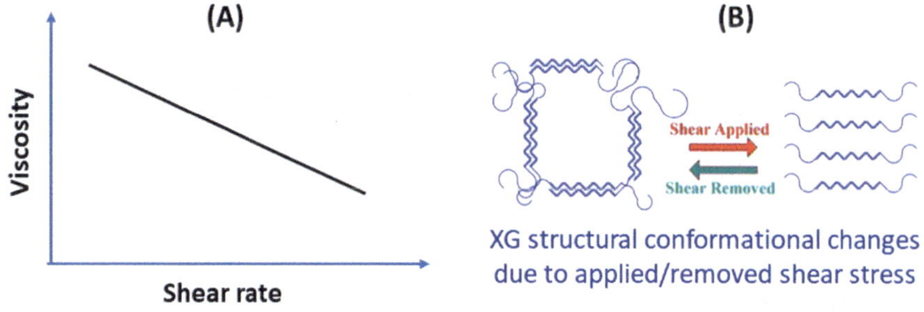

Fig. 3.4 Effect of shear rate/stress on **a** XG viscosity and **b** its structural conformational changes

the shear rate (0.11 to 0.022 1/s, respectively) (Chang et al. 2019). It has a unique shear-thinning behavior (Stokke and Christensen 1996) that can be observed in a 1% XG solution, whereby the viscosity at low shear rates was a factor of 5×10^1 to 1×10^5 times higher than for distilled water case; however, at higher shear rates only a factor of ten difference in viscosity was observed. It is to be noted that this pseudo-plastic behavior is maintained over wide ranges of temperature, salt concentrations and pH levels. This arises from the presence of an ordered rod-like (rigid) conformation, with a high persistence length (L_p) (Cuvelier and Launay 1986), which can self-associate creating domains of associate chains (aggregates) at low shear rates or at-rest (Fig. 3.4). On first applying the stress at high shear rates, the gel aggregates convert into solution; however, if the shear stress is applied again, the double-helical structure of the XG can be degraded into single chains. Notably, if the applied shear stress is removed, the XG can regain its original structure (Funahashi et al. 1987).

3.3.2.2 Temperature

It is known that the viscosities of XG solutions depend on the measurement temperature (T_M), where the viscosity is being measured, and on the XG dissolution temperature (T_D). Concerning the former, the viscosity decreases as T_M increases; however, this behavior is reversible within the temperature range of 10 to 80 °C. Regarding the later temperature (i.e., T_D), there are three zones that could be defined (see Fig. 3.5a): (i) from 20 to 40 °C, the viscosity declines as T_D increases; (ii) from 40 to 60 °C, the viscosity increases with increasing T_D; and (iii) for $T_D > 60$ °C, the viscosity declines as the temperature increased. Such behavior is directly linked to the conformational changes of the XG molecule, whereby there is a conformation shift from an ordered (low dissolution temperature) to a disordered (high dissolution temperature) state (Milas and Rinaudo 1979; Garcia-Ochoa and Casas 1994).

Because of the increase in the rate of reaction and the molecular mobility, increases in the measuring temperature negatively affect the XG network. The XG solutions exhibit non-Newtonian flow behavior at measuring temperatures below the ordered-disordered transition, where the XG helix in the solution may behave as a rigid

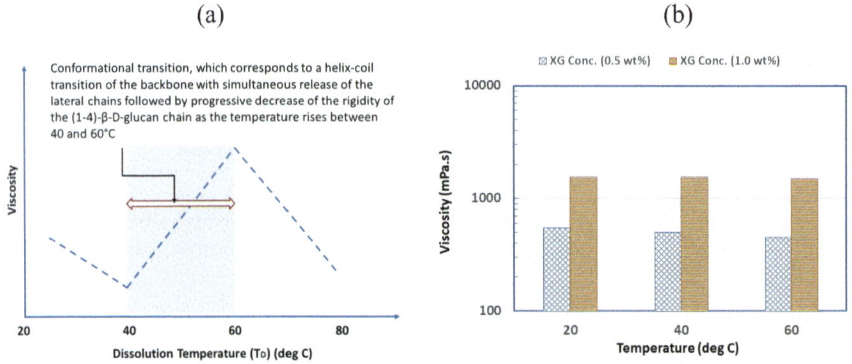

Fig. 3.5 Effects of temperature on XG viscosity: **a** conformational temperature-zone transitions; and **b** Viscosity performance of the commercial VANZAN-XG (data for Fig. 3.5b sourced from https://camsi-x.com/wp-content/uploads/2020/11/VANZAN_Xanthan_Gum.pdf)

rod. It has been reported that for 1% XG solution at 60 °C, a transition occurred from a partially ordered random broken helix conformation to disordered random coil conformation (Pelletier et al. 2001), which produced less viscosity and pseudoplasticity. However, the viscosity of a commercial VANZAN-XG was reported to be stable for a given XG concentration over a measuring temperature range from ambient temperature to 60 °C (see Fig. 3.5b). In another study, Capron et al. (1997) observed a decrease in molecular mass as a function of increasing measuring temperature for the XG native state (Fig. 3.6). The calculated mass per unit length of the polymer (i.e., M_L) corresponded with a double-stranded helix.

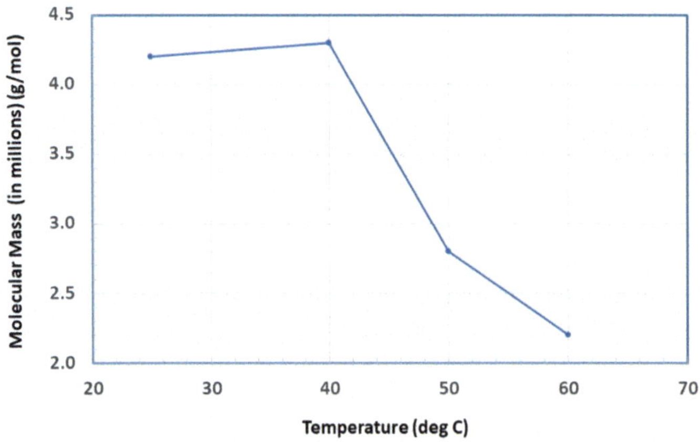

Fig. 3.6 Variations of molecular mass of XG in 0.01-M NaCl as a function of temperature (data from Capron et al. (1997))

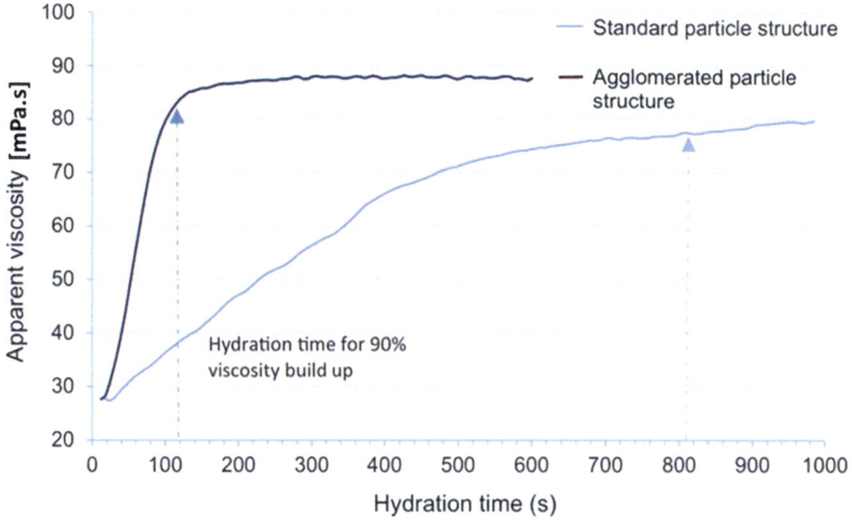

Fig. 3.7 Model curves of viscosity build-up for XG with two different particle structures added to distilled water at 0.3% (w/v) (Berninger et al. 2021) [This is an open access article under the terms of the Creative Commons Attribution-Non-Commercial-No Derives License, which permits use and distribution in any medium, provided the original work is properly cited, the use is non-commercial, and no modifications or adaptations are made.]

3.3.2.3 XG Particle Size (Granulation)

The particle-size range or granulation of XG strongly influences its hydrating and dispersing performance. A fine granulation results in a fast hydration and higher tendency towards forming lumps; however, coarser granulations hydrate slower. Figure 3.7 illustrates model curves of viscosity build-up for XG with two different particle structures, added to distilled water at 0.3% (w/v). The agglomerated grade reaches 9% of its maximum viscosity after about 115 s; while the standard grade reaches 90% of its maximum viscosity after about 810 s. It is to be noted that there is an agglomerated XG-based food thickener, which is widely used for the preparation of thickened beverages (Lee and Yoo 2020, 2021), that would allow for an easy dispersion without the formation of lumps.

3.3.2.4 Polarity of Fluids (Surface Tension)

Since water is a polar liquid, it is characterized by high surface tension due to the strong intermolecular interactions. Surface tension depends vastly on the concentration of XG, salinity, pH and temperature. For the XG concentration effect on the surface tension, Prud'homme and Long (1983) reported no changes in the surface activities (surface tension of 70.7 mN/m, which is equivalent to distilled water) for low concentration (0.1 wt%) XG aqueous solutions. However, as the XG concentration increased to 1%, a dramatic decrease in surface tension was observed (42.3 mN/m); i.e., about a 40% decrease from that obtained for low XG concentration.

For the salinity effect on the surface tension, Nedjhioui et al. (2009) observed a decrease in surface tension with increasing NaCl concentration in XG solutions, attributed to: (i) the NaCl screening effect on the negatively charged carboxyl groups of the XG chains; (ii) reduction of the electrostatic repulsive forces; (iii) favoring the formation of a compact helix conformation (Brunchi et al. 2014) because of the XG chains' self-aggregation (Secouard et al. 2005); (iv) since the non-polar XG chains trap the polar carboxyl groups in the helix conformation, their ability to form hydrogen bonds with water molecules is decreased because they are already formed between the XG chains.

The effect of pH and temperature variations on the surface tension can be evaluated using results shown in Fig. 3.8 for low XG concentration (0.02%) in water. At pH 5, the surface tension decreased from 67.1 mN/m at 28 °C to 62.9 mN/m at 45 °C, indicating that for low XG concentrations, the surface tension is not greatly affected by the increase in temperature. However, as pH is further decreased (i.e., from pH 5 to pH 3.5 and 2.5), greater reductions of the surface tension as a function of increasing temperature were observed. In view of the dependence of the surface tension on the structural conformation, Young et al. (1989) reported that compared to the helical conformation, the surface tension is higher for the coil conformation, which can be obtained in alkaline pH and high temperatures in dilute solutions. Furthermore, in concentrated saline solutions, the coexistence of both coil and helix conformations of XG chains might occur (Bejenariu et al. 2010).

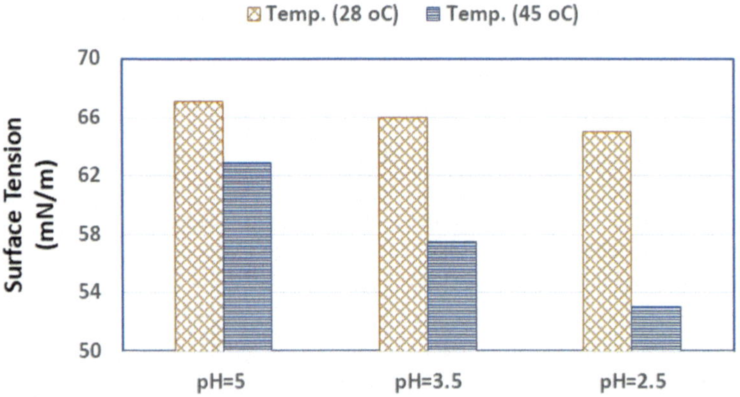

Fig. 3.8 Surface tension changes with temperature and pH for 0.02% XG concentration in water (data from Brunchi et al. (2016))

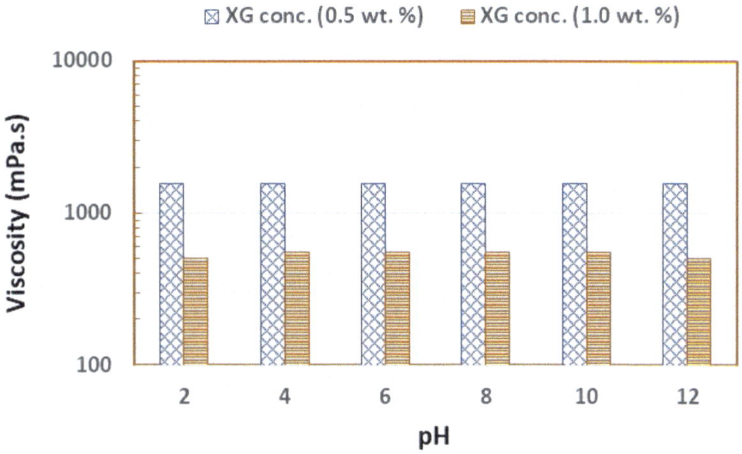

Fig. 3.9 Viscosity variations with pH for commercial VANZAN-XG solutions (data from https://camsi-x.com/wp-content/uploads/2020/11/VANZAN_Xanthan_Gum.pdf)

3.3.2.5 pH

The viscosity of the commercial VANZAN-XG remains nearly constant between pH 2 and pH 12, as shown in Fig. 3.9. Below pH 2 and above pH 12, the viscosity tends to decrease slightly. The pH of XG changes during its production, decreasing to pH 5, due to the presence of acid groups, with the pH of the broth increasing to 7–8 after 24 h, and to 8–9.5 after 48 h, depending on the temperature and agitation conditions. The structural conformation of the XG hydrogels produced by heating at 165 °C in the presence of citric acid (low pH) has higher crosslinking, higher gel content, homogeneous porous structure, and higher charge densities. On the other hand, those produced at 165 °C in the absence of citric acid (neutral pH) showed lower gel content, heterogeneous porous structure and nanofibrils (Mohammed et al. 2007; Bueno et al. 2013). The XG chains in networks produced with citric acid are coiled when expanded in water, but upon increasing the ionic strength, they undergo a conformational transition to the ordered state (Bueno and Petri 2014).

3.3.2.6 Electrolyte Concentration

The electrical conductivity of polyelectrolyte solution depends on the number of ions per unit volume and their mobility in the solution. Thus, for a certain XG concentration in distilled water, the electrical conductivity depends on the residual amounts of glucuronic and pyruvic acids (–OO^-), as well as the nature of the metallic cations (e.g., Na^+, K^+, Ca^{2+} and Mg^{2+}) that are linked to acidic residues and which originated from the salts employed in the XG production media, or were added after fermentation (Klaic et al. 2011; Khouryieh et al. 2007). For XG with low concentration

(0.01%) in distilled water solution, the reported values of conductivity varied from 0.019 mS/cm at 18 °C to 0.039 mS/cm at 45 °C due to ions mobility (Brunchi et al. 2016). Similarly, Jones et al. (1987) reported that in the presence and absence of salt, the conductivity of 0.25% XG solution increased with the temperature. Furthermore, in 0.01-M KBr salt solution, the changes in electrical conductivity of XG aqueous solution during the conformational transition were reported to be more pronounced than in salt-free solution.

Because of the presence of ionic species (solutes) in the fluid, solutes transport from fluid into XG networks occurs via the following mechanisms (Bueno et al. 2013): (i) capillary action; and (ii) Fick's law of diffusion, which is based on the concentration gradient between the concentration of ionic species in the diffused layer and the polymeric matrix. Notably, under both acidic and neutral conditions, XG hydrogels are stable; however, under alkaline conditions ester bonds are hydrolyzed (Bueno et al. 2013). The effects of different XG solution concentrations, in the presence of 0.75% and 3.5% NaCl, and 2.0% $CaCl_2$, on solution viscosity are demonstrated in Fig. 3.10. Generally, XG shows good resistance to high salinity, irrespective of the concentration of the salt, and the order of valency (monovalent or divalent).

The good resistance to high salinity is possible because XG is composed of chains that are conformed into different ordering depending on the degree of salinity and temperature (Fig. 3.1). The disordered form is favored at high temperature and/or at very low ionic strength and has been described as a single-stranded random coil (Milas and Rinaudo 1986), as well as a worm-like chain with a moderate persistence length (L_p) (Milas et al. 1995). The addition of NaCl and $CaCl_2$ induces the disorder—order conformational transition. The addition of salt affects the XG conformational transition through Coulomb interactions. Salt-induced ordering can be explained due to the alignment of trisaccharide sidechains with the main chain through hydrogen bonding (Kitagawa et al. 1985). A compact, double-stranded structure in 0.1-M NaCl was proposed to elongate to a less compact structure upon exposure to low salt concentration, and to dissociate into single-stranded chains at even lower salt concentrations (Lecourtier et al. 1986).

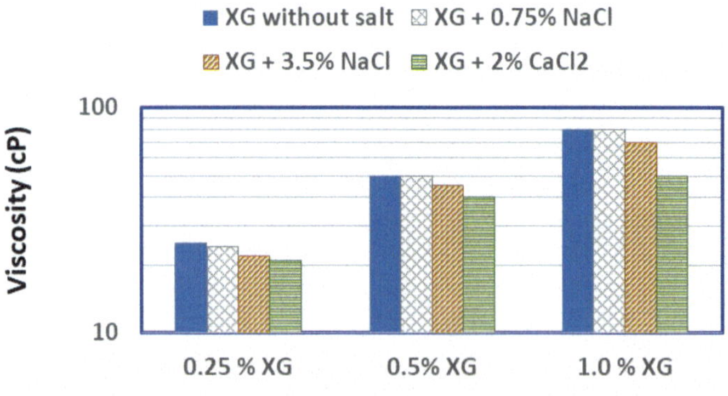

Fig. 3.10 Effects of XG, monovalent (Na^+) and divalent (Ca^{2+}) salt concentrations on the solution viscosity for a shearing rate of 100 rpm (data from Okechukwu Ezeh et al. (2021))

Notably, XG chains naturally carry Ca^{2+} ions, which stabilize the native helices' structures. The addition of Ca^{2+} ions to XG at 1:1 molar ratio might lead XG solution towards a gel-like state. However, at 2:1 molar ratio, the solution might go to a gel-like state due to the partial replacement of intermolecular site binding of Ca^{2+} ions by binding to individual carboxyl groups, decreasing the degree of complexation and network formation (Mohammed et al. 2007). Hence, the electrostatic interactions might lead to potential formation of a network structure between XG carboxylate groups and divalent cations (Milas et al. 1995).

The XG structural conformation can be further evaluated with the use of the results reported by Vega et al. (2015). They investigated XG (from Parafarm, N° 509, 2013, Argentina), with viscosity ranging between 1400 and 1600 Pa s, and pH of 6–8 for 0.1% XG solution in distilled water, as well as various monovalent and divalent salts with concentrations ranging from 5×10^{-6} to 0.1 M. The results indicated that in distilled water, XG tends to have a coil conformation structure; however, it is also highly dependent on XG concentration. With addition of salt at low concentration, the viscosity of the solution increases and the structure continues to be coiled. However, with salt concentration increase, the system tends to have a helix conformation structure. The salt concentration at which the system transitions from coil to helix conformation is highly dependent of the nature of the ionic species. For divalent cations, such as Ca^{2+} and Mg^{2+} the system change will occur at lower salt concentration than the monovalent ion such as Na^+ (see Fig. 3.11). Notably, divalent salts (such as Ca^{2+} and Mg^{2+}) can cause gelation and/or precipitation of XG at alkaline pHs (i.e., for pH > 10); however, trivalent salts (such as Al^{3+}, Fe^{3+} and Cr^{3+}) can cause gelation at acid and neutral pH levels.

With reference to the type of soluble salts, the flow behavior of solutions prepared with XG and $CaCl_2$ or $Ca(NO_3)_2$ at 1:1, 1:10 and 1:100 molar ratios indicated that: (i) the monovalent counter-ion type (chloride or nitrate) has no influence on the flow consistency or on the flow behavior; (ii) as the ionic strength increased (to 1:100), the flow consistency increased due to the formation of ordered state XG chains; and (iii) as the ionic strength increased (to 1:100), the solutions became less

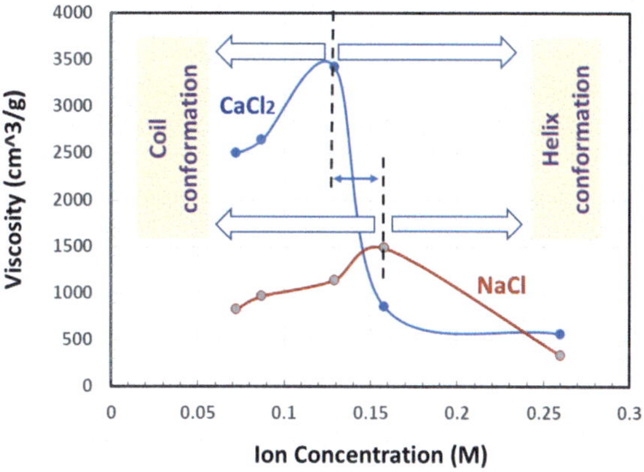

Fig. 3.11 Effects of ion concentrations and charges on XG viscosity (data from Vega et al. (2015))

pseudoplastic (Dario et al. 2011; Le and Turgeon 2013). However, for insoluble salts, fluids containing insoluble salts with large density values and strong hydration (e.g., $CaCO_3$ and $Ca_3(PO_4)_2$) showed high flow consistency because they act as weighting agents (Petri and Queiroz Neto 2010; Dario et al. 2011). High flow consistency is an important fluid property for drilling mud used in the petroleum industry.

3.4 Rheological Behavior in a Dynamic Shear Flow

3.4.1 Oscillatory Shear Testing and Theory

The preceding discussion was mainly related to the XG nonlinear rheological behavior in steady shear flow fields, which indicates that concentrated XG solutions exhibit a marked non-Newtonian shear-thinning flow behavior (Song et al. 2006). This kind of flow properties, in steady shear flow, is not capable of providing information regarding the relationship between the flow behavior and the microstructure of XG due to its nature-destructive technique that led to a structural breakdown of the biopolymer. In addition, since at high shear rates the deformation is very fast, the biopolymer does not have enough time to adjust to the applied shear stresses. To overcome such limitations, the XG's rheological behavior must be measured at small deformation rates compatible to field conditions. This is achieved with the use of a dynamic test, which is performed in oscillatory shear flow fields by applying a small amplitude of sinusoidal strain (or stress) and measuring the resulting stress (or strain), similar to the professional practices employed in the food, polymer and road-pavement (bitumen) industries.

In addition, this testing protocol is normally followed to ensure the material's response is within the linear viscoelastic region (Macosko 1994). The linear viscoelastic properties are obtained using small-amplitude oscillatory shear (SAOS) tests, while the nonlinear viscoelastic properties are obtained using the large-amplitude oscillatory shear (LAOS) tests (Fig. 3.12). SAOS testing allows one to: (i) measure the performance without experiencing microstructural damages to the tested biopolymer; (ii) relate the dynamic rheological parameters to the molecular structure of tested biopolymer; and (iii) calculate the linear viscoelastic properties, such as shear storage and loss moduli, and the tangent loss angle (Ferry 1980; Tschoegl 1989).

According to the theory of linear viscoelasticity, the complex modulus $G^*(\omega)$ is defined as the ratio of complex stress σ^* to complex strain γ^*; that is:

$$G^*(\omega) = \sigma^*(\omega)/\gamma^*(\omega) \tag{3.2}$$

where ω = angular frequency.

Also, the frequency-dependent complex modulus $G^*(\omega)$ can be expressed as follows:

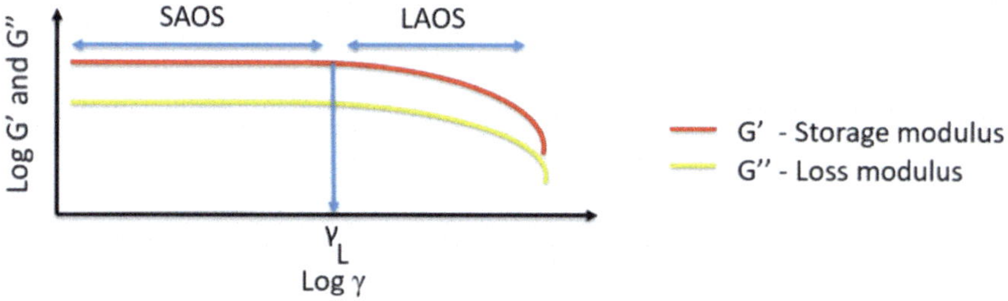

Fig. 3.12 Strain sweep test, showing linearity limit γ_L, small-amplitude oscillatory shear (SAOS), and large-amplitude oscillatory shear (LAOS) (adapted from Hyun 2011)

$$G^*(\omega) = K_1(\omega)^{a1} + K_2(i\omega)^{a2} \tag{3.3}$$

where i = SQRT(−1); K_1, K_2, a_1 and a_2 = fitting coefficients to be determined from the experimental data.

In addition, the complex modulus $G^*(\omega)$ can be expressed in terms of real and imaginary parts as given by:

$$G^*(\omega) = G'(\omega) + iG''(\omega) \tag{3.4}$$

where $G'(\omega)$ = shear storage (elastic) modulus (i.e., a measure of the energy stored within the system); and $G''(\omega)$ = shear loss modulus (i.e., a measure of the energy dissipated from the system, or its viscous behavior).

These moduli are mathematically expressed as follows:

$$G'(\omega) = K_1(\omega)^{a1}\cos((\pi/2) \times a_1) + K_2(\omega)^{a2}\cos((\pi/2) \times a_2) \tag{3.5}$$

$$G''(\omega) = K_1(\omega)^{a1}\sin((\pi/2) \times a_1) + K_2(\omega)^{a2}\sin((\pi/2) \times a_2) \tag{3.6}$$

The loss tangent, or tan(δ), which is a measure of the internal friction within the system, is expressed by the following ratio:

$$\tan(\delta) = G''(\omega)/G'(\omega) \tag{3.7}$$

However, based on the experimental results, the frequency dependence of the storage modulus $G'(\omega)$ and the loss modulus $G''(\omega)$ may be quantitatively described by a power law type relationship, as follows (Jampala et al. 2005; Kim and Yoo 2006):

$$G'(\omega) = K'(\omega)^{n'} \tag{3.8}$$

$$G''(\omega) = K''(\omega)^{n''} \tag{3.9}$$

where K', K'', n', and n'' = fitting coefficients to be determined from the experimental data—that is, K' and K'' are the intercepts and n' and n'' are the slopes in the bi-logarithmic plots of the shear storage and loss moduli against angular frequency, respectively.

The additional frequency-dependent material functions, such as the dynamic viscosity $\eta'(\omega)$, and the out-of-phase component of the complex viscosity $\eta''(\omega)$, can be obtained from the following relationships:

$$\eta'(\omega) = G''(\omega)/\omega \tag{3.10}$$

$$\eta''(\omega) = G'(\omega)/\omega \tag{3.11}$$

$$\eta^* = \left(G'^2 + G''^2\right)^{1/2}/\omega \tag{3.12}$$

Therefore, the above equations can be used to simulate the linear viscoelastic behavior of concentrated XG systems.

3.4.2 *Controlling Properties*

Much like the steady-state rheological behavior, the dynamic behavior of XG in aqueous solution depends on the XG type, grain size (granulation) and concentration, pH level, type of ionic species and concentration, temperature, and frequency. In the following, a discussion related to XG concentration, XG type, solution salinity, and ion type is presented.

3.4.2.1 XG Concentration

The following discussion regarding XG concentration is based on experimental data reported by Song et al. (2006). XG solutions with concentrations of 1, 2, 3, and 4 wt% were prepared by slowly adding the required amount of XG powder into a known volume of gently agitated distilled water medium filled in a glass container. The mixed solutions were maintained at room temperature, with constant stirring for 24 h. To prevent evaporation, the tops of the glass containers were sealed up using an airtight film. Then, a propeller-type variable-speed homogenizer was used to provide an additional agitation of each XG mixture for a period of 3–5 h, with a rotational speed of 300 rpm, until the XG was completely dissolved and the XG-distilled water solution was free from lumps. To ensure complete hydration of the XG polymer, the solutions were stored and undisturbed at room temperature for an additional 12 h before conducting the experiments. The dynamic frequency sweep tests were performed in SAOS flow fields at a fixed temperature (of 20 °C) for an angular frequency ranging from $\omega = 0.025$ to 100 rad/s, with a logarithmically increasing

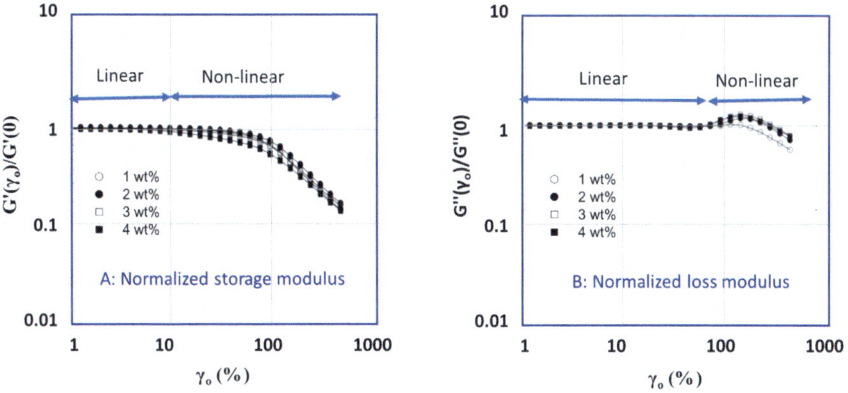

Fig. 3.13 Strain dependence of **a** the normalized shear storage modulus $G'(\gamma_0)/G'(0)$ and **b** the normalized shear loss modulus $G''(\gamma_0)/G''(0)$, at a fixed angular frequency of 1.0 rad/s, for aqueous XG solutions (data from Song et al. (2006))

scale at a constant strain amplitude of 0.01, which has been reported to lie within the linear viscoelastic region for high XG concentrations (Song et al. 2006).

The strain dependence of the normalized shear storage (elastic) modulus $G'(\gamma_0)/G'(0)$ and the normalized shear loss modulus $G''(\gamma_0)/G''(0)$ (i.e., both normalized considering a fixed angular frequency of $\omega = 1.0$ rad/s) for aqueous XG solutions with concentrations ranging from 1 to 4% is shown in Fig. 3.13. For the normalized storage modulus $G'(\gamma_0)/G'(0)$ (Fig. 3.13a), a linear region is shown in the bi-logarithmic plot for smaller strain range (i.e., $\gamma_0 < 10\%$), in which the storage modulus exhibits a constant value irrespective of the shear strain level. However, for strain levels increasing from 10 to 100%, the normalized storage modulus exhibits a decreasing nonlinear behavior. Then, a sharp decrease in storage modulus is observed with strain in excess of 100%, which indicates a strain-thinning behavior due to destruction of the XG internal structure (Isono and Ferry 1995).

At small strain of $\gamma_0 < 10\%$, the entanglement density did not change because of the maintenance of a balanced state between the rates of XG structural breakdown and rebuilding of new ones. The deformation exhibits a linear behavior (i.e., with a constant storage modulus), indicating that XG is responding in an elastic manner to external shear stresses. However, as the shear strain increases, the balance between the destruction and reformation of XG structures starts to decrease (non-equilibrium state), causing a reduction in entanglement density. Such a condition will lead to the observed nonlinear elastic behavior, as well as a decrease in the storage modulus.

For the normalized shear loss modulus $G''(\gamma_0)/G''(0)$ (Fig. 3.13b), a linear region is observed in the bi-logarithmic plot for strains smaller than 80%, whereby the loss modulus exhibits a constant value regardless of the strain level. However, for strain levels larger than 80%, the loss modulus exhibits a rise–fall response, increasing up to a certain strain (~150%) and subsequently decreasing below the constant value observed for $\gamma_0 < 80\%$. The latter behavior, known as a strain-overshoot phenomenon

(Hyun et al. 2002, 2003) and discussed further below, may arise due to the unique structural conformation of the XG polymer's long sidechains that contribute to the formation of relatively weaker structural complexes. Notably, in aqueous media at room temperature, the backbone of XG exists in a disordered state that can be highly extended due to an electrostatic repulsion between the charged groups on the side chains; hence, XG molecules may align and associate (due to hydrogen bonding) to form a weak structured material (Rochefort and Middleman 1987). However, when an external shear stress is applied and large strains occurred, the XG complex structure will resist the external imposed deformation up to a certain strain level; as such, the loss modulus increases. Then, the XG complex structure is broken down due to increased strain level above a critical strain value (of $\gamma_o = 150\%$), after which the XG chains align with the flow fields and the loss modulus decreases.

The reported results by Song et al. (2006) also indicated that the storage modulus $G'(\gamma_o)$ is always greater than the loss modulus $G''(\gamma_o)$ within the linear viscoelastic region; hence, XG behaves as an elastic material in this domain. Furthermore, the storage modulus shows nonlinearity at smaller strain levels than that of the loss modulus. However, at large strains ($> 200\%$), XG behaves as a viscous material due to the sharper decrease of the storage modulus (compared to that of the loss modulus) with increasing strain level.

The behaviors of large-amplitude oscillatory shear (LAOS) flow can be classified into four types (Fig. 3.14), which are as follows (Hyun et al. 2002, 2003):

(a) *Type I*, strain-thinning: Both the shear storage and loss moduli are decreased. This is explained by polymer-chain disentanglement and chain reorientation, or alignment of microstructures along the flow direction, which reduces local viscous drag.
(b) *Type II*, strain-hardening: Both the storage and loss moduli are increased. This is associated with the formation of complex microstructures.
(c) *Type III*, weak strain-overshoot: The storage modulus is decreased, whereas the loss modulus is first increased and then decreased. Such behavior could be attributed to the effective volume of temporal network structures (Tirtaatmadja et al. 1997), or to the destruction of microstructures developed during the oscillations and the reformation of the microstructures (Sim et al. 2003).
(d) *Type IV*, strong strain-overshoot: Both the storage and loss moduli are first increased and then decreased, reaching local maxima followed by local minima for both moduli (Hyun et al. 2006).

In view of the above characterization of the LAOS flow, the concentrated XG large strain flow behavior, shown in Fig. 3.13b, belongs to Type III behavior.

Furthermore, general material behavior in the LAOS flow fields can be employed to differentiate between strong and weak gels (Ross-Murphy et al. 1983). At large strains, strong gels rupture and fail akin to a solid material, which indicates that the complex modulus $G^*(\omega)$ is essentially strain-independent (i.e., linearly viscoelastic). However, weak gels flow without fracture and they exhibit a structured fluid-like behavior, indicating that the complex modulus is strain-dependent. In relation to XG,

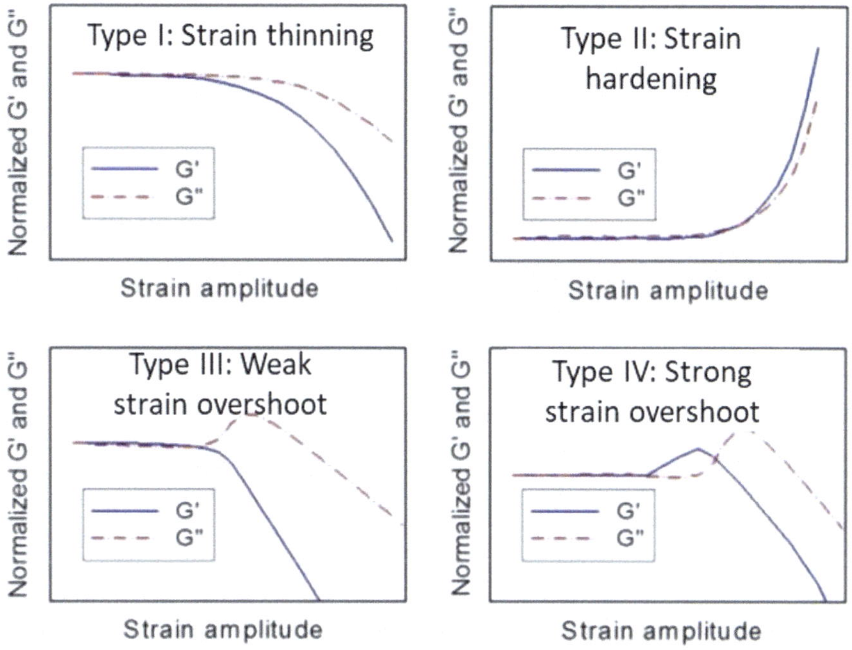

Fig. 3.14 Classification of large-amplitude oscillatory shear (LAOS) flow behavior (adopted from Hyun et al. (2002, 2003))

the experimental results showed that XG flows without fracture, such that concentrated XG can be classified as a weak gel, whereby its nonlinear viscoelastic behavior is mainly associated with a tendency to form aggregated structures that are broken down for the large strain condition.

The results shown in the bi-logarithmic plot of Fig. 3.15 (data from Song et al. (2006)) are for the storage modulus $G'(\omega)$, the loss modulus $G''(\omega)$, and loss tangent or $\tan(\delta) = G''(\omega)/G'(\omega)$, presented as functions of the angular frequency ω for aqueous XG solutions at 1 wt% concentration. As the angular frequency increases, both the storage and loss moduli are progressively increased. In addition, it is to be noted that the storage modulus is always greater than the loss modulus over the entire range of tested angular frequencies. This indicates that the linear viscoelastic behavior of the XG solutions is elastically dominated. Such behavior could be attributed to the presence of attractive interactions between the XG helices that formed by rigid XG macromolecules, which contributed to the formation of a weak three-dimensional network and a complex aggregation in solutions by means of physical entanglements and hydrogen bonding (Carnali 1992; Pal 1995).

The calculated power law parameters deduced from the shear storage and loss moduli against angular frequency results for aqueous XG solutions with different concentrations are shown in Table 3.2 (data from Song et al. (2006)). The K' values (for the storage modulus) are much larger than those of K'' (for the loss modulus) for

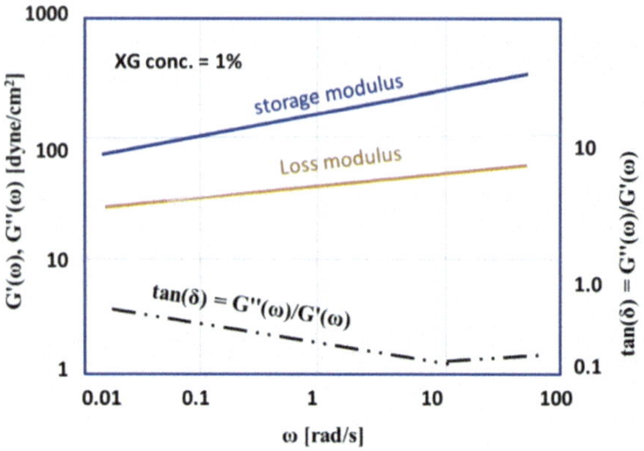

Fig. 3.15 Variations of the shear storage modulus $G'(\omega)$, loss modulus $G''(\omega)$, and loss tangent, tan(δ), as functions of the angular frequency ω for aqueous XG solutions at 1 wt% concentration (data from Song et al. (2006))

all XG solutions, indicating the dominance of the elastic behavior (storage modulus) over the viscous behavior (loss modulus). Also, the values of n' (for the storage modulus) and n'' (for the loss modulus) gradually decrease with increasing XG concentration, indicating that the frequency dependence of the linear viscoelastic functions becomes weaker for increasing XG concentration; i.e., the XG solutions become more elastic and can be characterized by a slower relaxation mechanism (Song et al. 2006).

With respect to the linear viscoelastic behavior in small-amplitude oscillatory shear (SAOS) flow fields, one can distinguish between strong and weak gels as follows: for $n' > 0$, the gel is known as physically weak, whereas for $n' = 0$ the gel is chemically strong (Ross-Murphy et al. 1983; Harrison et al. 1999; Jampala et al. 2005; Kim and Yoo 2006). For the concentrated XG solutions (Table 3.2), the n' values ranged between 0.13 and 0.18; hence, as demonstrated previously, this reaffirms that XG solutions do not form a stable (strong) gel, but instead exhibit a weak gel-like behavior.

It is worth noting that for intermediate and semi-dilute XG concentration regimes (i.e., $0.05\% < C_{XG} < 0.5\%$), both conformational ordering of the XG backbone (which is highly dependent on salt concentration) and intermolecular associations control the rheological properties of XG dilute solution (Rochefort 1987). However,

Table 3.2 Power law parameters of storage and loss moduli for aqueous XG solutions with different concentrations (calculated from data reported in Song et al. (2006))

XG concentration (wt%)	Storage modulus, $G'(\omega)$		Loss modulus, $G''(\omega)$	
	K' (Eq. 3.8) (dyn/cm^2/(rad/s)$^{n'}$)	n' (Eq. 3.10) ()	K'' (Eq. 3.9) (dyn/cm^2/(rad/s)$^{n''}$)	n'' (Eq. 3.11) ()
1	217	0.1842	65	0.1323
2	295	0.1723	84	0.1382
3	562	0.1450	131	0.1171
4	821	0.1337	183	0.1122

in high concentrated regime (i.e., $C_{XG} > 1\%$), stress nonlinearities and liquid–crystal type behavior become dominant (Lim et al. 1984).

3.4.2.2 Salt Concentration

As discussed earlier, XG has been characterized by its unusual rheological properties in the presence of salts. Normal polyelectrolyte behavior shows a viscosity decrease with the addition of salt because of charge screening that permits for a more compact structure. However, because of the salt-induced disorder → order structural transition, and the tendency for strong intermolecular associations of XG molecules, the rheological behavior is a function of the salt and XG concentrations. It has been reported that complete ordered XG conformation can be obtained for salt concentrations greater than 10^{-2} M, and that no further salt effects occur beyond this concentration level (Rochefort 1987).

The effects of both XG and salt concentrations on the shear storage (elastic) modulus can be evaluated using Fig. 3.16 (data from Rochefort (1987)). This figure illustrates the variations of the elastic modulus $G'(\gamma_o)$ normalized to the elastic modulus value at small shear strain $G'(0)$, with the strain applied at 1.0 rad/s frequency, investigating two XG concentrations with and without NaCl. Notably, the elastic modulus has been used here since it is the primary indicator of the XG gel-like structure. With these high concentrated XG solutions (0.25 and 0.5%), it is observed that 50% shear strain can be reached before XG structural breakdown begins, indicating that this is a reasonable strain level to adopt when performing frequency sweep experiments. At high strains for the same XG concentration, the structural breakdown is less for the high salt (0.5% NaCl) than that of deionized water (DW) solution. This could be attributed to the way the XG structure heals during shearing. In other words, for low salt concentrations, structure re-healing takes a long time, whereas in the case of high salt concentrations it takes extremely short time periods for re-association, indicating that the intermolecular associations are responsible for such rheological behavior of semi-dilute XG solutions (i.e., for $0.05\% < C_{XG} < 0.5\%$) (Rochefort 1987).

Notably, the interaction between Na^+ and COO^- screens the overall negative charges on the XG molecule causing: (i) a decrease in surface charge; (ii) a decrease of the repulsion forces acting between the XG molecules; (iii) destabilization of the XG gel suspension; (iv) XG molecules' entanglements and agglomeration; and (v) a compact structural conformation, which is arranged as a gel-like structure. This in turn: (i) results in a lower density of negative charges in the intra- and inter-chain (Galván et al. 2018); (ii) creates a larger conformational freedom of the isolated chain, which favors the pendant due to low entropy (Xu et al. 2013; Galván et al. 2018); (iii) contributes to the collapsing of the side chains down into the backbone, and forming an ordered conformation (Brunchi et al. 2014); (iv) leads to conforming a helix molecular conformation, as the side chains are folded in, and associated with the backbone, due to the formation of an ordered conformation (Zhong et al. 2013); (v) reduces the hydrodynamic diameter due to the formation of a compact

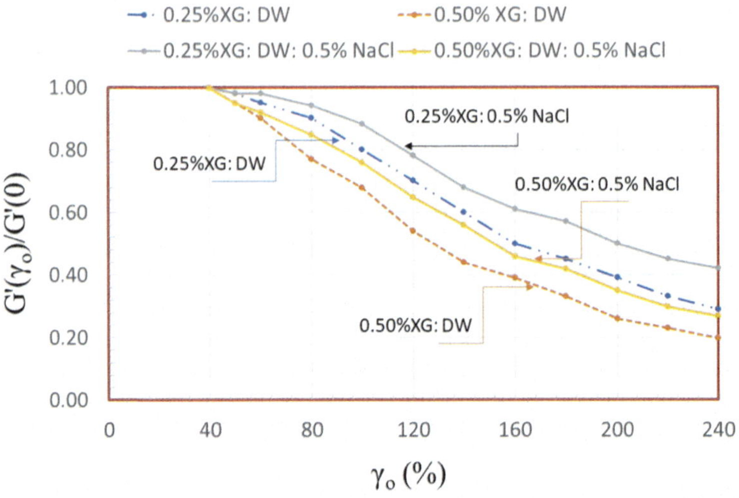

Fig. 3.16 Variations of the elastic modulus $G'(\gamma_o)$ normalized to the elastic modulus value at small shear strain $G'(0)$, with the strain applied at 1.0 rad/s frequency, investigating two XG concentrations of 0.25 and 0.5% with and without NaCl (data from Rochefort (1987)). *Note* DI, distilled water (solution)

conformation; and (vi) results in viscosity decrease under shearing (Brunchi et al. 2014).

3.4.2.3 XG Type

The effects of XG type on the viscoelastic nature of the solution and its dynamic properties (such as the shear storage G' and loss G'' moduli) have been evaluated by Ghebremedhin et al. (2020). They examined four different types of XG (from Jungbunzlauer Ladenburg GmbH, Germany): (i) XG1, an unmodified XG (M_w = 1.35×10^6 g/mol); (ii) XG2, which contained less pyruvate groups and had the least molecular weight ($M_w = 1.08 \times 10^6$ g/mol); (iii) XG3, which did not contain any acetate groups and possessed the highest molecular weight ($M_w = 1.68 \times 10^6$ g/mol); and (iv) XG4, which included calcium cations, since the pH level during the fermentation process was adjusted using $Ca(OH)_2$.

The viscoelastic properties of the four XGs, with 0.5 wt% concentration in water, were determined using a Discovery HR-3 rheometer (TA Instruments). The experiments were performed with a 25-mm diameter plate at a constant frequency of f = 1 Hz, constant temperature of T_M = 25 °C and a soaking time of 120 s. Then, the storage and loss moduli were determined as a function of the shear strain in the range of 0.001 to 1000%. This method is generally used to: (i) measure the storage G' and loss G'' moduli; (ii) describe the deformation behavior of polymers; (iii) determine the viscoelastic nature of the XGs that can be observed for values of $G' > G''$ in the linear viscoelastic range, which indicate a predominantly elastic behavior or solid-like structure due to network formation; and (iv) determine the sol–gel transition; hence the gelation point, which is the crossover point at $G' = G''$.

It was observed that for all four XGs tested, the shear storage and loss moduli exhibit a constant linear behavior, with fairly constant values of G' and G'' for up to 10% strain, followed by a decrease for further strain increase (Fig. 3.17; data from Ghebremedhin et al. (2020)). Moreover, the unmodified XG1, the XG4, which contained calcium ions, as well as the deacetylated XG3, show the same G' of about 5 Pa, and the same G'' of about 2 Pa. Since the results indicate that $G' > G''$, these three XGs exhibit properties of a gel type. For the pyruvate-poor XG (i.e., XG2), the storage modulus is about 4 Pa lower than those of the other three XGs, and is just below the loss modulus; hence, it has the weakest gel character among the investigated XGs.

In addition, XG1, XG2 and XG3 showed sol–gel temperature transition points (Fig. 3.18), where the experimental storage and loss moduli crossover (i.e., for $G' = G''$). The pyruvate-poor XG2 showed the lowest G' and G'' moduli, and a sol–gel transition occurring at a low temperature of about 18.1 °C. As such, with the rheometer tests performed at a constant T_M of 25 °C (i.e., > 18.1 °C), this explains the higher values of G'' (than G') measured for the XG2. For the deacetylated XG3, the moduli are higher than those obtained for XG2, and the sol–gel transition occurs at about 55.5 °C. The unmodified XG1 shows a sol–gel transition at a higher temperature of about 74.6 °C. The high moduli as well as the sol–gel transition point of XG1 could be ascribed to the nature of the interaction in the disordered polymer conformation via the non-specific entanglements.

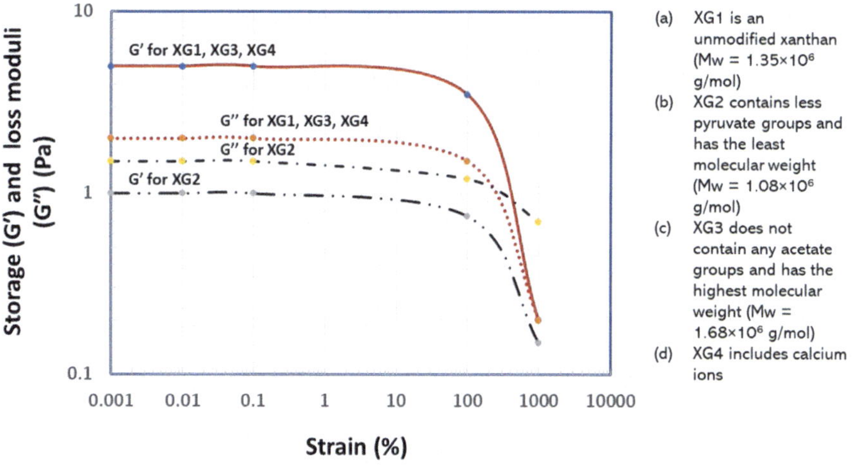

Fig. 3.17 Variations of the shear storage G' and loss G'' moduli measured at a constant temperature of 25 °C for four different XGs at 0.5 wt% concentration in water (data from Ghebremedhin et al. (2020))

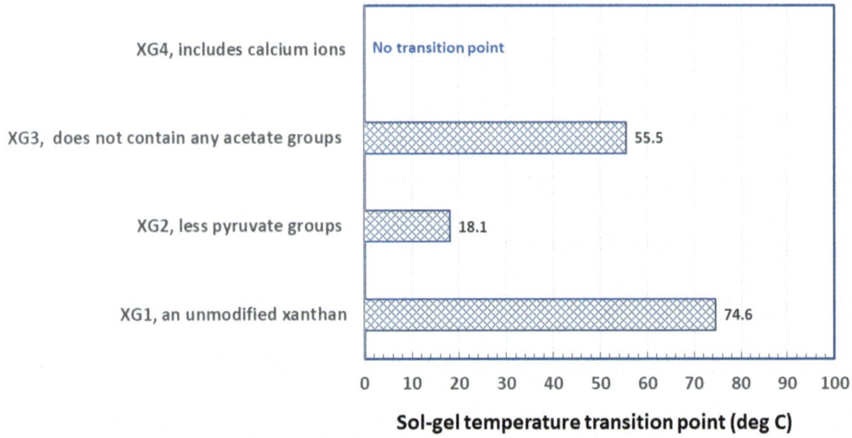

Fig. 3.18 Sol–gel temperature transition point of four different XGs at 0.5 wt% concentration in water (data from Ghebremedhin et al. (2020))

The XG2 showed the greatest temperature dependency of G' and G'', and had the lowest moduli values compared to the three other XG types tested, since: (i) it contained less pyruvate groups; hence, the number of negative charges, the repulsive forces between the polymer molecules, and the applied thermal energy to overcome these electrostatic repulsions, are all less; (ii) it had the least molecular weight; (iii) its molecules were less immobilized; and (iv) the repulsive forces at the negative side chains in the helix conformation of the polymer are high. The helix structure of XG3, which does not contain any acetate groups, is less stable due to the fewer number of acetyl residues. However, compared to XG1, XG3 is more soluble in water because it has more hydroxyl groups that form hydrogen bonds with the aqueous water solution.

Finally, XG4, which contained calcium cations, displayed the highest moduli, and did not exhibit a sol–gel temperature transition point. In addition, XG4 showed: (i) the smallest changes in the moduli with changes in temperature; (ii) nearly parallel curves for both moduli with temperature change; and (iii) a pronounced gel character, even at high temperatures, due to the formed bonds between the negatively charged XG4 molecules and the Ca^{2+} units, along with the crosslinking structure that contributed to formation of stiffer and larger molecule assemblies. Therefore, XG4 had the highest thermal stability among the tested XGs, which highlights the role of divalent cations on the stability of XG molecules.

3.4.2.4 Temperature

In view of the temperature effect, the results of Ghebremedhin et al. (2020) have indicated that for all four tested XGs there was an increase in the storage G' and loss G'' moduli with decreasing measurement temperature due to the reduced movement of the XG molecules as temperature decreased. In other words, during the cooling process, the dynamics of the XG molecules slow down; hence, they are slowly immobilized by random blocking as a result of the electrostatic repulsive forces between the negative charges. Notably, at high temperatures: (i) XG molecules are highly mobile, although their dynamic mobility is still limited due to the negative charges on their side chains; and (ii) due to the excessive negative charges, XG molecules might be still stiff even if they are not in the helical conformation. Accordingly, it seems that the coil–helix transition hardly influences the viscosity change of the XG.

As such, it can be noted that XG undergoes an order–disorder conformational transition upon heating, and the ordered conformation can be stabilized by adding salt (Milas and Rinaudo 1979). The molecular mechanism for this order–disorder conformational transition has not yet been fully established, and there is still discussion on whether the ordered form is a double- or single-stranded helix. Also, it is known that at higher concentrations, XG solutions exhibit gel-like behavior, which is attributed to weak intermolecular interactions between the ordered aggregates.

3.4.2.5 Cation Type

The effects of cation type and XG concentration on the rheological properties can be evaluated using the experimental data reported by Ma and Barbosa-Cánovas (1997). The investigated XG was produced in a pure-culture fermentation process, yielding a high-molecular-weight XG having a creamish color, with high stability to pH, salt and heat. XG concentrations of 0.5 and 1.0% were used with three types of cations, namely Na^+, Ca^{2+} and Fe^{3+}, which were in chloride forms; i.e., sodium chloride (NaCl), calcium chloride ($CaCl_2$) and ferric chloride ($FeCl_3{\cdot}6H_2O$). The results have indicated that the frequency sweep profiles for 0.5 and 1.0% XG gels in the absence or presence of Na^+, Ca^{2+}, and Fe^{3+} ions had similar patterns, except the magnitudes of both the shear storage and loss moduli were higher than that of deionized water (DW). Also, the frequency dependence of the storage modulus G' can be approximated by the following simple power law (Ma and Barbosa-Cánovas 1997):

$$G' = \alpha\omega^P \tag{3.13}$$

where the fitting coefficients (α and P) can be obtained from the bi-logarithmic plot of the storage modulus (G') versus frequency (ω).

From the experimental data for DW, NaCl, $CaCl_2$, and $FeCl_3$ solutions (Fig. 3.19; data from Ma and Barbosa-Cánovas (1997)), the respective values of P were deduced as 0.30, 0.22, 0.19, and 0.18 for 0.5% XG gels, and 0.24, 0.18, 0.12, and 0.14 for 1.0% XG gels. This clearly indicates the effects of both the ionic charge and XG concentration on the storage modulus. In other words, the value of P decreased should the XG concentration increase for the same ionic charge, or if the ionic charge increases for the same XG concentration. Furthermore, the addition of ionic species to XG solution decreases the slope of the elastic modulus with the frequency.

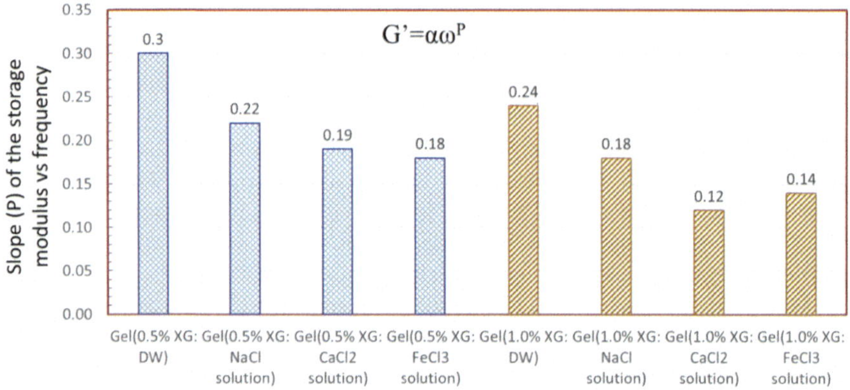

Fig. 3.19 Effects of ion type and XG concentration on the storage modulus (data from Ma and Barbosa-Cánovas (1997)). *Note* DW, deionized water

As discussed earlier, when the physical properties of XG solutions are measured under small-amplitude oscillatory shear, elastic (solid-like) properties (characterized by the storage modulus G') predominate over viscous (liquid-like) response (loss modulus G''), and both moduli exhibit only slight variations with frequency (ω). This behavior is typical of a gel rather than a solution. Thus, the XG solutions had a gel-like structure. Since the storage modulus is always greater than the loss modulus, and is relatively independent of shear frequency, the XG gel can be characterized as a weak gel. In addition, the role of the ionic species is to reduce intermolecular repulsion and promote network formation.

Furthermore, the role of ionic species valences (Na^+, Ca^{2+}, and Fe^{3+}) can be evaluated using data from Ma and Barbosa-Cánovas (1997) (see Fig. 3.20). For DW, NaCl, $CaCl_2$ and $FeCl_3$ solutions, the respective storage moduli were measured as 3.1, 8.3, 10, and 2700 Pa for 0.5% XG gels, and 2.6, 37, 42, and 13,000 Pa for 1.0% XG gels. The storage moduli of both Na^+ and Ca^{2+} were higher than that of 0.5% and 1.0% XG concentrations with DI; for Na^+, the increase was 2.7 to 14 times, and for Ca^{2+}, 3.2 to 16.2 times, for the two concentrations, respectively. However, the addition of Fe^{3+} increased the magnitude of the storage moduli by a factor of 870 and 5000 times greater than the storage moduli of 0.5% and 1.0% XG concentrations with DI, respectively. The increase of the storage moduli with addition of Na^+ and Ca^{2+} may be through the mechanism of a noncovalent cross-linking, in which different forces (e.g., Coulombic, dipole–dipole, van der Waals, charge transfer, and hydrophobic and hydrogen bonding interactions) are involved. However, Fe^{3+} may be involved in permanent chemical covalent bonds rather than physical cross-linking, since the presence of the ferric ion resulted in instantaneous gelation and a thermo-irreversible very strong gel (Ma and Barbosa-Cánovas 1997).

In addition to the storage and loss moduli, creep/recovery experiments provide valuable information about the XG gels' viscoelastic properties. Figure 3.21 shows a schematic representation for the creep behavior of XG gels for DI, NaCl, $CaCl_2$,

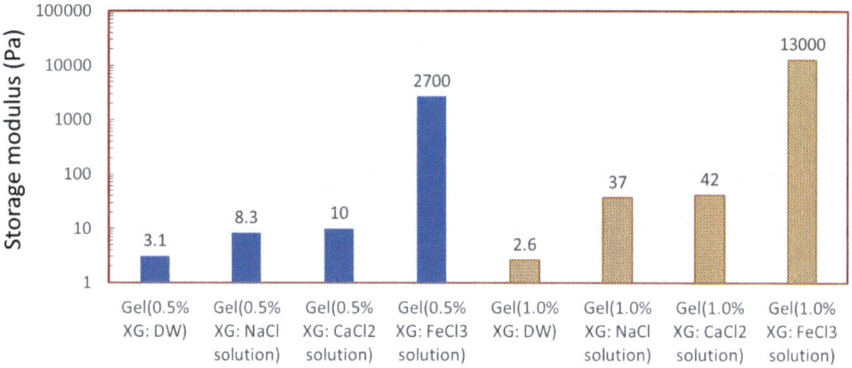

Fig. 3.20 Effect of ion type and XG concentration on the storage modulus measured at 1 Hz frequency (data from Ma and Barbosa-Cánovas (1997)). *Note* DW, deionized water

and $FeCl_3$ solutions. The compliance (J) shown on the vertical axis is the inverse of the loss modulus (i.e., $J = 1/G''$). In general, the creep behavior starts with a linear region, where the applied shear stress was small, and the creep/recovery response is independent of the applied stress. On achieving the peak compliance, this is followed by a nonlinear region, where a breakdown of structure occurs, and the creep response varies with the applied stress.

The variations of the peak compliance (calculated at the end of the linear region) with cation type and XG concentration are shown in Fig. 3.22 (data from Ma and Barbosa-Cánovas (1997)). With respect to XG gel with DI, the observed peak compliances for the XG gels with Na^+, Ca^{2+}, and Fe^{3+} cations were: (i) for 0.5% XG concentration, 4.8%, 1.2%, and 2×10^{-3}%, respectively; and (ii) for 1% XG concentration, 60%, 30%, and 1×10^{-3}%, respectively. The sharp decrease of the peak compliance occurring for the 0.5% XG gels indicates that the XG gel behavior is mainly viscous.

Fig. 3.21 Schematic representation of the creep/recovery behavior for XG gels with deionized water, NaCl, $CaCl_2$, and $FeCl_3$ solutions

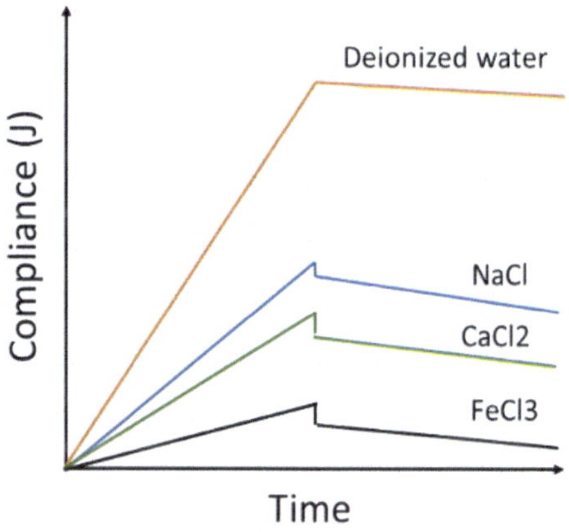

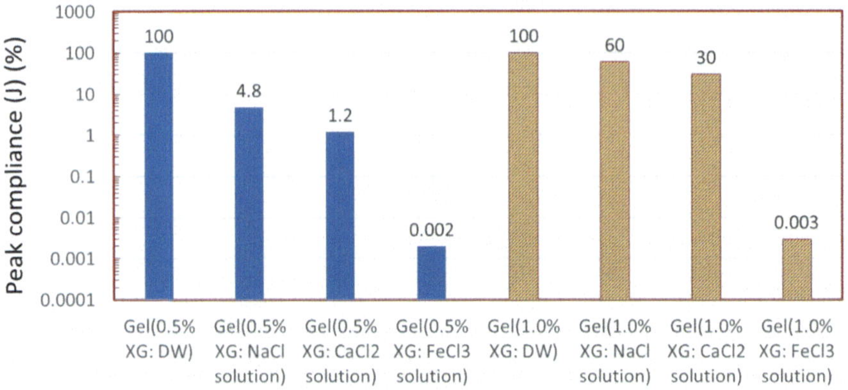

Fig. 3.22 Effects of ion type and XG concentration on the peak compliance *J* (data from Ma and Barbosa-Cánovas (1997)). *Note* DW, deionized water

However, for the 1% XG gels, the response became viscoelastic. Overall, the presence of cations contributed to a decrease of gel compliance due to an increase in the XG structural rigidity (Lynch and Mulvihill 1994). Notably, the presence of Fe^{3+} decreases the peak compliance by approximately three or four orders of magnitudes compared to that with Na^+ and Ca^{2+}; hence, the structural rigidity of the XG gel increased by Fe^{3+} addition.

In addition, Fig. 3.23 displays the elastic recovery (defined as the recoverable deformation at the end of recovery time expressed as a percentage of the total creep deformation) for different ion types and XG concentrations. The very small elastic recovery of 3.8% for the 0.5% XG gel with DI indicates that that the gel behaves as a viscous fluid. The higher elastic recovery of 27.8% measured for the 1.0% XG gel with DI was about seven times greater, indicating its behavior was more viscoelastic. Notably, the addition of Na^+, Ca^{2+}, and Fe^{3+} cations increased the elastic component in the viscoelastic properties as well as the elastic recovery for both XG concentrations.

3.5 XG Surface Charge and Interaction

The effect of XG type, as well as NaCl concentration, on the zeta potential of XG can be evaluated in view of the experimental data reported by Ghebremedhin et al. (2020) (see Fig. 3.24). They tested two different types of XG (from Jungbunzlauer Ladenburg GmbH, Germany): (i) XG1, which was an unmodified XG ($M_w = 1.35 \times 10^6$ g/mol); and (ii) XG2, which contained less pyruvate groups and had a lower molecular weight ($M_w = 1.08 \times 10^6$ g/mol). Zeta potentials were measured at an XG concentration of 0.001 wt% in differently concentrated NaCl solutions (0, 0.1, 0.5, 1, 5, 10, 50, 100, and 500 mM). As anionic polyelectrolytes, XG1 and XG2

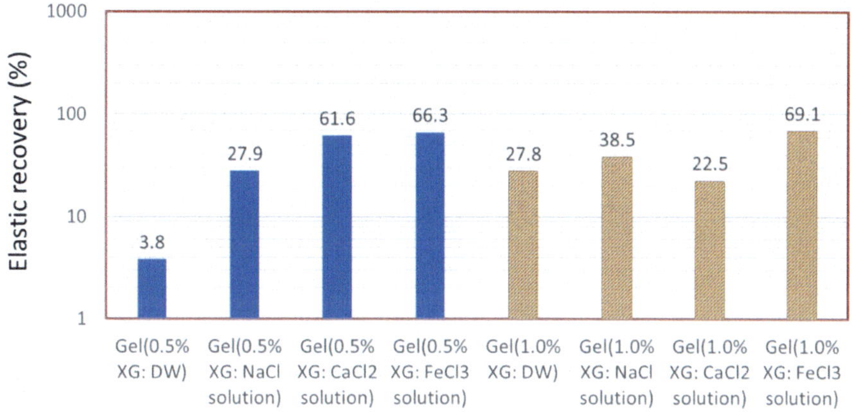

Fig. 3.23 Effects of ion type and XG concentration on the elastic recovery (data from Ma and Barbosa-Cánovas (1997)). *Note* DW, deionized water

showed a negative zeta potential. Figure 3.24 shows a clear increase in zeta potential (decrease in negative values) with increasing NaCl concentrations. With increasing shielding of the negative side chains by the sodium counter-ions, the XG polymers exhibit a zeta potential closer to zero.

The nature of the interaction of the active functional groups of the polysaccharide with metallic ions also affects the properties of polymers, e.g., polymer morphology and solution rheology. Some polysaccharides provide intramolecular binding sites for the cations, and their binding results in a considerable reduction of the polymer's

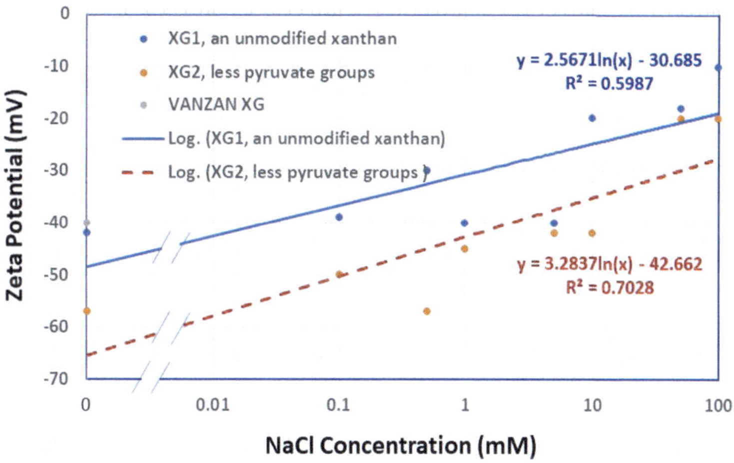

Fig. 3.24 Variations of zeta potential with NaCl concentration for various two types of XG polymers (data from Ghebremedhin et al. (2020))

hydrodynamic radius, and thus decreases the viscosity of the solution. Other polysaccharides form chelate-like complexes with metal cations via intermolecular cross-linking, which leads to a vast increase in the solution viscosity (Furth et al. 2008). Therefore, the interaction between XG and the bivalent heavy metal ions can be through forming intramolecular complexes between the side chains of the same XG molecule by: (i) the pyruvate units that are located at the outer ends of the side chains, as suggested by Bergmann et al. (2008) for the cases of Mn^{2+}, $Cd^{2+,}$ and Pb^{2+}; and (ii) the carboxylate groups of the adjacent glucuronic acid residues in the binding of the lighter bivalent cations, e.g., Ca^{2+} and Mg^{2+}, as suggested by Tako and Nakamura (1987); and (iii) the carboxylic groups of the adjacent glucuronic acid residues in the ligation of Cu^{2+}, $Co^{2+,}$ and Fe^{3+}, to form octahedral or square planar structures, as proposed by Ciesielski and Tomasik (2008).

3.6 XG Hydrogels Swelling and Strength

Hydrogels are three-dimensional networks of hydrophilic polymers, which can swell in the presence of water. Their ability to respond to external loading (e.g., pressure, temperature, pH, ionic strength, electric or magnetic fields) depends on the nature of the polymer chains. Understanding the swelling behavior of hydrogels in the presence of ions is important for a variety of science and engineering applications.

In presence of water, swelling occurs due to: (i) cation hydration, whereby cations with a very small radius or a large charge density have more influence on electron-pair acceptance of water, increasing its ability to hydrate electron donator groups, as carbonyls; (ii) specific interactions between ions and macromolecules and their first hydration shell, which involve the relative polarizabilities of the ions and the particle-to-particle interaction in the ion hydration layer; and (iii) the osmotic effect (concentration gradient), which drives water molecules out of the hydrogel structure, thereby causing gel contraction. Such ionic concentration gradient, which occurs inside the gel structure, could be developed because ions are totally excluded from intermediate and bound waters, and their transport inside the gel structure occurs only in the bulk phase (Wisniewski and Kim 1980). This in turn results in moving both the intermediate and bound waters to the bulk water, thereby enhancing stabilization of the system. Therefore, the polymer chains' dehydration and the resultant increase of hydrophobic interactions among them will enhance contraction/shrinking of the gel structure.

The swelling and strength behaviors of XG hydrogels can be evaluated based on the experimental data reported by Bueno et al. (2013). They investigated commercial XG (Kelzan®, CP Kelco, USA), with a pyruvate degree of 0.38, acetyl degree of 0.41, molecular weight of about 1.0×10^6 g/mol, and polymerization degree of 1072. XG films were produced by casting a 6 g/l XG aqueous solution in the absence or presence of citric acid at 0.3 g/l as a crosslinker. The solutions were homogenized with a stirrer at 18,000 rpm for 3 min, and then centrifuged for 5 min at 3600 rpm. Crosslinking was obtained by heating the dried films (with thickness of about 0.02 mm) at 165 °C for 7 min. Then, the sol fraction (i.e., the fraction of the polymer that was not part of the crosslinked network) was extracted with water at a temperature of about 70 °C, under gentle magnetic stirring, over a period of 24 h. The prepared un-crosslinked and acetic-acid-crosslinked XG hydrogels were then dried at 45 °C for 48 h before testing.

The swelling and strength (elastic modulus, stress at break, and elongation strain at break) properties of the three investigated XG gels are presented in Table 3.3. The gel content (G_c as %) and swelling degree at equilibrium with deionized water (i.e., S_{DW}) were calculated as follow:

$$G_c = \left[1 - \left(\frac{m_p - m_{dg}}{m_p}\right)\right] \times 100 \quad (3.14)$$

$$S_{DW} = \frac{m_w}{m_{dg}} = \frac{m_{sg} - m_{dg}}{m_{dg}} \quad (3.15)$$

where m_p = initial mass of polymer; m_{dg} = mass of dried gel; m_w = mass of water; and m_{sg} = mass of swollen water.

The results shown in Table 3.3 indicate that the dried acetic-acid-crosslinked XG had more gel content than the dried un-crosslinked XG. Furthermore, the density of XG films varied depending on the treatment method used. That is, the un-crosslinked XG hydrogel exhibited higher density compared to the dried un-crosslinked and acetic-acid-crosslinked XG hydrogels due to their irregular crosslinking points and increase of the gel content. This implies that the dried acetic-acid-crosslinked XG is slightly more rigid than the dried un-crosslinked XG, due to the higher homogeneity

Table 3.3 Swelling and strength properties of un-crosslinked, dried un-crosslinked and dried acetic-acid-crosslinked XG hydrogels (data from Bueno et al. (2013))

Property	XG hydrogel type		
	1	2	3
	Un-crosslinked (commercial XG, considering 6 g/l XG aqueous solution and no citric acid)	Dried un-crosslinked (gel type 1, dried at 45 °C for 48 h)	Dried acetic-acid-crosslinked (gel type 1 with citric acid after drying at 45 °C for 48 h)
Gel content, G_c (%)	–	20 ± 5	75 ± 15
Swelling degree at equilibrium, S_{DW}	–	150 ± 10	27 ± 3
Density (g/cm^3)	1.8	0.60	0.62
Elastic modulus (GPa)	3.1 ± 0.5	1.1 ± 0.1	1.9 ± 0.8
Stress at break (MPa)	40 ± 4	14 ± 4	18 ± 5
Strain at break (%)	3 ± 2	0.2 ± 0.1	0.10 ± 0.05

of the crosslinking. It has been established that the crosslinking density is inversely proportional to the swelling capacity (Peppas 1986). The dried un-crosslinked XG showed a higher swelling degree and lower gel content than the dried acetic-acid-crosslinked XG hydrogel; because no crosslinker was added and pure XG chains have limited number of groups that can be crosslinked.

The strength properties determined for both dried un-crosslinked and acetic-acid-crosslinked XG gels were smaller than those measured for un-crosslinked XG chains. The differences in the strength properties are likely due to the different packing densities in each sample arising from the different treatment methods. Irregular crosslinking points induce space formation between the XG chains, prompting a decrease in the packing density, elastic modulus and elongation strain (at breakage).

The swelling properties of XG hydrogels are known to be influenced by the solution pH (Thakur et al. 2011). Since dried un-crosslinked and acetic-acid-crosslinked XG hydrogels have acidic groups from O-acetyl (an acetyl group—normally CH_3CO attached to an oxygen atom from another functional group or molecule unit) and pyruvyl residues, or from the citric acid, they can be completely deprotonated at pH > 6. In addition, under basic medium, the unreacted hydroxyl groups are also deprotonated, and ester bonds undertake alkaline hydrolysis. Therefore, electrostatic repulsion and crosslink-disruption favor the increase in swelling ratio at equilibrium (i.e., S_{dpH}/S_{DW}, where S_{dpH} denotes the swelling degree with solutions at specific pH, and S_{DW} is the swelling degree with deionized water (DW)) observed for hydrogels in basic medium, as shown for pH 10 in Fig. 3.25 (data from Bueno et al. (2013)). Because it has more crosslinking to be hydrolyzed, dried acetic-acid-crosslinked XG hydrogel is more sensitive to changes in the pH level than dried un-crosslinked XG hydrogel. However, for pH 2 and pH 6.5, there is no significant effect of pH on the swelling ratios.

Bueno et al. (2013) also evaluated the influence of different ionic strength on the swelling of the XG gel, investigating differently prepared samples, as follows. First, the hydrogels were totally swollen in water, then they were immersed into different

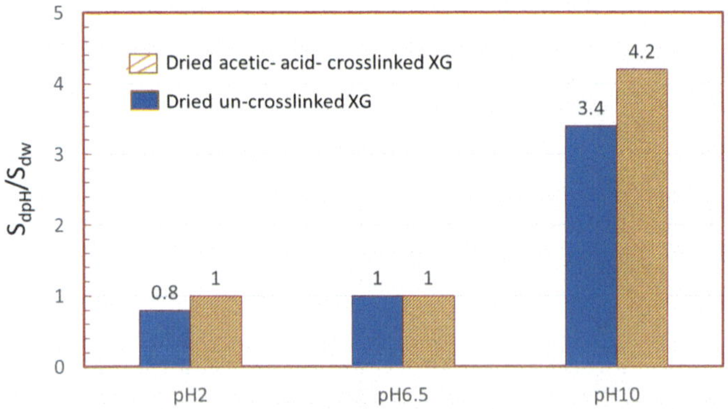

Fig. 3.25 Effects of pH on the swelling ratio (i.e., S_{dpH}/S_{DW}) of dried un-crosslinked and acetic-acid-crosslinked XG hydrogels (data from Bueno et al. (2013))

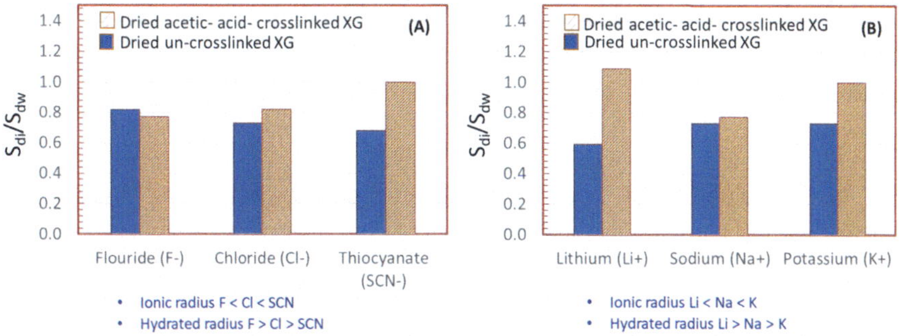

Fig. 3.26 Effects of monovalent ionic species on relative swelling of dried un-crosslinked and acetic-acid-crosslinked XG hydrogels: **a** anions; and **b** cations (data from Bueno et al. (2013))

chemical solutions, namely HCl at 0.1 mol/l, NaOH at 0.1 mmol/l, or LiCl, KCl, NaCl, NaSCN, and NaF at 1.0 mol/l. After 24 h, the hydrogels were removed from the medium, weighed and the relative swelling ratio (i.e., S_{di}/S_{DW}, where S_{di} denotes the swelling degree with a specific ion in the solution, and S_{DW} is the swelling degree with deionized water (DW)) was calculated.

The swelling ratio S_{di}/S_{DW} at equilibrium of dried un-crosslinked and acetic-acid-crosslinked XG hydrogels in the presence of monovalent anions (fluoride (F^-), chloride (Cl^-) and thiocyanate (SCN^-)) with Na^+ as the counter-ion, at 1.0 mol/l and pH 6.5, are shown in Fig. 3.26a. It was observed that for all investigated hydrogels, the gel shrunk/contracted (i.e., $S_{di}/S_{DW} < 1.0$), apart from the acetic-acid-crosslinked XG in the presence of the SCN^- anion because of the increase of the gel content and the electrostatic repulsion that together triggered swelling ($S_{di}/S_{DW} > 1$). In the presence of fluoride, contraction effects (reduction of diffused ion layers) were smaller because of the increase of solution pH to about 9 (alkaline), such that crosslinking is disrupted, and both charge density and hydrogel swelling are increased.

In addition, the contraction behavior of dried un-crosslinked and acetic-acid-crosslinked XG hydrogels was investigated in the presence of Li^+, Na^+ and K^+, with Cl^- as the common counter-ion, at 1.0 mol/l and pH 6.5 (see Fig. 3.26b). The dried un-crosslinked XG gels contracted more than the dried acetic-acid-crosslinked XG gels, especially in the presence of Li^+ (about 40%) due to osmotic effects, due to the relatively lower carbonyl content. Whereas the slight expansion behavior (about 10%) in the case of the dried acetic-acid-crosslinked XG hydrogel could be attributed to specific interactions between Li^+ and solvation layer of XG carbonyl or OH^- groups, which increases both hydration and swelling of the hydrogel.

3.7 Summary and Concluding Remarks

The nonlinear rheological behavior of XG solution in steady-shear flow fields was discussed in this chapter. XG exhibits a unique shear-thinning behavior, different from the generalized shear rate dependence that is known for disordered "random-coil" polysaccharides. Also, the linear rheological behavior of XG solution in dynamic shear flow fields was highlighted.

As summarized below, the rheological behavior of XG in aqueous solution depends on XG type, granulation and concentration, shear rate, dissolution and measurement temperatures, pH level, the monovalent and divalent metal ion concentration, and surface tension.

(a) The XG pseudoplastic and shear-thinning behaviors are maintained over a wide range of temperature, salt concentrations and pH levels. This can be related to the existence of an ordered rod-like (rigid) conformation with a high persistence length that can self-associate, creating domains of associate chains (aggregates) at low shear rates or at-rest. At high shear rates, however, the gel aggregates convert into solution and the double-helical structure of XG degrades into single chain. Nevertheless, if the applied pressure is removed, the XG will regain its original structure.
(b) XG solution viscosity highly depends on the dissolution temperature, i.e., the viscosity declines as dissolution temperature increases from 20 to 40 °C, while it increases with increasing dissolution temperature from 40 to 60 °C, and then declines again for dissolution temperatures greater than 60 °C. Such behavior is directly linked to the conformational changes of the XG molecule, whereby there is a conformation shift from an ordered (low dissolution temperature) to a disordered (high dissolution temperature) state.
(c) The particle size or granulation of XG strongly influences its hydrating and dispersing performance. A fine granulation results in a fast hydration with dispersion because of the higher tendency towards forming lumps, whereas coarser granulations hydrate slower with an easier dispersion.
(d) Surface tension depends strongly on polymer concentration, salinity, pH and temperature. At low XG concentrations (0.1 wt%), no changes occur in the surface activities. However, as XG concentration increased to 1.0 wt%, a dramatic decrease of surface tension was observed. Furthermore, surface tension decreases with increasing NaCl concentration in XG solutions due to NaCl screening of the negatively charged carboxyl groups, reducing electrostatic repulsion, and favoring the formation of a compact helix conformation because of self-aggregation. For the same temperature, surface tension decreases with decreasing pH of the solution. Whereas for the same pH, surface tension decreases with increasing temperature. Surface tension of XG in aqueous solutions is highly influenced by the nature of the XG conformational structure, i.e., surface tension is high for the coil conformation (which can be obtained in alkaline pH, and at high temperatures in dilute solutions) compared to that of

the helix conformation. Under both acidic and neutral conditions, XG hydrogels are stable; however, under alkaline conditions ester bonds are hydrolyzed.

(e) XG generally shows good resistance to high salinity, irrespective of the salt concentration and the order of valency (monovalent or divalent). The good resistance to high salinity is possible because XG is composed of chains that are conformed into different ordering, depending on the degree of salinity and temperature. As anionic polyelectrolytes, XG shows negative zeta potential with a clear increase occurring for increasing NaCl concentration. With increasing shielding of the negative sidechains by the sodium counter-ions, the XG polymers exhibit a zeta potential closer to zero.

(f) The structural conformation of XG can be interpreted from viscosity measurements by adopting the Mark–Houwink theory. However, new formula to account for the high-molecular-weight polymers, such as XG, and the chain stiffness of worm-like chains, should be developed.

The preceding discussion was mainly related to the XG's nonlinear rheological behavior in steady-shear flow fields, which are not capable of providing information regarding the relationship between the flow behavior and the microstructure of the XG due to its nature-destructive technique. To overcome this limitation, the XG's rheological behavior must be measured at small deformation rates, compatible to field conditions. This is achieved using the dynamic test, which is performed in oscillatory shear flow fields by applying a small amplitude of sinusoidal strain (or stress) and measuring the resulting stress (or strain), similar to the professional practices adopted in the food, polymer and road-pavement (bitumen) industries. This method is generally used to: (i) measure the shear storage G' and loss G'' moduli; (ii) describe the deformation behavior of the polymers; (iii) determine the viscoelastic nature of the XGs that can be observed for values of $G' > G''$ in the linear viscoelastic range, which indicate a predominantly elastic behavior or solid-like structure due to network formation; and (iv) determine the sol–gel transition, and hence the gelation point, which is the crossover point at $G' = G''$. The following points are notable in respect of the XG's small-strain rheological behavior:

(a) At small strains, the entanglement density does not change because of the maintenance of a balanced state between the rates of XG structural breakdown and rebuilding of new ones. This indicates that the XG is responding in an elastic manner to the external shear stresses, and the deformation exhibits a linear behavior (i.e., producing a constant storage modulus G'). As the strain increases, however, the balance between the destruction and reformation of XG structures starts to decrease (non-equilibrium state), leading to a reduction in entanglement density. Such a condition will lead to the observed nonlinear elastic behavior, and a decrease in the storage modulus.

(b) The results also indicate that the storage modulus is always greater than the loss modulus within the linear viscoelastic region; hence, XG behaves as an elastic material in this domain. In addition, the storage modulus shows nonlinearity at a smaller strain level than that of the loss modulus. However, at large strain,

XG behaves as a viscous material due to the sharper decrease of the storage modulus (compared to that of the loss modulus) with increasing strain level.

(c) Since the experimental results indicated that XG flows without fracture, concentrated XG can be classified as weak gel, whereby its nonlinear viscoelastic behavior is mainly associated with a tendency to form aggregated structures that are broken down for the large strain condition.

Additionally, the swelling and strength behavior/properties of XG hydrogels were discussed. Swelling occurs due to: (i) cation hydration, whereby cations with a very small radius or a large charge density have more influence on electron-pair acceptance of water, increasing its ability to hydrate electron donator groups, as carbonyls; (ii) specific interactions between ions and macromolecules and their first hydration shell, which involve the relative polarizabilities of the ions and the particle-to-particle interaction in the ion hydration layer; (iii) osmotic effect (concentration gradient), which drives water molecules out of the hydrogel structure, thereby causing gel contraction. Such ionic concentration gradient, which occurs inside the gel structure, could be developed because ions are totally excluded from intermediate and bound waters, and their transport inside the gel structure occurs only in the bulk phase. This in turn results in moving both the intermediate and bound waters to bulk water, thereby enhancing stabilization of the system. As such, the polymer chains' dehydration and the resultant increase of hydrophobic interactions among them will enhance contraction/shrinking of the gel structure.

3.8 Recommendations

It is of high importance to develop standard procedures for XG testing and establish rheological criteria for XG use in geotechnical and geo-environmental applications, akin to those developed by AASHTO (2016) for bitumen use in the road-pavement industry to evaluate both the rutting and the fatigue cracking resistances due to cyclic loadings.

For instance, in evaluating the rutting resistance of the Rolling Thin Film Oven (RTFO)-aged binder, the stiffness parameter $[G^*/\sin\delta]$ at any strain level should be greater than or equal to 2200 Pa in order to perform well against rutting (AASHTO 2016). A higher value of the stiffness parameter represents better resistance against the permanent deformation under a wheel load, with specific examples including:

(a) Rutting resistance of RTFO-aged Performance Grade (PG) 58–28 binder at 58 °C, $[G^*/\sin\delta] = 2500$ Pa ≥ 2200 Pa (shear strain ranging from 0.01 to 100%)

(b) Rutting resistance of RTFO-aged PG 64–28 binder at 64 °C, $[G^*/\sin\delta] = 3250$ Pa ≥ 2200 Pa (shear strain ranging from 0.01 to 100%)

(c) Multiple stress creep recovery (MSCR) test result for RTFO-aged PG 58–28 binder at 58 °C,

- Non-recoverable creep compliance (Jnr) @ 3.2 kPa^{-1} = 2.5 (for standard traffic condition, Jnr $\leq$ 5 kPa^{-1}; Heavy Traffic "H" Grade, Jnr $\leq$ 2 kPa^{-1})
- % recovery of 24%.

Furthermore, for fatigue cracking resistance evaluation, the stiffness parameter $[G^* \times \sin \delta]$ of pressure-aging-vessel-aged binders at different shear strain levels at an intermediate temperature should be less than the Superpave specified maximum critical value of 5×10^6 Pa (AASHTO 2016). For instance, the fatigue resistance of RTFO-aged PG 64–28 binder at 64 °C, $[G^* \times \sin \delta] = 1 \times 10^6 \leq 5 \times 10^6$ Pa (shear strain ranging from 0.01 to 100%).

References

American Association of State Highway Transportation Officials (AASHTO) R 28-12 (2016) Accelerated aging of asphalt binder using a pressurized aging vessel (PAV). Washington, DC

Arendt O, Kulicke WM (1998) Determination of the viscoelastic properties of a homologous series of the fermentation polymer xanthan gum. Die Angew Makromol Chem 259(1):61–67. https://doi.org/10.1002/(SICI)1522-9505(19981001)259:1%3c61::AID-APMC61%3e3.0.CO;2-V

Bejenariu A, Popa M, Picton L, Le Cerf D (2010) Effect of concentration, pH and temperature on xanthan conformation: a preliminary study before crosslinking. Rev Roum Chim 55(2):147–152

Bergmann D, Furth G, Mayer C (2008) Binding of bivalent cations by xanthan in aqueous solution. Int J Biol Macromol 43(1):245–251. https://doi.org/10.1016/j.ijbiomac.2008.06.001

Berninger T, Dietz N, Lopez OG (2021) Water-soluble polymers in agriculture: xanthan gum as eco-friendly alternative to synthetics. Microb Biotechnol 4(5):1881–1896. https://doi.org/10.1111/1751-7915.13867

Berth G, Dautzenberg H, Christenson BE, Harding SE, Rother G, Smidsrod O (1996) Static light scattering studies on xanthan in aqueous solutions. Macromolecules 29:3491–3500

Bradshaw IJ, Nisbet BA, Kerr MH, Sutherland IW (1983) Modified xanthan—its preparation and viscosity. Carbohydr Polym 3(1):23–38. https://doi.org/10.1016/0144-8617(83)90010-3

Brunchi CE, Morariu S, Bercea M (2014) Intrinsic viscosity and conformational parameters of xanthan in aqueous solutions: salt addition effect. Colloids Surf B 122:512–519. https://doi.org/10.1016/j.colsurfb.2014.07.023

Brunchi C-E, Bercea M, Morariu S, Avadanei M (2016, September 2016) Investigations on the interactions between xanthan gum and poly(vinyl alcohol) in solid state and aqueous solutions. Euro Polym J 84. https://doi.org/10.1016/j.eurpolymj.2016.09.006

Bueno VB, Petri DFS (2014) Xanthan hydrogel films: molecular conformation, charge density and protein carriers. Carbohydr Polym 101:897–904. https://doi.org/10.1016/j.carbpol.2013.10.039

Bueno VB, Bentini R, Catalani LH, Petri DFS (2013) Synthesis and swelling behavior of xanthan-based hydrogels. Carbohydr Polym 92(2):1091–1099. https://doi.org/10.1016/j.carbpol.2012.10.062

Capron I, Brigand G, Muller G (1997) About the native and renatured conformation of xanthan exopolysaccharide. Polymer 38(21):5289–5295. https://doi.org/10.1016/S0032-3861(97)00079-7

Carnali JO (1992) Gelation in physically associating biopolymer systems. Rheol Acta 31:399–412. https://doi.org/10.1007/BF00701120

Carriere CJ, Amis EJ, Schrag JL, Ferry JD (1993) Dilute-solution dynamic viscoelastic properties of xanthan polysaccharide. J Rheol 37:469–478

Chang I, Kwon YM, Im J, Cho GC (2019) Soil consistency and interparticle characteristics of xanthan gum biopolymer–containing soils with pore-fluid variation. Can Geotech J 56:1206–1213. https://doi.org/10.1139/cgj-2018-0254

Ciesielski W, Tomasik P (2008) Metal complexes of xanthan gum. Electron J Pol Agric Univ 11(2):25. http://www.ejpau.media.pl/volume11/issue2/art-25.html

Cuvelier G, Launay B (1986) Concentration regimes in xanthan gum solutions deduced from flow and viscoelastic properties. Carbohydr Polym 6(5):321–333. https://doi.org/10.1016/0144-8617(86)90023-8

Dario AF, Hortencio LMA, Sierakowski MR, Queiroz Neto JC, Petri DFS (2011) The effect of calcium salts on the viscosity and adsorption behavior of xanthan. Carbohydr Polym 84(1):669–676. https://doi.org/10.1016/j.carbpol.2010.12.047

Ferry JD (1980) Viscoelastic properties of polymers, 3rd edn. John Wiley & Sons, New York, USA

Funahashi H, Yoshida T, Taguchi H (1987) Effect of glucose concentration on xanthan gum production by *xanthomonas campestris*. J Ferment Technol 65(5):603–606. https://doi.org/10.1016/0385-6380(87)90124-5n

Furth G, Knierim R, Buss V, Mayer C (2008) Binding of bivalent cations by hyaluronate in aqueous solution. Int J Biol Macromol 42(1):33–40. https://doi.org/10.1016/j.ijbiomac.2007.09.001

Galván ZRN, Soares LdS, Medeiros EAA, Soares NdFF, Ramos AM, Coimbra JSdR, de Oliveira EB (2018) Rheological properties of aqueous dispersions of xanthan gum containing different chloride salts are impacted by both sizes and net electric charges of the cations. Food Biophys 13(2):186–197. https://doi.org/10.1007/s11483-018-9524-9

Garcia-Ochoa F, Casas JA (1994) Apparent yield stress in xanthan gum solution at low concentration. Chem Eng J 53(3):41–46. https://doi.org/10.1016/0923-0467(93)06043-P

Ghebremedhin M, Schreiber C, Zielbauer B, Dietz N, Vilgis TA (2020) Interaction of xanthan gums with galacto- and glucomannans. Part II: heat induced synergistic gelation mechanism and their interaction with salt. J Phys Mater 3(3):034014. https://doi.org/10.1088/2515-7639/ab9ac9

Graessley WW (1982) Entangled linear, branched and network polymer systems—molecular theories. In: Abe A et al (eds) Synthesis and degradation rheology and extrusion. Advances in polymer science, vol 47. Springer, Berlin, Heidelberg. https://doi.org/10.1007/BFb0038532

Harrison G, Franks GV, Tirtaatmadja V, Boger DV (1999) Suspensions and polymers: common links in rheology. Korea Aust Rheol J 11:197–218

Hyun K, Kim SH, Ahn KH, Lee SJ (2002) Large amplitude oscillatory shear as a way to classify the complex fluids. J Non Newt Fluid Mech 107(1):51–65. https://doi.org/10.1016/S0377-0257(02)00141-6

Hyun K, Nam JG, Wilhelm M, Ahn KH, Lee SJ (2003) Nonlinear response of complex fluids under LAOS (large amplitude oscillatory shear) flow. Korea Aust Rheol J 159(2):97–105

Hyun K, Nam J, Wilhelm M, Ahn K, Lee SJ (2006) Large amplitude oscillatory shear behavior of PEO-PPO-PEO triblock copolymer solutions. Rheol Acta 45(3):239–249. https://doi.org/10.1007/s00397-005-0014-x

Hyun K, Wilhelm M, Klein CO, Cho KS, Nam JG, Ahn KH, Lee SJ, Ewoldt RH, McKinley GH (2011) A review of nonlinear oscillatory shear tests: analysis and application of large amplitude oscillatory shear (LAOS). Progr Polym Sci 36(12):1697–1753, ISSN 0079-6700. https://doi.org/10.1016/j.progpolymsci.2011.02.002

Isono Y, Ferry JD (1995) Stress relaxation and differential dynamic modulus of polyisobutylene in large shearing deformations. J Rheol 29:273–280

Jamieson AM, Southwick JG, Blackwell J (1982) Dynamical behavior of xanthan polysaccharide in solution. J Polym Sci Polym Phys Ed 20(9):1513–1524. https://doi.org/10.1002/pol.1982.180200901

Jampala SN, Manolache S, Gunasekaran S, Denes FS (2005) Plasma-enhanced modification of xanthan gum and its effect on rheological properties. J Agric Food Chem 53(9):3618–3625. https://doi.org/10.1021/jf0479113

Jones SA, Goodall DM, Cutler AN, Norton IT (1987) Application of conductivity studies and polyelectrolyte theory to the conformation and order-disorder transition of xanthan polysaccharide. Eur Biophys J 15:185–191. https://doi.org/10.1007/BF00263683

Kim C, Yoo B (2006) Rheological properties of rice starch xanthan gum mixtures. J Food Eng 75(1):120–128. https://doi.org/10.1016/j.jfoodeng.2005.04.002

Khouryieh HA, Herald TJ, Aramouni F, Alavi S (2007) Intrinsic viscosity and viscoelastic properties of xanthan/guar mixtures in dilute solutions: effect of salt concentration on the polymer interactions. Food Res Int 40(7):883–893. https://doi.org/10.1016/j.foodres.2007.03.001

Kitagawa H, Sato T, Norisuye T, Fujita H (1985) Optical rotation behavior of xanthan in mixtures of water and cadoxen. Carbohydr Polym 5(6):407–422. https://doi.org/10.1016/0144-8617(85)90002-5

Klaic PMA, Nunes AM, da Silveira Moreira A, Vendruscolo CT, Ribeiro AS (2011) Determination of Na, K, Ca and Mg in xanthan gum: sample treatment by acid digestion. Carbohydr Polym 83(4):1895–1900. https://doi.org/10.1016/j.carbpol.2010.10.059

Le XT, Turgeon SL (2013) Rheological and structural study of electrostatic cross-linked xanthan gum hydrogels induced by β-lactoglobulin. Soft Matter 9(11):3063–3073. https://doi.org/10.1039/C3SM27528K

Lecourtier J, Chauveteau G, Muller G (1986) Salt-induced extension and dissociation of a native double-stranded xanthan. Int J Biol Macromol 8(5):306–310. https://doi.org/10.1016/0141-8130(86)90045-0

Lee H, Yoo B (2020) Agglomerated xanthan gum powder used as a food thickener: effect of sugar binders on physical, microstructural, and rheological properties. Powder Technol 362:301–306. https://doi.org/10.1016/j.powtec.2019.11.124

Lee H, Yoo B (2021) Agglomeration of galactomannan gum powders: physical, rheological, and structural characterizations. Carbohydr Polym 256:117599. https://doi.org/10.1016/j.carbpol.2020.117599

Lim T, Uhl JT, Prud'hornme RK (1984) Rheology of self-associating concentrated xanthan solutions. J Rheol 28:367–379. https://doi.org/10.1122/1.549757

Lynch MG, Mulvihill DM (1994) The influence of caseins on the rheology of iota-carrageenan gels. Food Hydrocolloids 8:312–329

Ma L, Barbosa-Cánovas GV (1997) Viscoelastic properties of xanthan gels interacting with cations. J Food Sci 62(6):1124–1128. https://doi.org/10.1111/j.1365-2621.1997.tb12227.x

Macosko CW (1994) Rheology: principles, measurements and applications. VCH Publishers, New York, USA

Milas M, Rinaudo M (1979) Conformational investigation on the bacterial polysaccharide xanthan. Carbohydr Res 76:189–196. https://doi.org/10.1016/0008-6215(79)80017-8

Milas M, Rinaudo M (1986) Properties of xanthan gum in aqueous solutions: role of the conformational transition. Carbohydr Res 158:191–204. https://doi.org/10.1016/0008-6215(86)84017-4

Milas M, Rinaudo M, Duplessix R, Borsali R, Linder P (1995) Small angle neutron scattering from polyelectrolyte solutions: from disordered to ordered xanthan chain conformation. Macromolecules 28(9):3119–3124. https://doi.org/10.1021/ma00113a014

Mohammed ZH, Haque A, Richardson RK, Morris ER (2007) Promotion and inhibition of xanthan 'weak-gel' rheology by calcium ions. Carbohydr Polym 70(1–2):38–45. https://doi.org/10.1016/j.carbpol.2007.02.026

Muller G, Lecourtier J (1988) Temperature-induced extension and dissociation of native xanthan. Carbohydr Polym 9(3):213–225, ISSN 0144-8617. https://doi.org/10.1016/0144-8617(88)90026-4

Nedjhioui M, Moulai-Mostefa N, Canselier JP, Bensmaili A (2009) Investigation of combined effects of xanthan gum, sodium dodecyl sulphate, and salt on some physicochemical properties of their mixtures using a response surface method. J Dispersion Sci Technol 30(9):1333–1341. https://doi.org/10.1080/01932690902735538

Norton IT, Goodall DM, Frangou SA, Morris ER, Rees DA (1984) Mechanism and dynamics of conformational ordering in xanthan polysaccharide. J Mol Biol 175(3):371–394. https://doi.org/10.1016/0022-2836(84)90354-1

Okechukwu Ezeh O, Ikiensikimama SS, Akaranta O (2021) Comparative analysis of the effects of monovalent and divalent ions on imported biopolymer-xanthan gum and locally formulated biopolymers-gum arabic and terminalia mantaly. J Energy Res Rev 9(2):33–46. https://doi.org/10.9734/jenrr/2021/v9i230229

Pal R (1995) Oscillatory, creep and steady shear flow behavior of xanthan-thickened oil-in-water emulsions. AIChE J 41:783–794. https://doi.org/10.1002/aic.690410405

Pastor MV, Costell E, Izquierdo L, Duran L (1994) Effects of concentration, pH and salt content on flow characteristics of xanthan gum solutions. Food Hydrocolloids 8(3–4):265–275

Pelletier E, Viebke C, Meadows J, Williams PA (2001) A rheological study of the order–disorder conformational transition of xanthan gum. Biopolymers 59(5):339–346. https://doi.org/10.1002/1097-0282(20011015)59:5%3c339::AID-BIP1031%3e3.0.CO;2-A

Peppas NAM (1986) Hydrogels in medicine and pharmacy fundamentals. CRC Press, Boca Raton

Petri DFS, Queiroz Neto JC (2010) Identification of lift-off mechanism failure for salt drill-in drilling fluid containing polymer filter cake through adsorption/desorption studies. J Petroleum Sci Eng 70(1):89–98. https://doi.org/10.1016/j.petrol.2009.10.002

Prud'homme RK, Long RE (1983) Surface tensions of concentrated xanthan and polyacrylamide solutions with added surfactants. J Colloid Interface Sci 93(1):274–276

Rinaudo M, Milas M (1978) Polyelectrolyte behaviour of a bacterial polysaccharide from *Xanthomonas campestris*: comparison with carboxymethyl cellulose. Biopolymers 17(11):2663–2678. https://doi.org/10.1002/bip.1978.360171113

Rochefort WE (1987) Rheology of xanthan gum: salt, temperature, and strain effects in oscillatory and steady shear experiments. J Rheol 31(4):337–369. https://doi.org/10.1122/1.549953

Rochefort WE, Middleman S (1987) Rheology of xanthan gum: salt, temperature, and strain effects in oscillatory and steady shear experiments. J Rheol 31(4):337–369

Ross-Murphy SB, Morris VJ, Morris ER (1983) Molecular viscoelasticity of xanthan polysaccharide. Faraday Symp Chem Soc 18:115–129. https://doi.org/10.1039/FS9831800115

Sato T, Norisuye T, Fujita H (1984a) Double-stranded helix of xanthan in dilute solution: evidence from light scattering. Polym J 16:341–350. https://doi.org/10.1295/polymj.16.341

Sato T, Kojima S, Norisuye T, Fujita H (1984b) Double-stranded helix of xanthan in dilute solution: further evidence. Polym J 16(5):423–429. https://doi.org/10.1295/POLYMJ.16.423

Sato T, Kojima S, Norisuye T, Fujita H (1984c) Double-stranded helix of xanthan: dimensional and hydrodynamic properties in 0.1 M aqueous sodium chloride. Macromolecules 17:2696–2700

Secouard S, Malhiac C, Grisel M (2005) The effect of solution surface tension on aroma compound release from aqueous xanthan solutions. Flavour Fragr J 21(1):8–12. https://doi.org/10.1002/ffj.1695

Sim H, Ahn K, Lee S (2003) Three-dimensional dynamics simulation of electrorheological fluids under large amplitudes oscillatory shear flow. J Rheol 47:879–895

Song KW, Kuk HY, Chang GS (2006) Rheology of concentrated xanthan gum solutions: oscillatory shear flow behavior. Korea-Aust Rheol J 18(2):67–81

Southwick JG, Lee H, Jamieson AM, Blackwell J (1980) Self-association of xanthan in aqueous solvent-systems. Carbohydr Res 84(2):287–295. https://doi.org/10.1016/S0008-6215(00)85558-5

Stokke BT, Christensen BE (1996) Release of disordered xanthan oligomers upon partial acid hydrolysis of double-stranded xanthan. Food Hydrocoll 10(1):83–89. https://doi.org/10.1016/S0268-005X(96)80058-0

Tako M, Nakamur S (1984) Rheological properties of deacetylated xanthan gum in aqueous media. Agric Biol Chem 48(12):2987–2993

Tako M, Nakamura S (1987) Rheological properties of Ca salt of xanthan in aqueous media. Agric Biol Chem 51(11):2919–2924

Thakur A, Wanchoo RK, Singh P (2011) Structural parameters and swelling behavior of pH sensitive poly(acrylamide-co-acrylic acid) hydrogels. Chem Biochem Eng Q 25(2):181–194

Tinland B, Rinaudo M (1989) Dependence of the stiffness of the xanthan chain on the external salt concentration. Macromolecules 22(4):1863–1865. https://doi.org/10.1021/ma00194a058

Tirtaatmadja V, Tam KC, Jenkins RD (1997) Rheological properties of model alkali-soluble associative (HASE) polymers: effect of varying hydrophobe chain length. Macromolecules 30:3271–3282

Tschoegl NW (1989) The phenomenological theory of linear viscoelastic behavior. Springer, Berlin, Heidelberg. https://doi.org/10.1007/978-3-642-73602-5

Vega ED, Vásquez E, Diaz JRA, Masuelli MA (2015) Influence of the ionic strength in the intrinsic viscosity of xanthan gum. An experimental review. J Polym Biopolymer Phys Chem 3(1):12–18. https://doi.org/10.12691/jpbpc-3-1-3

Viebke C (1998) Order–disorder conformational transition of xanthan gum. In: Dumitriu S (ed) Polysaccharides: structural diversity and functional versatility. Marcel Dekker, Inc. @@Ch 17, p 16, ISBN: 0–8247–5480–8

Whitcomb PJ, Macosko CW (1978) Rheology of xanthan gums. J Rheol 22:493–505. https://doi.org/10.1122/1.549485

Whitcomb PJ, Ek BJ, Macosko CW (1977) Rheology of xanthan gum solutions. In: Sandford PA, Laskin A (eds) Extracellular microbial polysaccharides, vol 45. ACS Symposium Series, Washington DC, pp 160–173. https://doi.org/10.1021/bk-1977-0045.ch012

Wisniewski S, Kim SW (1980) Permeation of water-soluble solutes through poly(2-hydroxyethyl methacrylate) and poly(2-hydroxyethyl methacrylate) crosslinked with ethylene glycol dimethacrylate. J Membr Sci 6:299–308. https://doi.org/10.1016/S0376-7388(00)82171-7

Wyatt NB, Liberatore MW (2010) The effect of counterion size and valency on the increase in viscosity in polyelectrolyte solutions. Soft Matter 6(14):3346

Wyatt NB, Gunther CM, Liberatore MW (2011) Increasing viscosity in entangled polyelectrolyte solutions by the addition of salt. Polymer 52(11):2437–2444

Xu L, Xu G, Liu T, Chen Y, Gong H (2013) The comparison of rheological properties of aqueous welan gum and xanthan gum solutions. Carbohydr Polym 92(1):516–522. https://doi.org/10.1016/j.carbpol.2012.09.082

Young S-L, Torres JA (1989) Xanthan: effect of molecular conformation on surface tension properties. Food Hydrocoll 3(5):365–377, ISSN 0268-005X. https://doi.org/10.1016/S0268-005X(89)80011-6

Zatz JL, Knapp S (1984) Viscosity of xanthan gum solutions at low shear rates. J Pharm Sci 73(4):468–471. https://doi.org/10.1002/jps.2600730410

Zhong L, Oostrom M, Truex MJ, Vermeul VR, Szecsody JE (2013) Rheological behavior of xanthan gum solution related to shear thinning fluid delivery for subsurface remediation. J Hazard Mater 244–245:160–170. https://doi.org/10.1016/j.jhazmat.2012.11.028

Zirnsak MA, Boger DV, Tirtaatmadja V (1999) Steady shear and dynamic rheological properties of xanthan gum solutions in viscous solvents. J Rheol 43(3):627–650. https://doi.org/10.1122/1.551007

Chapter 4
Xanthan Gum–Soil Interactions

4.1 Introduction

To the civil engineer, soil is any uncemented or weakly cemented accumulation of mineral grains formed by physical and/or chemical weathering of rocks, with the void spaces between the solids containing water (pore water) and/or air. Chemical weathering causes changes in mineral form of the parent rock leading to the formation of groups of colloidal crystalline particles (i.e., < 2-μm sized particles according to their civil engineering classification) known as clay minerals. Basic structural units of most clay minerals are a silicon–oxygen tetrahedron and an aluminum–hydroxyl octahedron, with the valency imbalances in both units resulting in net negative charges.

Clay-rich soils are known to have complex physicochemical properties, arising from the negatively charged, plate-like clay mineral particles that have a high specific surface (i.e., a high surface area to mass ratio). The negative charges cause cations present in the pore water to be attracted to the clay particles, such that the clay particles are surrounded by an adsorbed water layer. Xanthan gum (XG), on the other hand, is a hydrocolloid that is commonly employed as a thickener, stabilizer and emulsifier in the food, cosmetic, pharmaceutical, petroleum, agricultural, and engineering industries. The interaction between clay soils and XG in aqueous environments has been studied by several researchers to understand the potential of using XG for improving the geoengineering properties of clay soils. The interaction between clay soils and XG could be viewed from the effects of XG addition on the physical, chemical, and mechanical properties of clay soils.

Physical Properties: XG is known to have a strong affinity for water molecules (hydrophilicity), which makes it an effective water-binding agent. When added to clay soils, xanthan forms a gel-like structure that traps water, thereby increasing the water-holding capacity of the soil. This gel-like structure also reduces the porosity of the soil, which can improve the soil's ability to retain moisture, alkaline metal ions, and pollutants. The addition of xanthan to clay soils can also improve the

A. M. O. Mohamed et al., *Sustainability in Ground Improvement: The Case of Xanthan Gum Biopolymer*, Green Energy and Technology,
https://doi.org/10.1007/978-3-031-75313-8_4

soil structure, for instance, by promoting interparticle cohesion (for granular soil particles) and flocculation (for clay particles), thereby reducing water/wind-induced soil erosion.

Chemical Properties: Xanthan is anionic, and when added to clay soils, it interacts with the clay particles through electrostatic and van der Waals forces. The interaction between xanthan and clay particles leads to the formation of a more coherent, stable soil matrix, thereby enhancing soil stability/resilience against external loading and/or environmental factors.

Mechanical Properties: Xanthan has been shown to improve the mechanical properties of clay soils, including enhanced shear strength and reduced compressibility. When added to clay soils, xanthan acts as a lubricant, reducing the friction between the particles, and improves the flow properties of the clay soils. Xanthan also increases the viscosity of the clay soils, which can reduce the rate of soil settlement and improves the stability of clay soil structure.

Therefore, the interaction between clay soils and xanthan has been shown to improve the physical, chemical and mechanical properties of clay soils. Xanthan acts as a water-binding agent, stabilizer, and lubricant, which can improve soil structure, alleviate soil erosion, and improve the flow properties of the soil. However, further research is needed to optimize the use of xanthan for different soil types and/or under environmental conditions. This chapter discusses the mode of interactions between xanthan and clay minerals, in general, and for montmorillonite clay, kaolinite clay, and clay mineral oxide—xanthan suspensions, as well as for xanthan interactions with various nanometals.

4.2 Nature of the Interactions Between Clay Soils and Xanthan

The physical and chemical interactions between clay soils and xanthan are multifaceted and involve a variety of mechanisms. Below are some examples of the nature of these interactions.

4.2.1 Physical Interactions

Adsorption: Xanthan molecules can adsorb onto the surface of clay particles through (weak) van der Waals forces, which are the attractive forces between nonpolar molecules. The extent of adsorption depends on the concentration of xanthan (and the methodology used for its application to the soil; i.e., wet or dry mixing), the size of the clay particles, and the ionic strength of the soil suspension. The adsorption of xanthan onto clay particles can increase the stability of soil aggregates by reducing the repulsive forces between particles. For instance, in a study by Reddy and Varaprasad (2020), dispersive soils A and B, characterized by plasticity index values of 16 and 39%, dispersivity (measured by using a double hydrometer) of 35.2 and 68.3%, and free swelling index values of 25 and 60%, were treated by 1 wt% XG and cured for 90 days. The double hydrometer results showed that as curing time increased from 0 to 90 days, the erosion rate decreased from 0.068 to 0.0218 kg/s/m^2 and from 0.0505 to 0.0208 kg/s/m^2 for soil A and soil B, respectively.

Gelation: Xanthan molecules can form a three-dimensional network of polymer chains in the presence of water, which leads to the formation of a gel-like structure. When added to clay soils, xanthan molecules can interact with water molecules and form a gel-like structure that can (i) increase the soil's water-holding capacity, (ii) reduce soil erosion and settlement, and (iii) improve its nutrients retention capacity. For example, a study by Mendonça et al. (2021) found that the use of XG in soil treatment induces the partial filling of the soil voids and generation of additional links between the soil particles, which decreases the permeability coefficient and enhances the mechanical properties of the soil.

4.2.2 Chemical Interactions

Electrostatic interactions: Xanthan is anionic, and when added to clay soils, it can interact with the clay particles through electrostatic forces. The negative charges on the xanthan molecules can attract positively charged ions in the Diffused Double Layers (DDLs) of the negatively charged clay particle surfaces, thereby inducing clay particle flocculation and hence producing a more coherent soil matrix.

Hydrogen bonding: Xanthan molecules contain numerous hydroxyl groups, which can participate in hydrogen bonding with clay particles. Hydrogen bonding can improve the stability of soil aggregates and increase the shear strength of the soil. XG can also improve soil structure by binding the soil particles together, which can reduce the tendency of the soil to compress.

Chelation: Xanthan molecules can chelate metal ions in the soil, which can improve the soil's agricultural productivity and reduce the toxic effects of heavy metals. Xanthan can also maintain the availability of nutrients by preventing their leaching from the soil. Furthermore, the chelation of metal ions by xanthan can reduce soil pH, which can improve the solubility of some nutrients. For instance, in a study by Bergmann et al. (2008), the interaction between xanthan and selected bivalent cations (Ca^{2+}, Mg^{2+}, Mn^{2+}, Fe^{2+}, Cu^{2+}, Zn^{2+}, Cd^{2+}, and Pb^{2+}) was evaluated and the results showed that (i) xanthan chains (with molecular mass of 1.4×10^5 to 2.9×10^6 g/mol) reversibly bind Me^{2+} species in aqueous solution at pH 6; (ii) a single bivalent cation forms a complex that involves two disaccharide units of the main chain together with two side chains; (iii) interactions between Mn^{2+} and individual carbon positions of xanthan form the main binding sites, a structure of a chelate-like complex that involves the pyruvate units at the terminal ends of the side chains; (iv) a single bivalent cation is bound between the terminal ends of two side chains, leading to an intramolecular cross-link and a reduced hydrodynamic radius of the overall xanthan macromolecule; and (v) heavy metal ions (Cd^{2+} and Pb^{2+}) link stronger to the xanthan chain than lighter cations (Ca^{2+} and Mg^{2+}).

In summary, the physical and chemical interactions between clay soils and xanthan are multifaceted and depend on several factors. The nature of these interactions can vary depending on the concentration of xanthan (and the methodology used for its application to the soil; wet or dry mixing), the type and size of clay minerals present in the soil, and the environmental conditions. The interactions between clay soils and xanthan have the potential to improve soil properties, such as enhancing mechanical stability, water-holding capacity, and nutrient availability, and reducing leachability of toxic metal ions and groundwater pollution.

4.3 Interaction Mechanisms and Bonding

Chemical bonding is characterized as the bonding forces that hold the soil particles and the gelling agent together at their surfaces. At a microscopic level, the bonding depends mainly on the type of forces present at the interfaces between the soil particles and the biopolymer, namely ionic/electrostatic or covalent bonds (chemisorption), hydrogen bonding (strong polar attraction), and van der Waals forces (physical absorption). Short-range bonds (i.e., ionic/electrostatic and covalent) have the highest bonding energies and form strong bonding. Long-range bonds, such as van der Waals forces, which describe the interaction between dipoles in the bulk pore fluid within the soil aggregates, are the weakest bonds. In addition to these types of bonding, an increase in the concentration of the gelling polymer leads to increases in (i) the thickness of the gelling material that extends over the surfaces of the soil particles, as well as the soil pore voids, and (ii) the viscosity and molecular weight of the gelling polymer that withstand the crystallization of the biopolymer macromolecule chain. In turn, these contribute physically to an increase in the strength of the ties that connect adjacent soil particles together, as well as over the soil pore voids.

Since xanthan is an anionic polysaccharide, it partakes in hydrogen bonding and cationic bridging with clay particles (Azimi et al. 2024). When xanthan is dissolved in water, the carboxylic acid ($R–COOH^-$) dissociates into carboxylate ion ($R–COO^-$) and hydrogen ion (H^+). The monovalent H^+ ion acts as a bridge between COO^- on the XG tail and OH^- on the clay particle surfaces (cationic bridging). Furthermore, the $R–COO^-$ may link with cations (particularly those of divalent types) present in the DDLs of the clay particles, resulting in the clay–XG biopolymer network. Since cohesionless soils carry no particle charges, these types of bonding (hydrogen and electrostatic attraction) are not present (Mohamed and Paleologos 2018).

It is known that clay crystals exhibit two very distinct adsorption sites: crystal edges displaying pH-dependent surface charge because of abundant surface hydroxyl groups (≡Si–OH, ≡Al–OH), and basal (001) planes with permanent negative surface charge caused by isomorphic ionic substitutions in tetrahedral (e.g., $Al^{3+} \rightarrow Si^{4+}$) or octahedral layers (e.g., $Mg^{2+} \rightarrow Al^{3+}$). Some organic compounds can be additionally adsorbed inside the interlayer spaces of certain clay types, such as bentonite

(Mohamed and Antia 1998; Mohamed and Paleologos 2018). Edge hydroxyl groups comprise a highly versatile bonding environment in which carboxyl groups (from XG) can bind in three major ways:

(a) As inner-sphere complexes, via strong covalent bonds between carboxylic C–O and surface metal atoms (−OH exchanged with −COOH);
(b) As outer-sphere complexes, by either (i) electrostatic forces between surface (de)protonated hydroxyls and carboxylic groups, or (ii) hydrogen bonding between oxygen and hydrogen atoms in various configurations.

Binding of xanthan molecules onto clay particle surfaces is highly influenced by multivalent ions. Inorganic cations can complex −COOH groups and significantly enhance adsorption of xanthan molecules onto negatively charged surfaces via cationic bridging (with Ca^{2+} acting more efficiently under alkaline conditions with deprotonated $-COO^{-}$). Adsorption of xanthan molecules onto clay minerals considerably alters the surface properties of the clay particles, which in turn significantly affects the aggregation of clay crystals, changing the structure of the soil matrix. The degree of aggregation depends also on the structure of the xanthan molecules; that is, the mutual positions of reactive functional groups (e.g., −COOH).

Therefore, the mode of interaction is expected to be (see Fig. 4.1): (i) as the number of contact points between the xanthan and soil particle surfaces increases, significant binding interactions of (de)protonated carboxylic group ($-COO^{-}$) (from xanthan) with hydroxyls (from water) will occur leading to water bridging; and (ii) due to the presence of exchangeable multivalent cations in the soil pore water, cationic bridging will occur between both the xanthan and the clay particle surfaces.

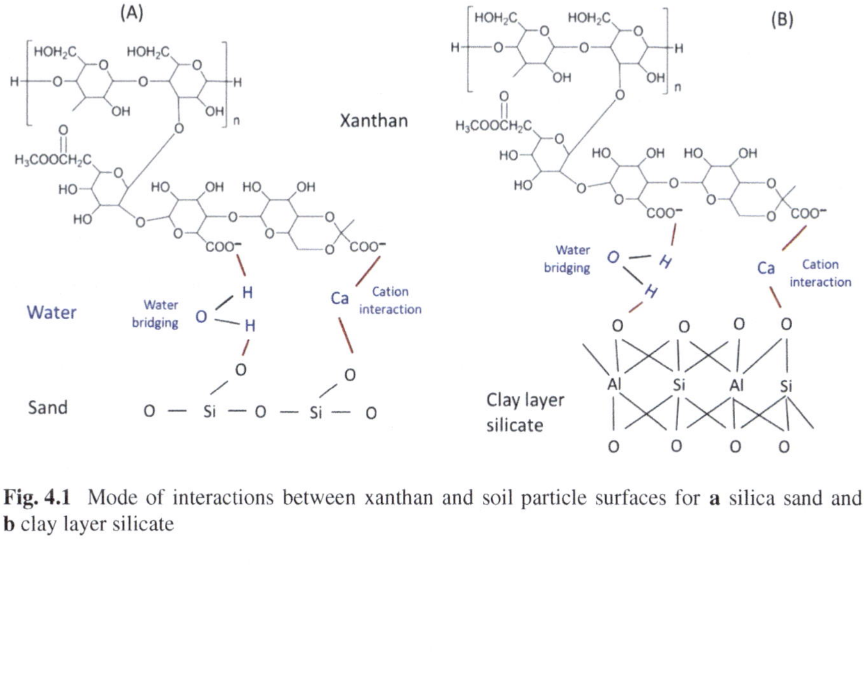

Fig. 4.1 Mode of interactions between xanthan and soil particle surfaces for **a** silica sand and **b** clay layer silicate

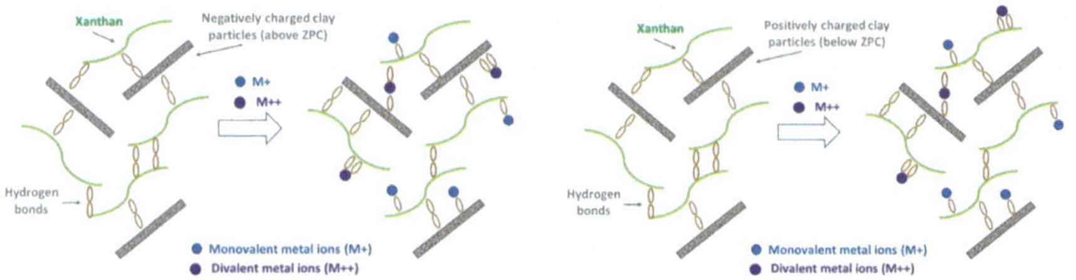

Fig. 4.2 Possible interaction mechanisms of XG, clay particles, and metal ions. *Note* ZPC, zero point of charge

In addition, the main interaction mechanism between xanthan and the clay particle surfaces is via hydrogen bonding, which makes the molecular chain of the clay–xanthan system very stretched in aqueous suspension; hence, the fluidity of the aqueous suspension can be controlled. However, in the presence of monovalent metal ions (e.g., Na^+ and K^+) in the soil pore fluid, cations inhibit the electrostatic repulsion between carboxyl ions present in the xanthan; hence, the hydrogen bonding between clay particle surfaces and xanthan is broken, making the xanthan molecule fold (Fig. 4.2).

However, when salt (monovalent cations or divalent cations) is added to the clay–xanthan suspension, the viscosity of the suspension will be reduced. The divalent cations would have greater influences on the viscosity than monovalent cations, because divalent cations will interact more strongly with the polar groups on xanthan molecules. As a result, the following ensue: (i) the xanthan molecules lose their internal repulsion and contract/shrink the coils; (ii) the hydrogen bonding between the xanthan polymer bonds and the interacting water molecules is reduced; (iii) the network structure of extended xanthan macromolecules is destroyed; (iv) the effective diameter of the xanthan molecules in suspension is reduced; and (v) the overall suspension viscosity is decreased.

In case of positively charged clay particle surfaces, i.e., at low pH values (below the Zero Point of Charge, ZPC), the interaction between xanthan and the clay particle surfaces is enhanced due to electrostatic attraction, and the bonding strength increases; as such, the strength of the aggregated flocs is enhanced. However, in the presence of metal ions, the adsorption sites on the xanthan surfaces are decreased, and the surface charge of the clay colloidal particles is reversed. Consequently, there is (i) an increase of the repulsive forces between the clay particles and the xanthan molecules; (ii) a decrease in the strength of the bonding; (iii) a decrease in the potential adsorption bridging between the clay particles and xanthan molecules; and (iv) an overall decrease in the suspension viscosity.

The carboxyl groups on the xanthan chains will form hydrogen bonds with the hydroxyl groups on the surfaces of the clay particles. In addition, the hydrogen ions in the hydrophilic group of the xanthan chains will replace the alkali metal ions present on the surfaces of the clay particles. Consequently, the thickness of the DDL of the clay particles will be reduced, and with the increase of the attraction energy between clay particles, their flocculation and hence soil aggregation will be promoted. In the meantime, because of the presence of hydrophobic long chains in the xanthan molecules, a network structure of the xanthan can be formed, the nature of the network depends on the xanthan concentration. As xanthan concentration increases, the xanthan structure transforms from coiled to coiled-disordered, to helix in single or double form, and finally to aggregated helices, which will promote flocculation as well as aggregation of the overall soil–xanthan structure. The resultant xanthan network structure (hydrogels) will be formed on the soil particle surfaces as well as in the void spaces via diffusion, infiltration, and entanglement, leading to the development of bonding forces between the soil particles. These intricate interactions between the xanthan hydrogels, pore water, and soil particles can lead to the formation of a three-dimensional hydrogel network connecting the soil particles, thereby providing a soil-reinforcing effect in the xanthan-stabilized soil (Chang and Cho 2012; Muguda et al. 2020; Mendonça et al. 2021; Kumar et al. 2023). On drying, the hydrogel formations connecting the soil particles change their physical nature from a "rubbery" to a "glassy" state, which has a significant influence on further improving the soil strength (Ayeldeen et al. 2016). The physical integrity and the chemical activity of the hydrogels, which are dependent on the chemical properties of the xanthan, control the overall strength development of xanthan-stabilized soils (Reddy and Varaprasad 2020).

However, on drying and re-saturation, the outer chains of the xanthan absorb and hold water that fills up the pore voids in the soil matrix. This makes the hydrogels swell before starting their dilution, which is the process where the hydroxyl groups of the hydrogel dissociate and dissolve in the water. The amount of water held by the hydrogels and the time taken for complete dilution depend on the intrinsic chemical properties of the xanthan. To delay or avoid the dilution of the xanthan hydrogels, cross-linking with other biopolymer (i.e., a process where the outer chains of the xanthan are made to interact with other polymer chains, to enhance the physical integrity and mechanical properties of the hydrogel) is recommended. For instance, Zhang et al. (2023) evaluated the synergistic interaction and gelling kinetics between XG and Locust Bean Gum (LBG) at different mass ratios (XG/LBG of 9:1, 7:3, 5:5, 3:7, 1:9) and found that the mixtures of XG and LBG induced gel formation, and the strongest gel structure was found for the mixture of 3:7 (XG/LBG) according to yield stress and storage modulus experimental results.

4.4 Mode of the Interaction Between Xanthan and Montmorillonite Clay

Xanthan is a natural polysaccharide that can adopt extended conformational structures due to its flexible nature. Montmorillonite clay, on the other hand, is a type of layered silicate mineral with a high specific surface and net negative charge. When these two materials come into contact, several interaction mechanisms can occur between them, including physical adsorption, electrostatic interactions, and hydrogen bonding.

(a) *Physical adsorption* occurs when xanthan molecules adhere to the surface of the montmorillonite clay particles through van der Waals forces. This type of interaction is primarily dependent on the surface area of the clay particle and the size of the xanthan molecule. The adsorption process can be influenced by factors, such as pH, temperature and ionic strength.
(b) *Electrostatic interactions* occur when the negatively charged clay particle surface interacts with the positively charged regions of the xanthan molecule. The carboxyl groups of xanthan can form ionic bonds with the surfaces of the clay particles, leading to a stable interaction. This type of interaction is highly dependent on the pH and ionic strength of the xanthan-clay suspension. In addition, the degree of ionization of the xanthan molecule can also affect the strength of the interaction.
(c) *Hydrogen bonding* occurs when the hydroxyl groups of xanthan form hydrogen bonds with the surfaces of the clay particles. This type of interaction is highly dependent on the temperature and the moisture content of the system.

Examples of the interaction between xanthan and montmorillonite clay can be found in the literature. For instance, a study conducted by Hua et al. (2020) showed that the montmorillonite-XG hydrogel networks are closely associated through hydrophobic interactions, hydrogen bonds, and ionic coordination interactions between $-COO^-$ of xanthan chains and Fe^{3+} ions. The hydrogel could effectively dissipate energy and obtain excellent fatigue-resistance, and self-healing abilities through the interactions between different crosslinking agents. In addition, the hydrogels present pH-dependent swelling behaviors, attributed to the presence of $-COO^-$ from the xanthan chains.

In another study by Chang et al. (2018), the liquid limit of XG-montmorillonite with deionized water (DW) gradually decreased with an increase in XG content, which implies XG mainly aggregates montmorillonite particles rather than affecting the viscosity of the pore fluid. The overall electrical sensitivity reduction with XG treatment implies inter-particle aggregation to be the main phenomenon of clays at low XG content. However, the behavior of XG-montmorillonite was governed by pore fluid chemistry when brine and kerosene were used. With brine, counter-ions decreased the liquid limit of montmorillonite significantly, while it tended to increase with higher XG contents. However, for kerosene, the liquid limit did not show noticeable changes with XG-montmorillonite treatment.

Therefore, the interaction mechanisms between the extended conformational structure of xanthan and montmorillonite clay can involve physical adsorption, electrostatic interactions, and/or hydrogen bonding. These interactions can be influenced by several factors, including pH, temperature, and ionic strength.

The effect of pH on the mode of interaction between the extended conformational structure of xanthan and montmorillonite clay is mainly due to the degree of ionization of the xanthan functional groups, such as carboxyl and hydroxyl groups, and the surface charge of the clay particles. At different pH values, the clay particle surface charge and the ionization of functional groups in xanthan can vary, leading to changes in the mode of interaction.

At low pH values, the carboxyl groups in xanthan are protonated, leading to a decrease in the negative charge of the xanthan molecule. Consequently, the electrostatic interactions between the negatively charged surfaces of the clay particles and xanthan are weakened, and the physical adsorption becomes more dominant. In other words, physical adsorption occurs due to the van der Waals forces, which are not dependent on the charge. Therefore, at low pH values, the mode of interaction between xanthan and clay particles is primarily physical adsorption.

Whereas at high pH values, the carboxyl groups in xanthan are deprotonated, leading to an increase in the negative charge of the xanthan molecule. The negative charge of xanthan enhances the electrostatic interaction between xanthan and the clay particles, and the electrostatic interaction becomes dominant over physical adsorption. In other words, at high pH values, the mode of interaction between xanthan and the clay particles is primarily electrostatic interaction.

4.5 Mode of the Interaction Between Xanthan and Kaolinite Clay

Kaolinite clay is a negatively charged mineral comprising alternating silica–gibbsite sheets that are held together relatively strongly by hydrogen bonding, while xanthan is a negatively charged biopolymer with a flexible chain structure. Much like montmorillonite, kaolinite clay can interact with xanthan through physical entanglement and electrostatic interactions. In other words, when mixed together, the long, flexible chains of xanthan can physically entangle with the platy layers of kaolinite clay, creating a stable suspension. The electrostatic repulsion between the negatively charged clay particles and xanthan molecules can also contribute to the stability of the suspension. Examples of the use of xanthan with kaolinite clay in various industries include:

(a) Oil and gas industry: Xanthan is often added to drilling fluids to provide viscosity and suspension properties, while kaolinite clay is used to control the rheological properties of the fluid. The combination of kaolinite clay and xanthan helps to maintain the stability of the drilling fluid, preventing settling and ensuring consistent flow properties.

(b) Food processing industry: Xanthan is often used as a thickener and stabilizer in food products, while kaolinite clay can be used as an anti-caking agent. When mixed together, the combination of kaolinite clay and xanthan can help to stabilize emulsions and prevent the separation of ingredients in food products.

(c) Personal care products: Xanthan is often used in personal care products (e.g., lotions and creams) as a thickener and stabilizer, while kaolinite clay can be added as a rheology modifier. The combination of these two ingredients can help to create a stable and smooth texture in the product.

(d) Paints and coatings: Kaolinite clay is commonly used as a filler in paints and coatings to improve their properties, such as opacity, brightness, and durability. Xanthan can be added as a rheology modifier to control the flow and viscosity properties of the paint. The combination of kaolinite clay and xanthan can help to improve the overall performance and stability of paint products.
(e) Agriculture: Xanthan is used in agriculture as a soil stabilizer and water retention agent, while kaolinite clay is used as a carrier for pesticides and herbicides. The combination of kaolinite clay and xanthan can help to improve the adhesion and coverage of the pesticide on the plant surface.
(f) Paper industry: Kaolinite clay is used in the paper industry as a coating pigment to improve the printability and smoothness of the paper surface. Xanthan can be added as a binder to improve the strength and adhesion of the coating. The combination of kaolinite clay and xanthan can help to produce high-quality paper with improved printability and surface properties.

Overall, the interaction between kaolinite clay and xanthan can be beneficial in a variety of industries and applications, helping to improve product performance, stability and properties.

The surface charge of kaolinite clay particles is pH dependent, meaning that their interaction with other molecules, such as xanthan, can vary depending on the pH of the system, with various examples given above. At low pH values (below the ZPC at pH 4.2 (Yong et al. 1992; Mohamed and Antia 1998; Mohamed and Paleologos 2018)), kaolinite clay has a net positive charge, while at high pH values (above the ZPC), it has a net negative charge.

When kaolinite clay is mixed with xanthan, the interaction between the two materials will depend on the relative charges of the kaolinite clay and the xanthan. If the pH of the system is such that the clay has a net negative charge, while xanthan also carries a net negative charge, there will be electrostatic repulsion between the two materials leading to flocculation. On the other hand, if the pH of the system is such that the clay has positive charges, while xanthan carries a net negative charge, there will be electrostatic attraction between the two materials. This can lead to increased interaction and enhanced stability of the system. Therefore, the pH of the system can play an important role in determining the interaction between kaolinite clay and xanthan, and can ultimately affect the stability and properties of the system. It is important to carefully consider the pH conditions of any system in which these two materials are used together.

At low pH values, below the ZPC of kaolinite clay, the surfaces of the clay particles are positively charged. Xanthan is a polyanionic biopolymer with a negatively charged backbone. Therefore, in a mixture of kaolinite clay and xanthan at low pH, there will be electrostatic attraction between the positively charged kaolinite clay and the negatively charged xanthan molecules. This can lead to the formation of intermolecular aggregates or complexes between the two materials.

The exact structure of the kaolinite clay–xanthan mixture at low pH will depend on the relative concentrations of the two materials, as well as the pH and ionic strength of the system. It is possible that the xanthan chains can physically entangle with the platy layers of kaolinite clay, creating a stable suspension of the two materials. Alternatively, the xanthan molecules may adsorb onto the surface of the kaolinite clay particles, coating the particles and potentially affecting their properties. Therefore, the structure of the kaolinite clay and xanthan mixture at low pH will depend on the specific conditions of the system, including the pH, concentration, and ionic strength of the suspension.

As an example, in a study by Kwon et al. (2023), the impact of xanthan on kaolinite fabrics and the ensuing changes in the compaction behavior and shear resistance of kaolinite soils were evaluated. The analysis revealed that (i) xanthan changed kaolinite fabrics into face-to-face associations; (ii) formation of xanthan bridges between kaolinite particles, resulted in the change in fabrics and subsequently improved the resistance of kaolinite to external forces; (iii) as xanthan content increased, the maximum dry density decreased, and the undrained shear strength increased; (iv) the viscous xanthan hydrogels produced a higher optimal moisture content and increased resistance to shear force; and (v) xanthan affects the mechanical properties of kaolinite through changing kaolinite fabrics (up to 0.5% of the xanthan-to-kaolinite mass ratio) and absorbing pore-fluids (excess xanthan over 0.5% of the xanthan-to-kaolinite mass ratio).

In another study by Kang et al. (2019), the effects of various biopolymers on the settling characteristics of kaolinite and fly ash suspensions revealed that cationic biopolymers induced fabric changes in kaolinite through bridging and charge neutralization, resulting in an increased settling velocity. Moreover, in a study by Shakeel et al. (2022), the rheological properties of kaolin suspensions modified with different biopolymers (XG, sodium carboxymethyl cellulose (Na-CMC), potato starch, chitosan, and apple fiber) were evaluated by varying the biopolymer type and content, and clay content. The results showed that both the viscosity and complex modulus, $G^*(\omega)$, were significantly influenced by adding different biopolymers into the kaolin suspensions. The viscosity of the modified suspensions increased by three to four orders of magnitude by adding 1 wt% XG or 5 wt% Na-CMC and by six orders of magnitude by adding 5 wt% apple fiber. Likewise, the complex modulus of modified clay suspensions was higher than the pure kaolin suspension at low XG or Na-CMC content (0.1 wt%).

The structure of kaolinite–XG mixtures at low pH levels is influenced by various factors, including the pH, concentration, and ionic strength of the solution, and the electrostatic interactions between the constituent materials. The specific structure of the mixture will vary depending on these factors, but typically involves the formation of intermolecular aggregates or complexes that improve the stability and rheological properties of the suspension. The mode of the interaction at low pH levels is governed to the electrostatic attraction between the positively charged kaolinite clay particles and the negatively charged xanthan molecules, which led to the formation of intermolecular complexes.

At high pH levels, the mode of interaction between kaolinite and XG changes due to the change in the surface charge of the kaolinite particles. In other words, kaolinite particles become more negatively charged at high pH levels, which increases the electrostatic repulsion between the particles. Xanthan molecules, which are anionic in nature, can be adsorbed onto the surfaces of the kaolinite particles through electrostatic interactions, hydrogen bonding, and van der Waals forces.

The adsorption of xanthan onto the surfaces of the kaolinite particles can result in the formation of a gel-like network that can increase the viscosity and yield stress of the mixture. The extent of adsorption and the rheological properties of the kaolinite–XG mixture depend on the concentration of XG, pH, and ionic strength of the suspension. For instance, in a study by Chang et al. (2018), the liquid limit of XG-kaolinite mixture peaked at 0.5 wt% XG content, and it decreased until a content of 1.0 wt% was reached, which implies the simultaneous hydrogel formation in pores and particle aggregation induced by XG.

4.6 Mode of Interaction Between Xanthan and Clay Mineral Oxides in Suspension

Mineral oxides (e.g., silica, alumina, and titanium dioxide) have been found to interact with xanthan through electrostatic and steric mechanisms. The specific mode of interaction depends on various factors, such as pH, ionic strength, and concentration of the system. For instance, at low pH, the interaction between mineral oxides and xanthan is mainly electrostatic, with the negatively charged xanthan adsorbing onto the positively charged surface of the mineral oxide. At high pH, however, the interaction becomes more steric in nature, with xanthan molecules forming a protective layer around the mineral oxide surface to prevent aggregation.

Studies have shown that the presence of XG can significantly affect the rheological and surface properties of mineral oxide suspensions. For example, XG has been found to improve the stability and dispersibility of silica and alumina suspensions, as well as enhance the viscoelastic properties of titanium dioxide suspensions.

4.6.1 Interaction with Silica

In a study by Saravanan and Subramanian (2016), the adsorption of xanthan molecules onto silica indicated that the adsorption density is negligible throughout the investigated pH region (acidic (pH 2) to alkaline (pH 11)), which is like the results reported by Gebhardt and Fuerstenau (1983). The nil adsorption can be attributed to the electrostatic repulsion between silica and the xanthan, because of the high negative charge associated with both. The zeta potential of silica suspension as a function of pH in the absence and presence of different xanthan concentrations is shown in Fig. 4.3a (data from Saravanan and Subramanian (2016)). In the absence of XG, the zeta potential values of silica are negatively charged throughout the investigated pH range, with the minimum zeta potential value of ~28 mV at pH 2 gradually increased to a maximum value of ~74 mV at pH 11. The addition of different concentrations

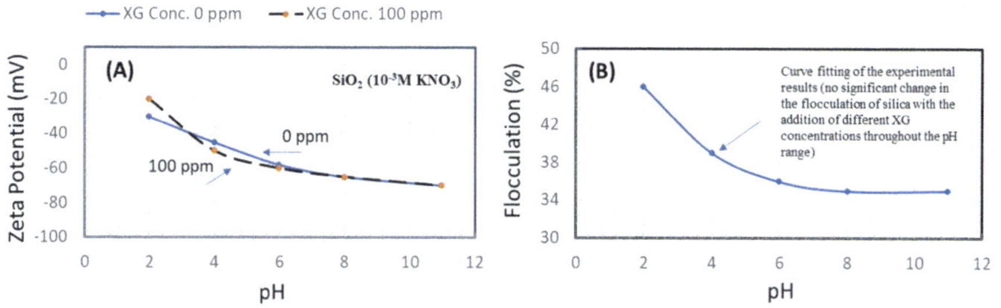

Fig. 4.3 **a** Zeta potential variations with XG concentrations and pH of XG-silica mixture; and **b** flocculation of XG on silica with XG concentration and pH environment (data from Saravanan and Subramanian (2016))

of XG hardly made any effect on the zeta potential values of silica throughout the pH range. This in turn explains the observed negligible adsorption.

The flocculation of silica as a function of pH in the absence and presence of different XG concentrations is shown in Fig. 4.3b. In the absence of the XG, the flocculation of silica particles decreases as the pH increases, which agrees with the electrokinetic behavior, wherein the negative zeta potential values increased as the pH was increased. This can result in the increased electrostatic repulsion between the particles, leading to a reduction in the particles' aggregation. Consequently, the percentage flocculation values of silica are decreased with the increase in pH. Furthermore, there is no significant change in the flocculation of silica with the addition of different XG concentrations throughout the investigated pH range.

In another study by Dário et al. (2011), the adsorption of xanthan onto Si/SiO_2 surfaces in the presence of calcium salts ($CaCl_2$, $Ca(NO_3)_2$, $CaSO_4$, $CaCO_3$, and $Ca_3(PO_4)_2$) was investigated by ellipsometry and atomic force microscopy (AFM) imaging. The adsorbed layer of xanthan onto Si/SiO_2 consisted of two regions: (i) a thin acid resistant sublayer, where xanthan chains were like highly entangled fibers, and (ii) a thick upper layer, whose morphology was calcium salt dependent.

It is worth noting that the O-acetyl and pyruvyl residues in the xanthan side chains provide conditions for complexation with bivalent cations. Several experimental techniques, such as conductometric, viscometric titrations and nuclear magnetic resonance spectroscopy, revealed that a single bivalent cation forms a complex, which involves two disaccharide units of the main chain and two carboxylate groups at side chains of the xanthan, leading to intramolecular cross-linking and chains contraction. Moreover, heavy metal ions (Cd^{2+} and Pb^{2+}) presented stronger binding to the xanthan chain than lighter cations (Ca^{2+} and Mg^{2+}) (Bergmann et al. 2008).

Moreover, the gel-like behavior of xanthan was investigated as a function of temperature and $CaCl_2$ concentration. Solutions passed through states of maximum and minimum gel-like character at 100 and 200% stoichiometric equivalence of Ca^{2+}, which were ascribed to the partial replacement of intermolecular site binding of calcium ions by binding to individual carboxyl groups, maximizing the degree of complexation and network formation (Mohammed et al. 2007). The use of Ca^{2+} ions as chelating agents was applied for xanthan chains and for other charged polysaccharides, such as alginate (Draget et al. 1994) and hyaluronic acid (Furth et al. 2008).

4.6.2 Interaction with Alumina

In a study by Saravanan and Subramanian (2016), the changes in the surface chemical properties of alumina and silica suspensions consequent to the addition of XG were evaluated. The adsorption density of xanthan for alumina was found to be maximum at pH 2, whereas nil adsorption was found for silica throughout the investigated pH range. The ZPC of alumina and silica was found to be located at pH 9.1 and likely less than pH 2, respectively. The addition of xanthan shifted the ZPC of alumina toward more acidic pH values, in proportion to the concentration of xanthan added, with a concomitant change in the zeta potentials to more electronegative or less electropositive values. The addition of xanthan beyond a certain concentration (> 50 ppm) increases the flocculation of alumina suspension to a higher value (~92%) in the pH range of 2–9. The electrokinetic and flocculation characteristics of the silica were not altered much by the addition of xanthan, akin to the adsorption result described above. Moreover, the Fourier Transfer Infrared (FTIR) spectral investigations provided evidence in support of hydrogen bonding and chemical interaction forces between alumina and xanthan. Maximum separation efficiency of ~93% for alumina-silica synthetic mixture was observed in the selective flocculation tests.

The adsorption density of XG for alumina (Al_2O_3) as a function of pH, at three different initial XG concentrations, is shown in Fig. 4.4a (data from Saravanan and Subramanian (2016)). The results indicated that (i) for all initial XG concentrations, the adsorption densities of XG decrease steadily from acidic environment (pH 2) to alkaline environment (pH 10); (ii) the adsorption density of XG for alumina increases with the increase in the initial XG concentration; and (iii) the maximum adsorption densities of 0.3, 0.8 and 1.4 mg/m^2 for initial XG concentrations of 50, 100 and 200 ppm, respectively, occurred at pH 2.

The electrokinetic behavior of alumina suspension as a function of pH in the absence and presence of different concentrations of XG is shown in Fig. 4.4b. In the absence of the XG, the ZPC of alumina is found to be located at pH 9.1. Whereas in the presence of (anionic) XG, the ZPC value of the original alumina surface shifted toward the acidic pH region with the increase in the XG concentrations (i.e., pH of 6.8 and 2.3 for XG concentrations of 1 and 50 ppm, respectively). The addition of XG, not only produces a decrease in the positive zeta potential values of alumina, but also charge reversal occurred, with the magnitude of the negative zeta potential values increasing with the increase in the XG concentration, as a function of pH.

In view of the mode of the interaction, the adsorption of xanthan onto alumina is governed by the hydrogen bonding and electrostatic forces. Notably, the number of hydroxyl groups present on the surface of alumina decreases as the pH increases. At the same time, alumina is highly positively charged in the acidic region, and its

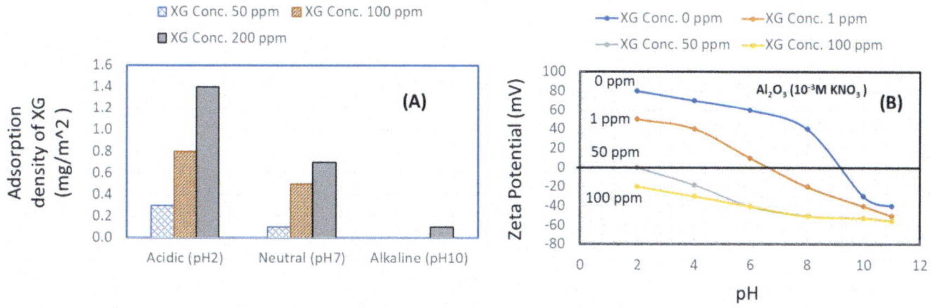

Fig. 4.4 **a** Adsorption density of XG on alumina with XG concentration and pH environment; and **b** zeta potential variations with XG concentrations and pH of XG-alumina mixture (data from Saravanan and Subramanian (2016))

charge gets reversed beyond pH 9.1 (see Fig. 4.4b). Since xanthan is an anionic polysaccharide, the higher adsorption of negatively charged xanthan molecules onto alumina, in the acidic pH region, through the combination of hydrogen bonding (the interaction of carboxylic group of xanthan and surface OH groups of alumina) and electrostatic interaction forces (the interaction of positively charged alumina and negatively charged xanthan) can be expected.

Flocculation of XG-alumina mixtures as functions of XG concentration and pH environment is shown in Fig. 4.5 (data from Saravanan and Subramanian (2016)). In the absence of the XG, the trend of flocculation as a function of pH follows the surface charge characteristics of alumina particles. The flocculation is minimum (~6%) at pH 2, and it increases gradually to a maximum value of ~92% at pH 9, thereafter again decreases to reach a value of ~40% at pH 11 without XG addition. The higher flocculation at pH 9 (near to the ZPC of alumina) can be attributed to the net surface charge becoming zero, which leads to the higher aggregation of alumina. On the other hand, the existence of high positive and negative surface charge of alumina in the extreme acidic and basic pH regions, respectively, lead to the enhanced electrostatic repulsion between the alumina, and hence the observed low flocculation in these regions. The addition of XG to the alumina suspension increases the flocculation in proportion with the concentration of the XG, in the pH range of 2–9. Beyond 50 ppm XG, the percentage flocculation attains a maximum value of ~92%, in the pH range of 2–9. In the pH range 9–11, the presence of XG, irrespective of its concentration, has not significantly altered the magnitude of flocculation of the alumina suspension, relative to the flocculation values obtained in the absence of the XG.

The increase in the flocculation of alumina with the increase in the XG concentration can be attributed to the availability of a greater number of xanthan chains for the enhanced bridging between alumina particles. The redispersion of the particles beyond pH 9 may be attributed to electrostatic repulsion between the negatively charged alumina suspension and the anionic xanthan.

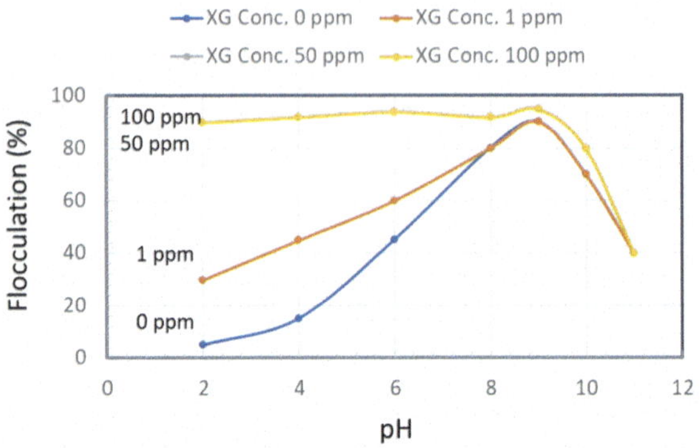

Fig. 4.5 Flocculation of XG on alumina with XG concentration and pH environment (data from Saravanan and Subramanian (2016))

4.6.3 Interaction with Calcium Carbonate

Bentonite-free water-based mud (BFWM) is frequently used to drill the pay-zone of reservoirs since bentonite deformability contributes to pore throat plugging (Burrows et al. 2004; Huang et al. 2016). Polymers (e.g., XG, starch, and polyanionic cellulose) and $CaCO_3$ particles are used to substitute for bentonite and impart the desired rheological and filtration properties of the drilling fluid (Hamed and Belhadri 2009; Whatley et al. 2019). However, due to the potential mobility of $CaCO_3$ and plugging of the pores (Whatley et al. 2019), nanoparticles (NPs) have been used to enhance the rheological properties, such as thinning of the drilling fluids and tightening of the filter cakes to prevent reservoir damage (Barry et al. 2015; Li et al. 2016; Bayat et al. 2018; Parizad et al. 2018; Perween et al. 2018; Smith et al. 2018). Studies have demonstrated the effectiveness of using SiO_2NPs to improve the properties of a BFWM filter cake and enhance the interactions between the filter cake constituents (Srivatsa and Ziaja 2012; Clavijo et al. 2019; Vargas et al. 2019).

In a study by Clavijo et al. (2021), the interactions among the constituents of a filter cake consisting of SiO_2NPs, BFWM, $CaCO_3$, and XG were evaluated. The main functional groups participating in the interaction process with BFWM are (i) for SiO_2NPs, O–H stretching, and symmetric and asymmetric vibrations of Si−O; (ii) for XG, −OH groups, C−H stretching for methyl groups, C=O stretching, C−O of enols, and C−H; and (iii) for $CaCO_3$, −OH groups, C−O, and C=O. The overall interaction between these constituents is through hydrogen bonding with the −OH groups.

To evaluate the effectiveness of SiO_2NPs to BFWM, Clavijo et al. (2021) investigated the hydraulic permeability of the BFWM filter cake, and the results showed an appreciable reduction in the permeability for SiO_2NP contents ranging between 0.01 and 0.1 wt% (Fig. 4.6). However, in comparison with the control filter cake material (i.e., 0 wt% SiO_2NP), the 0.3 wt% SiO_2NP content produced only a small decrease in permeability. Whereas SiO_2NP content up to 0.1 wt% reduced the permeability by 42–72% relative to the control. Upon adsorption onto the $CaCO_3$ particles, the SiO_2NPs reduced the electrostatic repulsion among the deposited $CaCO_3$ particles,

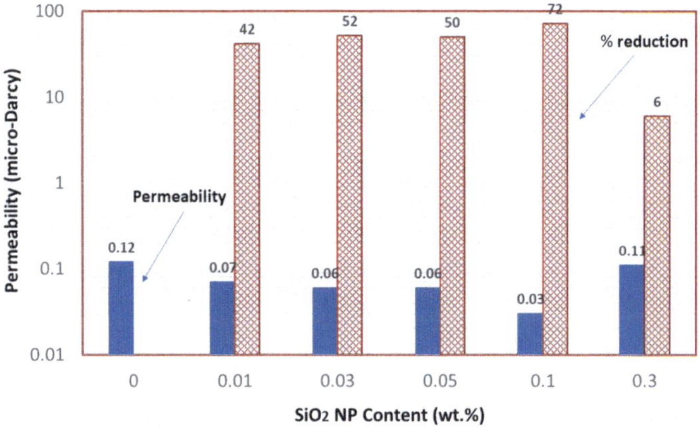

Fig. 4.6 Permeability changes with SiO_2NP content of bentonite-free water-based mud mixed with 0.14 wt% XG and 4 wt% $CaCO_3$ (data from Clavijo et al. (2021))

favoring particle attraction and more compact filter cake (Clavijo et al. 2019); hence reducing the permeability.

Considering the zeta potential of $CaCO_3$ at pH 10 is −28.60 mV (Mahmoud et al. 2016), which falls at the limit of the unstable region, the SiO_2NPs are likely to interact through their silanol (Si−OH) and hydroxyl (−OH) groups with the C−O groups of the $CaCO_3$ via hydrogen bonding, thus enabling reduction of the negative charge contributed by surface $CO_3{}^{2-}$. On the other hand, XG is more likely to react with the positive charges on Ca^{2+} surface (Backfolk et al. 2002).

Overall, the results of using a filter cake consisted of SiO_2NPs, BFWM, $CaCO_3$, and XG have shown that (i) SiO_2NPs improved the morphology of the filter cake; (ii) the dominant interaction was governed by sandstone and SiO_2NPs, followed by $CaCO_3$ and XG; and (iii) the SiO_2NPs reduced the repulsion energy among the $CaCO_3$ surfaces, promoting a tighter filter cake that hindered mud filtration through it.

In another study by Mandre and Panigrahi (1997), the selective flocculation of complex sulfide ores was done using cellulose xanthate, and the results demonstrated the effectiveness of removing Pb and Zn from the ore.

Similar results, in using xanthan (anionic polysaccharide) to flocculate calcium carbonate, were reported in a study by Friend and Kitchener (1973). They used anionic polyacrylamide to selectively flocculate calcite from its mixtures of quartz-rutile, and alumina-quartz. The results demonstrated that (i) the flocculation of calcite by the anionic polyacrylamides occurs even when the mineral and the polymer carry the same sign of charge; (ii) the strength of adsorption is almost independent of degree of ionic character, but the anionic polyacrylamides have greater bridging power because of their extended dimensions; (iii) the flocs can, however, be broken by adding sodium tripolyphosphate; and (iv) flocculation can be prevented by providing sufficiently strong negative zeta potential materials.

4.6.4 Interaction with Nanometal Oxides

At low concentrations, XG has been used as a stabilizer of NPs in aqueous solutions for thermal recovery of heavy oil (Hamedi Shokrlu and Babadagli 2011) and for drilling fluids (Ponmani et al. 2014; William et al. 2014). Also, other studies on the effect of NPs on the rheological behavior of XG solutions have been reported (Comba et al. 2011; Xue and Sethi 2012; Attia and Musa 2015; Kennedy et al. 2015). For example, in a study by Saha et al. (2018), the effect of hydrophilic silica NPs on the stability of crude oil-XG emulsions was evaluated, and the results indicated that the addition of the silica NPs improved the stability of the emulsions at all XG concentrations investigated (1000–5000 ppm). Nevertheless, higher emulsification and better stability periods were obtained at 5000 ppm of XG and 0.3 wt% of silica NPs due to the adsorption of the NPs at the oil–water interface, which reduced the interfacial tension, the oil droplet sizes, and the size distribution. Moreover, the NPs changed the wettability of the system from intermediately oil-wet to water-wet, and improved the oil recovery by between 18 and 20% for temperatures of 30 and 80 °C, respectively.

In another study by Corredor et al. (2019), the effect of TiO_2, Al_2O_3, $Fe(OH)_3$ and SiO_2NPs on the performance of XG solutions to enhance heavy oil recovery was evaluated. The NPs were first dispersed in deionized water (DW) at 0.2 wt% and ultrasonicated for 1 h. Sodium dodecyl sulfate (SDS) was added at 0.1 wt%, and the dispersions were stirred for 30 min. Then, XG at 0.4 wt% was added and stirred for 1 h. Finally, NaCl was added to the sample to achieve a concentration of 0.3 or 1.0 wt%.

It is to be noted that the colloidal stability of the NP dispersions can be predicted by the magnitude of the zeta potential. Dispersions with zeta potential values greater than +30 mV or less than −30 mV typically have a high degree of stability (ASTM 1985). Whereas NP dispersions with a low zeta potential value will eventually agglomerate under the effect of interparticle attractions. Accordingly, most of the nanofluids (NPs dispersed in DW) would potentially be unstable, as reported in Table 4.1. The mechanisms causing the instability of these NPs are related to the hydrogen bonding between the silanols groups of the NPs.

It was noted by Corredor et al. (2019) that the silica-nanofluid was stable despite its low zeta potential value (−17.5 mV), suggesting that the hydration layer formed between the silanols groups on the silica surface, and the water molecules (through hydrogen bonding) prevent the agglomeration of the SiO_2NPs. Also, effective steric hindrance leads to net repulsive inter-particle forces, despite the lack of charges on the silica surface (Colic et al. 1998). However, the hydration effect was not observed for the other NPs (i.e., Al_2O_3, TiO_2, and $Fe(OH)_3$), as discerned from the thermogravimetric analysis, showing that the weight loss associated with the desorption of physisorbed water (up to 400 °C) for SiO_2NPs was 25.7%, while the weight losses for Al_2O_3, TiO_2, and $Fe(OH)_3$ were 0.49, 3.78, and 5%, respectively. Similarly, the instability of Al_2O_3 and TiO_2 in water was reported by Hendraningrat and Torsæter (2014).

Table 4.1 Zeta potential and pH values of xanthan gum (XG) nanopolymer suspensions at 20 °C (data from Corredor et al. (2019))

		NaCl Concentration (wt.%)					
		0		0.2		0.3	1
NP type	Dispersion medium	Zeta potential (mV)	pH	Zeta potential (mV)	pH	pH	pH
SiO_2	DW	−17.5	6.89	–	–	–	–
	DW–SDS	−25.1	6.73	–	–	–	–
	XG–SDS	−65.7	6.62	−56.3	6.44	6.73	6.89
TiO_2	DW	−25.3	6.95	–	–	–	–
	DW–SDS	−28.8	6.78	–	–	–	–
	XG–SDS	−53.8	7.04	−46.4	6.87	7.41	8.15
Al_2O_3	DW	16.4	6.75	–	–	–	–
	DW–SDS	−26.9	6.54	–	–	–	–
	XG–SDS	−67.7	6.32	−59.5	6.04	6.48	6.64
$Fe(OH)_3$	XG	–	–	–	–	10.5	10.62

Note DW and SDS, deionized water and sodium dodecyl sulfate, respectively

The zeta potential for Al_2O_3NPs in DW water is +16.4 mV at pH 6.75, which is relatively high, consistent with the fact that this pH is far from their pH at zero point-of-charge (pH @ ZPC = 9.1) (Singh et al. 2005). When alumina is dispersed in water, it behaves like a basic oxide which consumes H^+ ions. Thus, it possessed positive surface charges (Singh et al. 2005). Upon adding SDS, the zeta potential changes sign (−26.9 mV), which could be explained by the adsorption of SDS onto the surface of the positively charged NPs as a bilayer, with the sulfate headgroups exposed to the solution (Husein 2017). Formation of a surfactant bilayer, and associated charge reversal of the alumina particles with anionic SDS surfactant, was previously reported by Somasundaran and Fuerstenau (1966). In general, the addition of SDS increased the negative magnitude of the zeta potential values of the NPs, but did not improve their stability.

With the addition of XG, the NPs-XG suspensions became stable (Table 4.1), with zeta potential values of less than −30 mV, due to (i) the XG chains adsorbed onto the NPs' surfaces inducing steric repulsion; and (ii) the increase in the viscosity of the dispersion medium (Pomogailo and Kestelman 2005). The adsorption of XG chains onto the surfaces of the SiO_2, $Fe(OH)_3$, and TiO_2 NPs likely occurs through hydrogen bonding between –OH groups on the NPs' surfaces and the carboxylic groups on the side chains of the XG, and on the positively charged Al_2O_3NPs through electrostatic attraction forces.

The viscosity measurements reported by Corredor et al. (2019) showed that (i) the XG polymer solutions and nanopolymer suspensions exhibited shear-thinning behavior due to the uncoiling and partial alignment of the XG chains at the high shear rate; (ii) the suspension with high concentration of XG (4000 ppm) and 0.3 and 1 wt%

NaCl exhibited slightly higher viscosity than the salt-free solution, which is consistent with Wyatt and Liberatore (2009, 2010), whereby below a critical XG concentration of ~2000 ppm, the viscosity decreases upon addition of salt. The addition of counter-ions (Na^+) contributes to neutralizing the electrostatic interactions between charges along the backbone of the XG chain, which allows the chain to fold, causing a reduction in the hydrodynamic size of the macromolecules. Since the XG chains decrease in size, the number of interactions with neighboring chains also decreases. For polymer concentrations exceeding the critical XG concentration, the interaction among the XG chains increases, and its effect on the viscosity of the solution becomes more important than the negative effect caused by the decrease in the hydrodynamic size of the macromolecules.

For the salt-free nanopolymer suspensions, the increment in viscosity can be attributed to the interaction between –OH groups on the surface of the NPs with the carboxylic groups on the trisaccharide side chains of the XG through hydrogen bonding (Kennedy et al. 2015). When NaCl is added to these nanopolymer suspensions, the hydrophobic interactions between SDS micelles, NPs and the XG chains increase, due to the screening of the electrostatic repulsion between them (Khan et al. 2008; Nedjhioui et al. 2009). The repulsion forces between the attached SDS micelles and SDS-coated NPs will prevent the coiling of the XG chains, facilitating the formation of a tridimensional network with other NP–XG–SDS complexes, increasing the viscosity.

The reduction in viscosity of XG-SDS solution caused by the addition of $Fe(OH)_3$, Al_2O_3, and TiO_2 NPs suggests high adsorption of polymer molecules at the NP surface. In other words, high polymer adsorption on the NPs' surfaces reduces the viscosity of the polymer solution, because it reduces the polymer concentration in the liquid phase and diminishes the bridging of polymer molecules with the NPs. The reduction in the polymer viscosity was noted at all salinities.

4.7 XG Interaction with Heavy Metals in Suspension

In a study by Mandre and Panigrahi (1997), the sulfide minerals from complex sulfide ore containing 6.73% Zn, 2.1% Pb, and 5.5% Fe were concentrated by selective flocculation using XG. The zeta potential of the virgin sulfide ore is shown in Fig. 4.7a (data from Mandre and Panigrahi (1997)). The sulfide ore shows a negative zeta potential throughout the pH range, and the negative potential increases from −46 to −185 mV with increase in pH. The zeta potential of treated sulfide ore with XG concentration of 1 wt% and pH of 11 is shown in Fig. 4.7b. The results indicated that the increase in the flocculant dosage increases (i.e., less negative) the zeta potential, giving rise to the adsorption of the XG. Moreover, for the recovery of heavy metals from solution, the results indicated that: (i) the metal grades were enhanced from 2 to 3.3% for Pb and from 6 to 10.57% for Zn, with percent recovery of 92.71 and 95.90 for Pb and Zn, respectively; and (ii) the calculated selectivity index, which

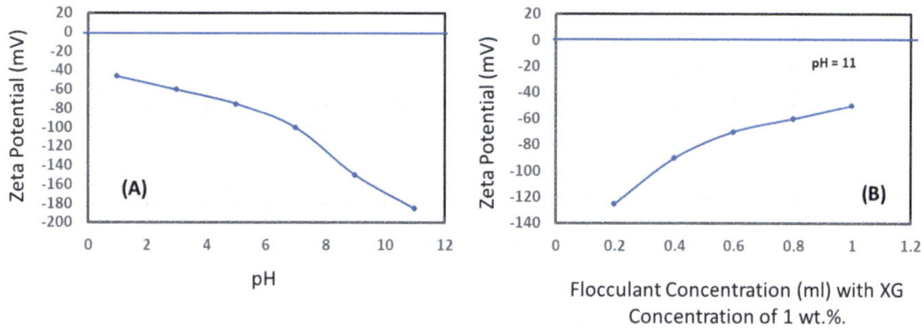

Fig. 4.7 **a** Zeta potential of sulfide ore with pH variations; and **b** zeta potential of sulfide ore concentrations with XG concentration of 1 wt% and pH 11 (data from Mandre and Panigrahi (1997))

measures the effectiveness of the flocculation process, indicates that the value for Zn (of 0.36) is higher than that of Pb (of 0.30).

The nature of the interaction between XG and bivalent heavy metal ions can be through forming intramolecular complexes between the side chains of the same XG molecule, such as: (i) the pyruvate units that are located at the outer ends of the side chains for the cases of Zn^{2+} and Pb^{2+}; (ii) the carboxylate groups of the adjacent glucuronic acid residues in the binding of lighter bivalent cations (e.g., Ca^{2+}), and (iii) the carboxylic groups of the adjacent glucuronic acid residues in the ligation of Cu^{2+}, Co^{2+}, and Fe^{3+}, and forming octahedral or square-planar structures.

It is to be noted that the nature of the interaction between the active functional groups of the polysaccharide and the metallic ions affects the properties of biopolymers. Some polysaccharides provide intramolecular binding sites for cations, and their binding causes a considerable reduction of the biopolymer's hydrodynamic radius and hence decreases the viscosity of the solution. Other polysaccharides form chelate-like complexes with metal cations via intermolecular cross-linking, leading to a notable increase in the solution viscosity. The ability to create cross-linking networks enables the biopolymers to encapsulate the bivalent heavy metal ions. The cross-linking of biopolymers also enhances the mechanical properties of soils and the durability of earthen construction materials. Since the polysaccharide chains bind both the metallic ions and soil particles, they can stabilize metals in subsurface soils by trapping them in stable complexes, while enhancing their geoengineering properties.

4.8 Summary and Concluding Remarks

The physical and chemical interactions between clay soils and XG are multifaceted and involve a variety of physical and chemical interaction mechanisms. The physical interaction can be in the form of adsorption, whereby the XG molecules adsorb onto the surfaces of the clay particles through weak van der Waals forces. The extent of adsorption depends on the concentration of XG, the type and size of the clay particles, and the ionic strength of the soil suspension. The adsorption of XG onto the clay particles increases the stability of soil aggregates by reducing the repulsive forces acting between clay particles, which can help to prevent soil erosion.

In addition, the physical interaction can be in the form of gelation, involving the XG molecules forming a three-dimensional network of biopolymer chains in the presence of water, leading to the formation of a gel-like structure. This gel-like structure increases the water-holding capacity of the soil pore voids and reduces soil deformation. Finally, the physical interaction can be in the form of lubrication, whereby the XG molecules reduce interparticle friction in the soil matrix, thereby improving the flow properties of the soil, and reducing the energy required for soil compaction operations.

The chemical interactions can be in the form of electrostatic interactions, whereby the XG molecules interact with the clay particles through electrostatic forces. The negative charges on the XG molecules attract positively charged ions present in the DDLs of the clay particles, leading to the availability of alkaline metal ions and nutrients to plants, and reduces leachability of toxic chemicals to groundwater. Furthermore, the chemical interaction can be in the form of hydrogen bonding, whereby the XG molecules participate in hydrogen bonding with the clay particles via their hydroxyl groups. The hydrogen bonding improves the stability of soil aggregates, improves soil structure and increases shear strength. Finally, chemical interaction can be in the form of chelation, whereby the XG molecules chelate metal ions in the soil, thus improving the soil's agricultural productivity, and reducing the toxic effects of heavy metals on the groundwater. The chelation of metal ions by XG can also reduce the soil pH, which can improve the solubility of some nutrients.

In summary, the physical and chemical interactions between clay soils and XG are multifaceted and depend on several factors. The nature of these interactions can vary depending on the concentration of XG, the type and size of clay minerals present in the soil, and external environmental conditions. The interactions between clay soils and XG have the potential to improve the soil properties, such as mechanical stability, enhanced water-holding and nutrients retention capacities, and reduced leachability of toxic metal ions and groundwater pollution.

References

ASTM (1985) American society for testing and materials. Methods of test for zeta potential of colloids in water and wastewater; ASTM D4187–82. American Society for Testing and Materials, West Conshohocken, PA, USA

Attia AM, Musa H (2015) Effect of sodium magnesium silicate nanoparticles on rheology of xanthan gum polymer. Int J Sci Eng Res 6:1349–1364

Ayeldeen MK, Negm AM, El Sawwaf MA (2016) Evaluating the physical characteristics of biopolymer/soil mixtures. Arab J Geosci 9:371

Azimi M, Soltani A, Mirzababaei M, Jaksa MB, Ashwath N (2024) Biopolymer stabilization of clayey soil. J Rock Mech Geotech Eng https://doi.org/10.1016/j.jrmge.2023.12.020

Backfolk K, Lagerge S, Rosenholm JB, Eklund D (2002) Aspects on the interaction between sodium carboxymethylcellulose and calcium carbonate and the relationship to specific site adsorption. J Colloid Interface Sci 248:5–12

Barry MM, Jung Y, Lee J-K, Phuoc TX, Chyu MK (2015) Fluid filtration and rheological properties of nanoparticle additive and intercalated clay hybrid bentonite drilling fluids. J Pet Sci Eng 127:338–346

Bayat AE, Moghanloo PJ, Piroozian A, Rafati R (2018) Experimental investigation of rheological and filtration properties of water-based drilling fluids in presence of various nanoparticles. Colloids Surf A 555:256–263

Bergmann D, Furth G, Mayer C (2008) Binding of bivalent cations by xanthan in aqueous solution. Int J Biol Macromol 43(3):245–251. https://doi.org/10.1016/j.ijbiomac.2008.06.001

Burrows K, Carbajal D, Kirsner J, Owen B (2004) Clay-free synthetic-based fluid results in zero barite sag. Offshore 64:88

Chang I, Cho GC (2012) Strengthening of Korean residual soil with β-1,3/1,6-glucan biopolymer. Constr Build Mater 30:30–35

Chang I, Kwon Y, Im J, Cho G-C (2018) Soil consistency and inter-particle characteristics of xanthan gum biopolymer containing soils with pore-fluid variation. Can Geotech J cgj-2018–0254–. https://doi.org/10.1139/cgj-2018-0254

Clavijo JV, Roldán LJ, Valencia L, Lopera SH, Zabala RD, Cárdenas JC et al (2019) Influence of size and surface acidity of silica nanoparticles on inhibition of the formation damage by bentonite-free water-based drilling fluids. Part I: nanofluid design based on fluid-nanoparticle interaction. Adv Nat Sci Nanosci Nanotechnol 10:045020

Clavijo JV, Moncayo-Riascos I, Husein M, Lopera SH, Franco CA, Cortés FB (2021) Theoretical and experimental approach for understanding the interactions among SiO_2 nanoparticles, $CaCO_3$, and xanthan gum components of water-based mud. Energy Fuels 35(6):4803–4814

Colic M, Fisher ML, Franks GV (1998) Influence of ion size on short-range repulsive forces between silica surfaces. Langmuir 14:6107–6112

Comba S, Dalmazzo D, Santagata E, Sethi R (2011) Rheological characterization of xanthan suspensions of nanoscale iron for injection in porous media. J Hazard Mater 185:598–605

Corredor LM, Husein MM, Maini BB (2019) Effect of hydrophobic and hydrophilic metal oxide nanoparticles on the performance of xanthan gum solutions for heavy oil recovery. Nanomaterials 9(1):94. https://doi.org/10.3390/nano9010094

Dário AF, Hortêncio LMA, Sierakowski MR, Neto JCQ, Petri DFS (2011) The effect of calcium salts on the viscosity and adsorption behavior of xanthan. Carbohydr Polym 84(1):669–676. https://doi.org/10.1016/j.carbpol.2010.12.047

Draget KI, Braek GS, Smidsrod O (1994) Alginic acid gels—the effect of alginate chemical-composition and molecular-weight. Carbohydr Polym 25:31–38

Friend JP, Kitchener JA (1973) Some physico-chemical aspects of the separation of finely-divided minerals by selective flocculation. Chem Eng Sci 28(4):1071–1080, ISSN 0009-2509. https://doi.org/10.1016/0009-2509(73)80010-7

Furth G, Knierim R, Buss V, Mayer C (2008) Binding of bivalent cations by hyaluronate in aqueous solution. Int J Biol Macromol 42:33–40

Gebhardt JE, Fuerstenau DW (1983) Adsorption of polyacrylic acid at oxide/water interfaces. Colloids Surf 7(3):221–231, ISSN 0166-6622. https://doi.org/10.1016/0166-6622(83)80048-1

Hamed SB, Belhadri M (2009) Rheological properties of biopolymers drilling fluids. J Pet Sci Eng 67:84–90

Hamedi Shokrlu Y, Babadagli T (2011) Transportation and interaction of nano and micro size metal particles injected to improve thermal recovery of heavy oil. In: Proceedings of the SPE annual technical conference and exhibition, Denver, CO, USA, 30 October–2 Nov. 2011, vol 3, pp 2257–2268

Hendraningrat L, Torsæter O (2014) Metal oxide-based nanoparticles: revealing their potential to enhance oil recovery in different wettability systems. Appl Nanosci 5:181–199

Hua D, Gao S, Zhang M, Ma W, Huang C (2020) A novel xanthan gum-based conductive hydrogel with excellent mechanical, biocompatible, and self-healing performances. Carbohydr Polym 247:116743, ISSN 0144-8617. https://doi.org/10.1016/j.carbpol.2020.116743

Huang W, Xie B, Qiu Z, Jiang L (2016) Bentonite-free water-based drilling fluid with high-temperature tolerance for protecting deep reservoirs. Chem Technol Fuels Oils 51:652–662

Husein M (2017) Preparation of nanoscale organosols and hydrosols via the phase transfer route. J Nanopart Res 19

Kang X, Xia Z, Chen R, Sun H, Yang W (2019) Effects of inorganic ions, organic polymers, and fly ashes on the sedimentation characteristics of kaolinite suspensions. Appl Clay Sci 181:105220. https://doi.org/10.1016/j.clay.2019.105220

Kennedy J, Kent K, Brown J (2015) Rheology of dispersions of xanthan gum, locust bean gum and mixed biopolymer gel with silicon dioxide nanoparticles. Mater Sci Eng C 48:347–353

Khan MY, Samanta A, Ojha K, Mandal A (2008) Interaction between aqueous solutions of polymer and surfactant and its effect on physicochemical properties. Asia Pac J Chem Eng 3:579–585

Kumar MA, Moghal AAB, Vydehi KV, Almajed A (2023) Embodied energy in the production of guar and xanthan biopolymers and their cross-linking effect in enhancing the geotechnical properties of cohesive soil. Buildings 13:2304. https://doi.org/10.3390/buildings13092304

Kwon YM, Chang I, Cho GC (2023) Xanthan biopolymer-based soil treatment effect on kaolinite clay fabric and structure using XRD analysis. Sci Rep 13(1):11666. https://doi.org/10.1038/s41598-023-38844-w

Li MC, Wu Q, Song K, De Hoop CF, Lee S, Qing Y et al (2016) Cellulose nanocrystals and polyanionic cellulose as additives in bentonite water-based drilling fluids: rheological modeling and filtration mechanisms. Ind Eng Chem Res 55:133–143

Mahmoud O, Nasr-El-Din HA, Vryzas Z, Kelessidis VC (2016) In nanoparticle-based drilling fluids for minimizing formation damage in HP/HT applications. In: SPE international conference and exhibition on formation damage control, society of petroleum engineers

Mandre NR, Panigrahi D (1997) Studies on selective flocculation of complex sulphides using cellulose xanthate. Int J Miner Process 50(3):177–186, ISSN 0301-7516. https://doi.org/10.1016/S0301-7516(97)00013-6

Mendonça A, Morais PV, Pires AC, Chung AP, Oliveira PV (2021) A review on the importance of microbial biopolymers such as xanthan gum to improve soil properties. Appl Sci 11(1):170. https://doi.org/10.3390/app11010170

Mohamed AMO, Antia H (1998). Geo-environmental engineering. Elsevier, Amsterdam, p 707. ISBN: 0-444-89847-6

Mohamed AMO, Paleologos EK (2018) Fundamentals of geo-environmental engineering: understanding soil, water, and pollutant interaction and transport. Butterworth-Heinemann, Elsevier, USA, p 708. Imprint eBook ISBN: 9780128051450; Paperback ISBN: 9780128048306

Mohammed ZH, Haque A, Richardson RK, Morris ER (2007) Promotion and inhibition of xanthan 'weak-gel' rheology by calcium ions. Carbohydr Polym 70:38–45

Muguda S, Lucas G, Hughes PN, Augarde CE, Perlot C, Bruno AW, Gallipoli D (2020) Durability and hygroscopic behavior of biopolymer stabilized earthen construction materials. Constr Build Mater 259:119725

Nedjhioui M, Moulai-Mostefa N, Canselier JP, Bensmaili A (2009) Investigation of combined effects of xanthan gum, sodium dodecyl sulphate, and salt on some physicochemical properties of their mixtures using a response surface method. J Dispers Sci Technol 30:1333–1341

Parizad A, Shahbazi K, Tanha AA (2018) SiO_2 nanoparticle and KCl salt effects on filtration and thixotropical behavior of polymeric water-based drilling fluid: with zeta potential and size analysis. Results Phys 9:1656–1665

Perween S, Beg M, Shankar R, Sharma S, Ranjan A (2018) Effect of zinc titanate nanoparticles on rheological and filtration properties of water-based drilling fluids. J Pet Sci Eng 170:844–857

Pomogailo AD, Kestelman VN (2005) Metallopolymer nanocomposites. Springer, Berlin/Heidelberg, Germany
Ponmani S, William JK, Samuel R, Nagarajan R, Sangwai JS (2014) Formation and characterization of thermal and electrical properties of CuO and ZnO nanofluids in xanthan gum. Colloids Surf A Physicochem Eng Aspects 443:37–43
Reddy JJ, Varaprasad BJS (2020) Effect of xanthan gum on internal erosion characteristics of dispersive soils—as an ecofriendly additive. Int J Adv Sci Technol 29(7s):4716–4724
Saha R, Uppaluri RVS, Tiwari P (2018) Silica nanoparticle assisted polymer flooding of heavy crude oil: emulsification, rheology, and wettability alteration characteristics. Ind Eng Chem Res 57:6364–6376
Saravanan L, Subramanian S (2016) Surface chemical properties and selective flocculation studies on alumina and silica suspensions in the presence of xanthan gum. Miner Eng 98:213–222. https://doi.org/10.1016/j.mineng.2016.08.022
Shakeel A, Ali W, Chassagne C, Kirichek A (2022) Tuning the rheological properties of kaolin suspensions using biopolymers. Colloids Surf A Physicochem Eng Aspects 654:130120. https://doi.org/10.1016/j.colsurfa.2022.130120
Singh BP, Menchavez R, Takai C, Fuji M, Takahashi M (2005) Stability of dispersions of colloidal alumina particles in aqueous suspensions. J Colloid Interface Sci 291:181–186
Smith SR, Rafati R, Sharifi Haddad A, Cooper A, Hamidi H (2018) Application of aluminum oxide nanoparticles to enhance rheological and filtration properties of water-based muds at HPHT conditions. Colloids Surf A 537:361–371
Somasundaran P, Fuerstenau DW (1966) Mechanisms of alkyl sulfonate adsorption at the alumina-water interface 1. J Phys Chem 70:90–96
Srivatsa JT, Ziaja MB (2012) In an experimental investigation on use of nanoparticles as fluid loss additives in a surfactant–polymer based drilling fluid. In: IPTC 2012: international petroleum technology conference, one petro, pp cp-280–00226
Vargas J, Roldán LJ, Lopera SH, Cardenas JC, Zabala RD, Franco CA et al (2019) In effect of silica nanoparticles on thermal stability in bentonite free water-based drilling fluids to improve its rheological and filtration properties after aging process. Offshore Technology Conference Brasil, Offshore Technology Conference
Whatley L, Barati R, Kessler Z, Tsau J-S (2019) In water-based drill-in fluid optimization using polyelectrolyte complex nanoparticles as a fluid loss additive. SPE International Conference on Oilfield Chemistry, Society of Petroleum Engineers
William JK, Ponmani S, Samuel R, Nagarajan R, Sangwai JS (2014) Effect of CuO and ZnO nanofluids in xanthan gum on thermal, electrical and high-pressure rheology of water-based drilling fluids. J Pet Sci Eng 117:15–27
Wyatt NB, Liberatore MW (2009) Rheology and viscosity scaling of the polyelectrolyte xanthan gum. J Appl Polym Sci 114:4076–4084
Wyatt NB, Liberatore MW (2010) The effect of counterion size and valency on the increase in viscosity in polyelectrolyte solutions. Soft Matter 6:3346–3352
Xue D, Sethi R (2012) Viscoelastic gels of guar and xanthan gum mixtures provide long-term stabilization of iron micro- and nanoparticles. J Nanopart Res 14:1–14
Yong RN, Mohamed AMO, Warkentin BP (1992) Principles of contaminant transport in soils. Elsevier, p 327
Zhang H, Yan Z, Xie F, Tian Y, Ai L (2023) Rheological properties and kinetics of gelation of binary polymers between xanthan gum and locust bean gum. Polymers (Basel) 15(23):4604. https://doi.org/10.3390/polym15234604

Chapter 5
Stability of Xanthan Gum–Soil Cohesive Networks

5.1 Introduction

As discussed in Chaps. 2 and 3, Xanthan Gum (XG) is a high-molecular-weight polysaccharide consisting of a repeating pentasaccharide unit, composed of glucose, mannose and glucuronic acids, and which are linked by β(1,4)-glycosidic bonds. The anionic carboxyl and hydroxyl groups on the XG molecule make it highly hydrophilic, and it can interact with various compounds, including water and metal ions. XG has excellent water-holding capacity, thickening ability and stabilizing properties, making it widely used in various applications, including agriculture, food processing, pharmaceutical, environmental and geotechnical engineering, and petroleum industries. In recent years, XG has also been increasingly investigated as a soil stabilizer to improve the physical and mechanical properties of soils.

Clay soils are composed of colloidal-sized mineral particles, such as silicates, aluminates and iron oxides, which have a high specific surface area (SSA) and a net negative charge due to the presence of exchangeable cations, such as Ca^{2+}, Mg^{2+}, and K^{+}. The high SSA and negative charge of clay particles make them highly reactive with various inorganic and organic compounds, including XG. When XG is added to clay soils, it can interact with the negatively charged clay particle surfaces through electrostatic forces. In other words, the anionic carboxyl and hydroxyl groups on the XG molecules allows them to interact with the positively charged metal ions (e.g., Ca^{2+} and Mg^{2+}) present in the Diffused Double Layer (DDL) of clay particles. This interaction leads to the formation of a stable and cohesive network of XG molecules on the surface of the clay particles that can improve soil structure (mainly through inducing clay particle flocculation) and its water-holding capacity.

The formation and stability of the cohesive network created by XG can be affected by various factors, including the concentration and molecular weight of XG, clay mineralogy, soil texture and cation exchange capacity (CEC), and environmental conditions, such as pH and temperature. Several studies have investigated the nature of the cohesive bonds developed between XG and clay mineral soils. For example,

A. M. O. Mohamed et al., *Sustainability in Ground Improvement: The Case of Xanthan Gum Biopolymer*, Green Energy and Technology,
https://doi.org/10.1007/978-3-031-75313-8_5

in a study by Chang et al. (2015), XG fibers were found to interact directly with the charged surfaces of clayey particles, while forming XG matrices that resemble a hard plastic between uncharged particles. Therefore, the strengthening effect of XG was shown to have the greatest effect for well-graded soils with fine particles. The results also found that with varying concentrations of XG, the strengthening effect had leveled off at higher concentrations, and was highly dependent on soil type, hydration level (e.g., moisture content), XG content, and mixing method.

The above discussion highlights the fact that the interactions between XG and soil particles are complex, and can have important implications for soil physical, chemical and mechanical properties, microbial activity and nutrient availability. Understanding the mechanisms and factors that govern the XG–soil interactions can help optimize the use of XG as a soil stabilizer, thereby maximizing its benefits for geotechnical and geo-environmental engineering applications.

This chapter discusses the consistency of XG-stabilized soils in view of the following factors: (a) soil type (cohesionless, cohesive with and without swelling characteristics); (ii) soil pore-fluid chemistry; (iii) XG physicochemical properties (size, initial structural conformation, concentration, viscosity, molecular mass quantity and distribution, shear rate and temperature); and (iv) the environmental conditions that the XG-stabilized soils will be subjected to in terms of variations in temperature and pH (i.e., from acidic to alkaline conditions). In addition, characterization and evaluation of the strength of the XG–soil cohesive bonds at both the macro- and micro-scales are highlighted.

5.2 Pore Fluid Viscosity

5.2.1 XG–Deionized Water (DW)

The viscosity of the soil pore-fluid is an important parameter affecting the nature of the interaction between the soil clay surfaces, XG and ionic species. In addition, the shear rate applied to XG-stabilized clay soils affects the viscosity of XG in the soil pore fluid, which can be evaluated in view of the data reported by Lee et al. (2021). They studied the shear-rate-dependent viscosity response of the XG (CAS:11138-66-2; Sigma Aldrich) investigating three different XG solutions, i.e., for different XG to deionized water (DW) mass ratios (or M_{XG}/M_{DW}). Their results showed that a drastic reduction in viscosity occurred with shear rate increase (Fig. 5.1), which can be anticipated to affect the shearing behavior of the XG-stabilized soil.

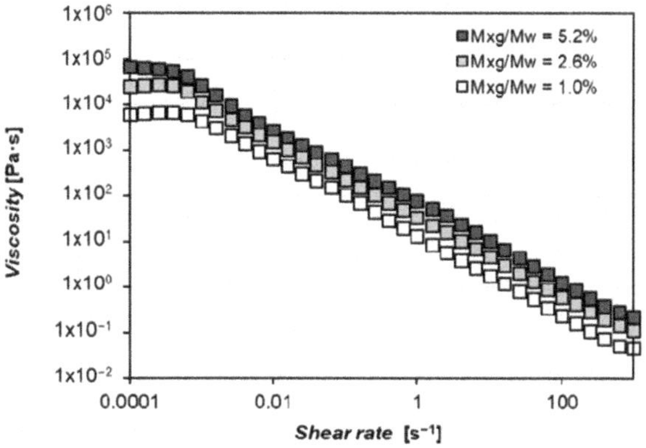

Fig. 5.1 Shear-rate-dependent viscosity reduction of XG hydrogels at different XG to water mass ratios (i.e., M_{XG}/M_{DW}) (Copyright by Lee et al. 2021. Open access article distributed under the terms and conditions of the Creative Commons Attribution (CC BY) license (https://creativecommons.org/ licenses/by/4.0/))

5.2.2 *XG–Soil Natural Pore Fluid*

Changes in the dynamic viscosity of extracted natural soil pore fluids as a function of XG concentration are illustrated in Fig. 5.2. The tested soil had the following characteristics: (i) chemical composition of 60.51% silica, 19.6% alumina, 8.94% ferrous oxide, 3.39% calcium oxide, 2.2% potassium oxide, 2.11% magnesium oxide, 1.12% titanium oxide, 1.05% sodium oxide, and small traces (< 1%) of other oxides (e.g., phosphorous pentoxide, lead oxide and barium oxide); (ii) mineralogical composition of kaolinite, dickite, palygorskite, sepiolite, allophane, imogolite and pargasite; and (iii) grading composition of 1.15% gravel, 58.85% sand and 40% fines comprising of silt and clay. As demonstrated in Fig. 5.2, at a low XG concentration of 0.5%, the viscosity of the extracted pore fluid is low (356 cP) and increases in a nonlinear manner with increasing XG concentration (to 4100 cP for 3% XG). However, at 5% XG concentration, the pore fluid forms a thick gel. Such an increase in the pore fluid viscosity affects the soil's geotechnical properties, related to compaction, compression (consolidation), strength, and water flow. Notably, the reported study did not indicate the shear rate at which the viscosity tests were conducted.

To compare the reported results by Sujatha et al. (2020) and Lee et al. (2021) for XG in soil natural pore fluid and DW, respectively, let us consider the viscosities of XG in DW solution at an XG/water ratio of $M_{XG}/M_w = 2.6\%$ (equivalent to an XG/dry soil ratio of 0.5 wt%, with an initial soil moisture content reported as 19.23%). From Fig. 5.1, the viscosities would be 1.0×10^4, 1.0×10^3 and 3.0×10^1 Pa.s for 0.001, 0.01 and 1.0 (1/s) shear rates, respectively. According to Sujatha et al. (2020), however, the corresponding viscosity would be 356 cP for 0.5 wt% XG in the soil pore fluid (see Fig. 5.2). Comparing these values (1 Pa.s = 1000 cP), one finds a considerable difference due to the shear rate, as well as the pore fluid concentration, such that it is of vital importance to specify the shear rate and the nature of the pore fluid used in the experiments.

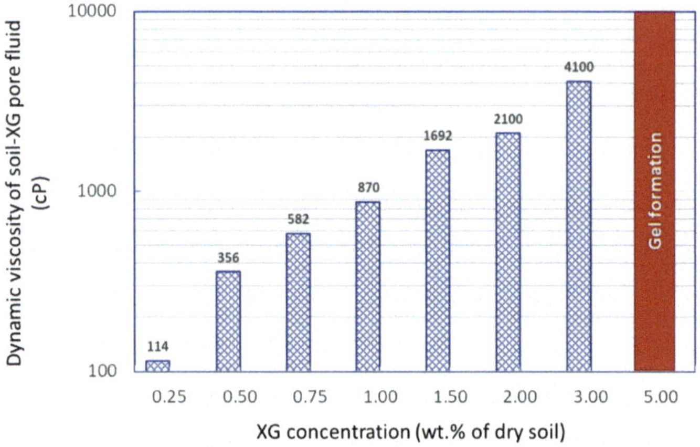

Fig. 5.2 Dynamic viscosity of the XG–natural soil pore-fluid mixture (data from Sujatha et al. (2020)

5.2.3 Solubility of XG in Deionized Water (DW)

The viscosity results shown in Fig. 5.3 were obtained for biopolymer solutions prepared at different XG concentrations (i.e., 0.25, 0.5, 0.75, 1.0, 1.25, 1.5, and 2% to the total volume of the DW used). The XG had an initial ash content of 10.25%, a natural moisture content of 12.89%, and contained carbon (53.18%), oxygen (41.16%), Na (1.58%), Si (0.06%), Cl (0.74%), and Ca (3.27%). In addition, the 1% XG in DW solution had a pH of 8.2, and a molecular weight of 2×10^6 g/mol. The appropriate amount of XG was dissolved in 50 ml of DW, agitated using a magnetic stirrer at 600 rpm for periods ranging from 0.5 to 1.0 h, in order to attain homogeneous XG solutions.

Based on the Fourier Transfer Infrared (FTIR) analysis, the XG exhibited various kinds of chemical bonds, such as O–H stretching (3398 cm^{-1}), C–H stretching (2922 cm^{-1}), attributed to the stretching of CH_2 or CH_3, –CC– stretching (2162 cm^{-1}), C=O stretching (1726 cm^{-1}), axial deformation of C=O of enols (1614 cm^{-1}), C–H deflection angle (1415 cm^{-1}), COO– asymmetric stretching (1372 cm^{-1}), C–O stretching (1274 cm^{-1}, 1062 cm^{-1}), strong C–H bending (900–700 cm^{-1}), and C–Br stretching (607 cm^{-1}).

Referring to Fig. 5.3, the results highlighted that (i) for the same shear rate, as XG concentration increases, the viscosity increases; (ii) for the same XG concentration, as the shear rate increases, the viscosity decreases due to the thinning phenomenon of the XG.

Furthermore, Kumar and Ramani (2021) have indicated that at an XG concentration of 1.25%, the biopolymer was completely dissolved in DW; however, at higher XG concentrations, the XG did not dissolve because of its very high viscosity that contributed to the formation of thick gels. Therefore, the temperature of the XG aqueous water solutions were increased from ambient room temperature to 80 °C

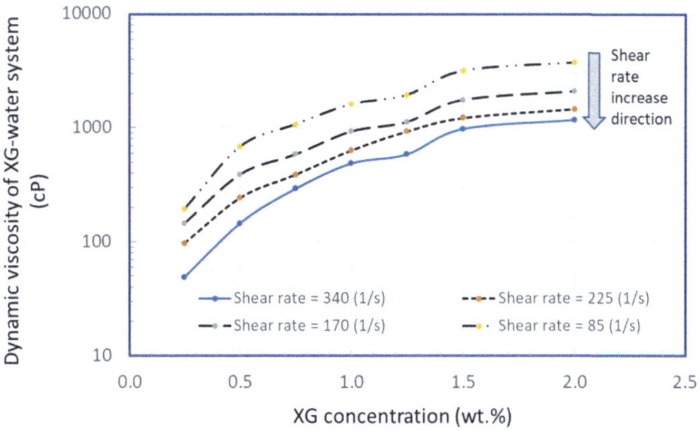

Fig. 5.3 Variations of dynamic viscosity (for different shear rates) with XG concentration in deionized water (data from Kumar and Ramani (2021))

for the preparation of solutions at higher XG concentrations. Notably, at such high temperature, the conformational structure of XG changes, and its mode of interaction with clay mineral soils varies. Such information is of great benefit because: (i) when dry mixing at ambient temperature is practiced, a homogenous mixing will not occur at higher XG concentrations; (ii) when dealing with natural soils, the actual pore fluid must be utilized in the determination of the XG viscosity because of its impact on the XG structural conformation; and (iii) when reporting experimental results, it is important to describe the XG concentration, the composition of the soil pore fluid, the shear rate, and the measurement temperature.

5.2.4 Pore Fluid Concentration and Clay Mineralogy

To appreciate the changes of the viscosity due to the potential presence of XG in the system, Fig. 5.4 illustrates the changes in viscosity of Na-kaolinite (at 0.01-M NaOH) and Na-montmorillonite in water at 25 °C, without XG. The viscosity of Na-kaolinite varies from 0.91 to 1.10 cP for kaolinite concentration increasing from 1.2% to 8.7%. Whilst, for Na-montmorillonite, the viscosity varies from 0.95 to 4.77 cP for montmorillonite concentrations of 0.1% and 2.2%, respectively. This implies that at low XG concentration of 0.25%, the XG–DW aqueous system viscosity (Fig. 5.3) is about 114 times greater than that of Na-kaolinite at 8.7% concentration, and 24 times that of Na-montmorillonite at 2.2% concentration. Considering once again the data in Fig. 5.3, an additional increase of XG to 1.0% in the DW aqueous system would likely lead to an increase of 870 times for Na-kaolinite, and of 183 times for Na-montmorillonite.

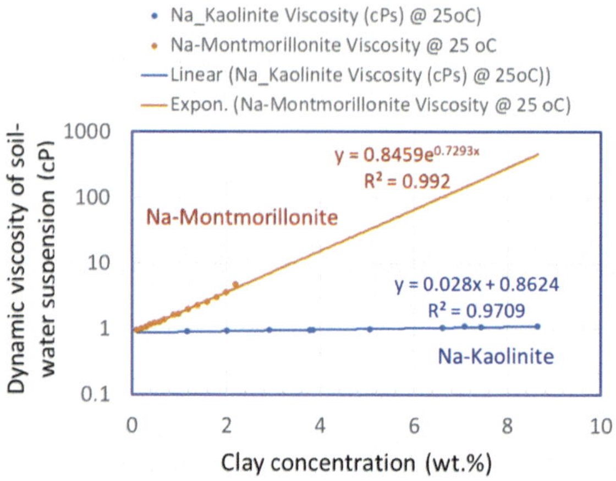

Fig. 5.4 Viscosity variations of Na-kaolinite and Na-montmorillonite in water with 0.01-M NaOH at 25 °C, without XG (data from Kasperski et al. (2011))

5.2.5 Temperature

The effects of temperature on the viscosity of a bentonite–XG mixture was evaluated by Akpan et al. (2020), as shown in Fig. 5.5. The viscosity decreases as the shear rate and temperature increased; i.e., the XG biopolymer underwent shear as well as thermal thinning. The viscosity changes at small shear rates are different from those occurring at high shear rates because of the initial presence of an ordered rod-like (rigid) XG conformation with a high persistence length (L_p), which can self-associate, creating domains of associate chains (aggregates) at low shear rates, or at-rest.

However, at high shear rates, the gel aggregates are converted into solution, and the single/double helical structure of XG are degraded into single chains. As discussed in Chap. 3, as temperature increases beyond the dissolution temperature of 60 °C, the viscosity declines because of the XG conformation shift from an ordered (low-dissolution temperature) to a disordered (high dissolution temperature) state (see Fig. 5.6).

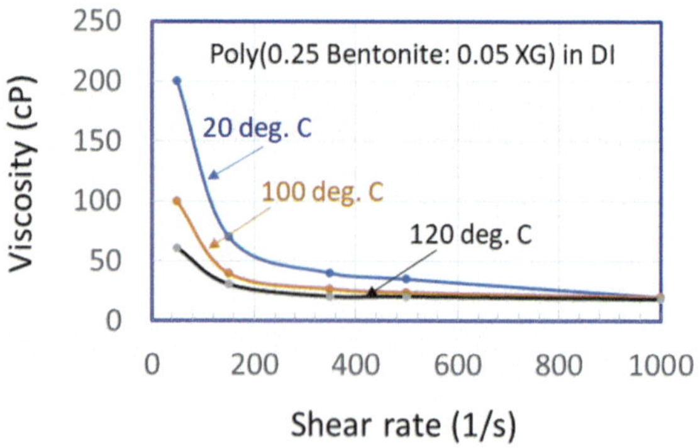

Fig. 5.5 Effects of temperature on viscosity with shear rate for poly(0.25 bentonite:0.05 XG) in deionized water (DW) (data from Akpan et al. (2020))

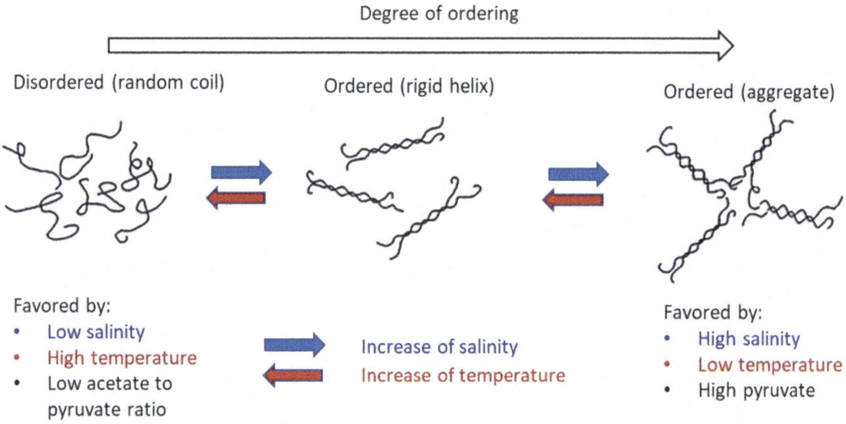

Fig. 5.6 Schematic showing XG structural ordering with changes in environmental conditions

5.2.6 Salt Concentration

The effects of salt concentration on the viscosity of bentonite–XG mixtures was also evaluated by Akpan et al. (2020). As shown in Fig. 5.7, with addition of NaCl in solution, the viscosity increased. As discussed in Chap. 3, the addition of NaCl to XG with high concentrations (> 0.2%) reduces the hydrodynamic diameter of the XG molecules while simultaneously increasing the viscosity of the XG gel. As such, the XG chains are contacting each other through hydrogen bonds, and hence forming denser XG gel solution. The increase of viscosity could be explained as follows. Without NaCl addition, the electrostatic repulsion between the XG chains tends to keep the XG chains at a certain distance from each other. With the addition of NaCl, repulsion forces are screened, making it possible for the XG chains to come closer in contact with each other and form hydrogen bonds. The development and propagation of the hydrogen bonds makes it difficult for the XG chains to move past one another in solution, thereby leading to an increase in the viscosity. Therefore, the establishment of hydrogen bonding and increased interaction have a larger effect on the viscosity of XG gel solutions than the decrease in the hydrodynamic diameter.

5.3 Emulsion Stability

The stability of XG-treated clay soils can be evaluated in view of the fluid consistency, flow behavior, aggregate surface areas, dispersivity, plasticity indices and water potential, as described below.

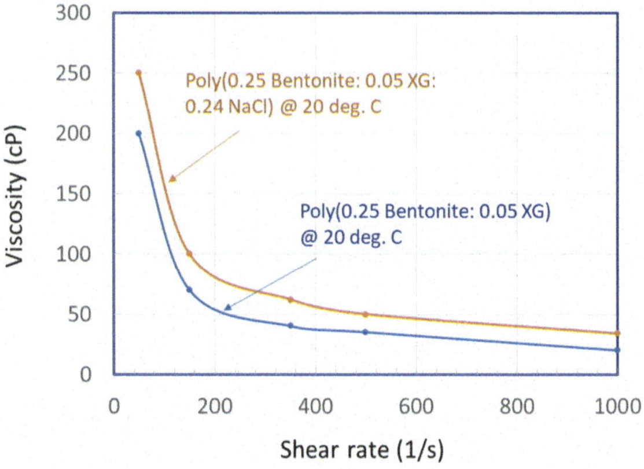

Fig. 5.7 Effects of salt concentration on viscosity with shear rate for poly(0.25 bentonite:0.05 XG) mixture (data from Akpan et al. (2020))

5.3.1 Fluid Consistency and Flow Behavior

Viscometers are generally used to evaluate the fluid consistency and flow behavior. For example, Akpan et al. (2020) used the fully automated Model 1100 viscometer to: (i) measure the shear stress (τ), shear rate (γ), apparent viscosity, low shear-rate viscosity, yield point (Y_P) and plastic viscosity (P_V); and (ii) estimate the flow behavior index (n), fluid consistency index (f_{ci}) and gel strength of formulated drilling muds at 25 °C. The software analysis models that are associated with common viscometers are based on the power-law model (n and f_{ci}), the Bingham plastic model, and the three parameter Herschel–Bulkley model. These models are briefly described below.

Herschel–Bulkley is a three-parameter model (Mitchell et al. 2011) that can be represented as follows:

$$\tau = \tau_y + f_{ci}\gamma^n \tag{5.1}$$

where τ is the shear stress; τ_y is the yield shear stress; γ is the shear rate; f_{ci} is the fluid consistency index; and n is the flow behavior index.

Fluid types can be characterized by using the Herschel–Bulkley model: (i) when $n < 1$, the fluid is yield-pseudoplastic; (ii) when $n > 1$, the fluid is dilatant; (iii) when $\tau_y = 0$ and $n < 1$, the fluid is pseudoplastic; (iv) when $n = 1$, the fluid is plastic; and (v) when $\tau_y = 0$ and $n = 1$, the fluid is Newtonian.

In addition, the fluid consistency index (f_{ci}) is employed as a measure of the drilling fluid mixture's thickness. It decreases with increasing temperature (Demirdal et al. 2007), and is an important parameter in designing the transport requirements of a drilling fluid employed in the petroleum industry. The flow behavior index (n) is used as a measure of the extent of shear thinning (i.e., with viscosity decreasing under shear strain); in other words, the smaller the value of n, the higher the shear thinning behavior.

The two parameter $\tau{:}\gamma$ power-law model defines a fluid as follows (Mitchell et al. 2011):

$$\tau = \eta\gamma^n \tag{5.2}$$

where η is the viscosity for a shear rate of 1 (1/s), expressed in lb/100 ft^2; and n is the flow behavior index.

The Bingham plastic model defines the shear stress–shear rate ($\tau{:}\gamma$) relationship of a fluid by two parameters as follows:

$$\tau = Y_\mathrm{p} + P_\mathrm{v}\gamma \tag{5.3}$$

where Y_p is the yield point; and P_v is the plastic viscosity.

When the shear stresses exceed the yield point, the viscosity of a Bingham plastic fluid becomes equal to the plastic viscosity.

The variations of the fluid consistency index and the flow behavior index for an XG-stabilized bentonite, both in water and in saline environment, are shown in Fig. 5.8. Addition of salt (i.e., for the drilling mud type M2 suspension) increased the fluid consistency index because of the viscosity increase; however, its shear thinning behavior, characterized by the flow behavior index values (of 0.2), remains the same.

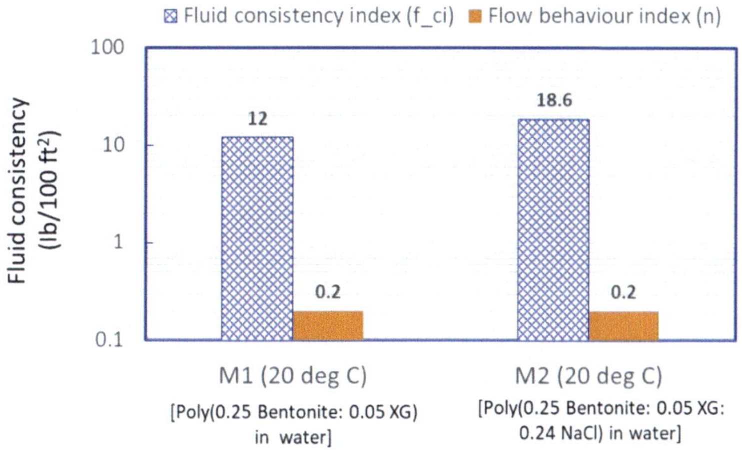

Fig. 5.8 Effects of salt on the fluid consistency and flow behavior indices of poly(0.25 bentonite:0.05 XG) mixture (data from Akpan et al. (2020))

5.3.2 Aggregate Surface Area

To shed light onto the nature of the soil–XG microstructural development upon addition of XG to pure clay minerals, Latifi et al. (2016) measured the external surface area (ESA) values of both untreated and XG-stabilized kaolinite and bentonite clays for curing periods ranging from 0 to 90 days (Fig. 5.9a). The results indicated that the addition of XG decreased the ESA of both stabilized clays due to aggregation. For bentonite, the ESA decrease from its initial value of 27.85 to 13.6 m^2/g after 90 days of curing (i.e., ESA decreased by about 51%); whilst, for kaolinite, it decreased from 17.14 to 5.36 m^2/g considering the same curing period (ESA decrease of about 69%). The rate of decrease for kaolinite was greater than that for bentonite because of the difference in the XG concentration used for the two soil materials (i.e., 1% for bentonite and 1.5% for kaolinite). Notably, their reported ESA value of 17.14 m^2/g for kaolinite (i.e., its SSA considering the pure clay material) is within the expected range (of 10–20 m^2/g) reported in the literature (Mohamed and Antia 1998; Mohamed and Paleologos 2018).

However, for bentonite (montmorillonite), a value of 27.85 m^2/g as ESA (SSA) is substantially below that reported in the literature (SSA varying from 600 to 800 m^2/g; 80% of which is due to internal surfaces). Irrespective of these discrepancies, the relative changes in the ESA of the kaolinite and bentonite clays due to XG addition, as well as XG curing, are remarkable. Similar results were reported by Hamza et al. (2023) for a 1.5% XG-treated natural high-plasticity clay soil (Fig. 5.9b), which showed a time-dependent decrease in the ESA by about 60% from its initial value after 60 days. This clearly indicates the continual XG-stabilized soil microstructural aggregation change with curing time.

It is worth noting that Na-bentonite is balanced with Na^+; however, due to isomorphous substitutions, it exhibits negative charges. The interlayers are not bonded, and water can easily enter between layers, as illustrated in Fig. 5.10. The distance of separation of the layers on hydration can be controlled by using certain organic liquids rather than water. Montmorillonite saturated with glycerol shows layer spacing of

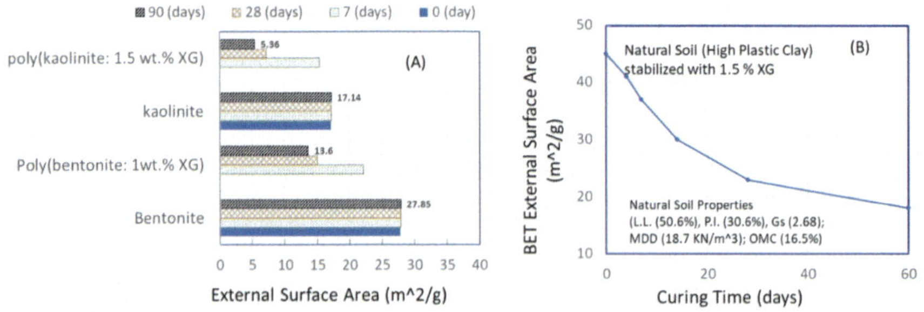

Fig. 5.9 Changes in external surface area (ESA) of aggregate due to treatment with XG: **a** kaolinite and bentonite clays (data from Latifi et al. (2016)); **b** natural clay with high plasticity (data from Hamza et al. (2023))

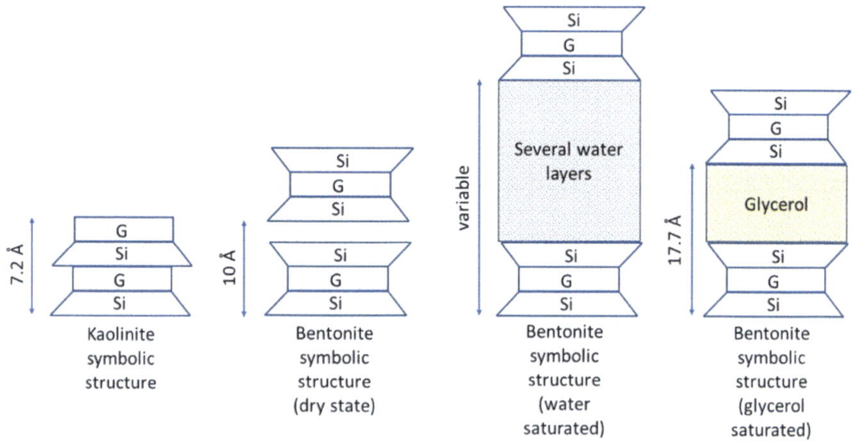

Fig. 5.10 Schematic illustration of kaolinite and montmorillonite structures

17.7 Å (i.e., 1.77 nm), of which 10 Å is the thickness of the layer and 7.7 Å of the glycerol. Hence, with the introduction of XG as a stabilizing agent, the XG will enter between the layers and control their spacing, depending on the status of the XG structural conformation (i.e., ordered (helix) or disordered (coil)).

However, since the XG structure varies with its concentration, different structures with variable viscosity and molecular mass prevail, such that the resultant interlayer spacings will vary (see Fig. 5.11). Also, the XG could be present as an amorphous material covering the external surfaces of the soil particles and filling the pore spaces. In addition, because of the XG structure occurring as either coiled, disordered-coiled, single helix, double helices or aggregated helices depending on its concentration, the nature of its interaction and the resultant XG–montmorillonite structure will vary from dispersed (at low concentration) to highly aggregated (at high concentration). Such arrangements have been reflected in the measured external surface area, as illustrated in Fig. 5.9a. Similar interaction mechanisms and structural formation will occur for XG–kaolinite, except no intercalation between kaolinite layers occurs, since they are fixed with an interlayer spacing of 7.2 Å.

5.3.3 Dispersivity

The effect of XG addition on the dispersion property of XG-stabilized soils can be evaluated in view of the experimental results reported by Joga and Varaprasad (2020) for an XG-stabilized natural dispersive soil from India. The tested soil was composed of 29% sand, 23% silt and 48% clay, with consistency limits properties of LL = 68%, PL = 29% and PI = 39%. It achieved a maximum dry unit weight (γ_{dmax}) of 16.8 kN/m^3 for standard Proctor-compaction at its optimum moisture

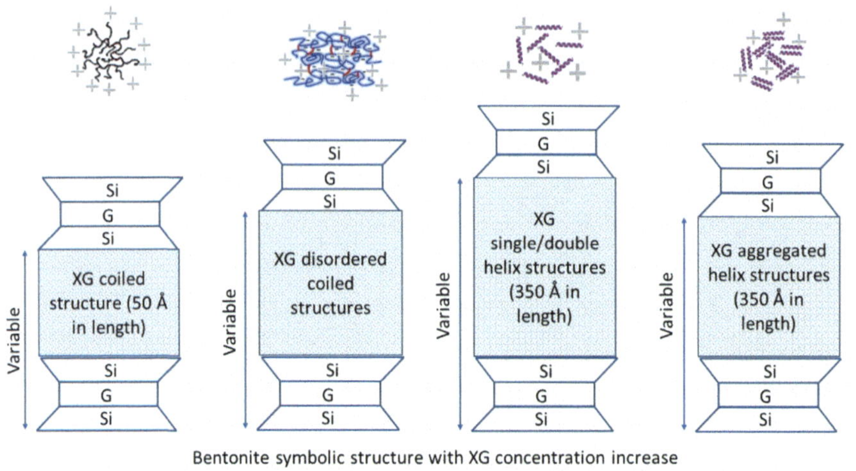

Fig. 5.11 Schematic illustration of montmorillonite structures intercalated by XG at varying concentrations (and hence imposing different structures)

content (OMC) of 19.2%, an unconfined compressive strength (UCS) of 360 kPa, and a hydraulic conductivity of 5.36 × 10^{-8} m/s (for the latter two properties, the soil was tested at its OMC). In addition, the soil pore fluid consisted of Na^+(174.56 meq/l), K^+(13.6), Ca^{2+}(36.9) and Mg^{2+}(14.53), with a total dissolved solids of 239.53 meq/l, sodium adsorption ratio (SAR) of 34.4%, electrical conductivity of 21.5 mS/cm, and pH of 8.2. The soil had a dispersion ratio of 69.5 (i.e., > 50), characterizing the soil as being dispersive (Elges 1985).

The different dispersivity assessment methods carried out were the crumb test, cylindrical dispersion test, and double hydrometer analysis. The crumb test, as per ASTM D 6572, employed distilled water as the submersion solution, with the results indicating that XG addition (1.0% XG) reduced the dispersivity of the soil, changing its classification from dispersive to non-dispersive.

The cylindrical dispersion test (Fig. 5.12), whereby the cylindrical soil specimen was submerged in a jar container, is used to assess the time required to reach, ultimately, the softened state of the stabilized soil. The results indicated that, in its natural state, the test specimen of dispersive soil collapsed, settling at the bottom of the jar, while a dark cloudy color suspension was seen above the collapsed soil, and extended to the water surface. However, upon the addition of 1% XG to the soil, the cylindrical test-specimen became stable as a column, while fully submerged in water for periods of 7, 14, and 28 days. Such behavior is due to the reduction of the repulsive forces, and the formation of aggregated soil–XG structures.

It is worth noting that the XG (an anionic biopolymer) interacts with the soil in different ways, such as: (i) interaction with divalent cations and forming a cross-linking network through hydrogen bonds and cation bridging (in favor of inducing clay particle flocculation); (ii) impacting on the thickness of the diffused ion layer due to the presence of XG and the different types of cations in different concentrations;

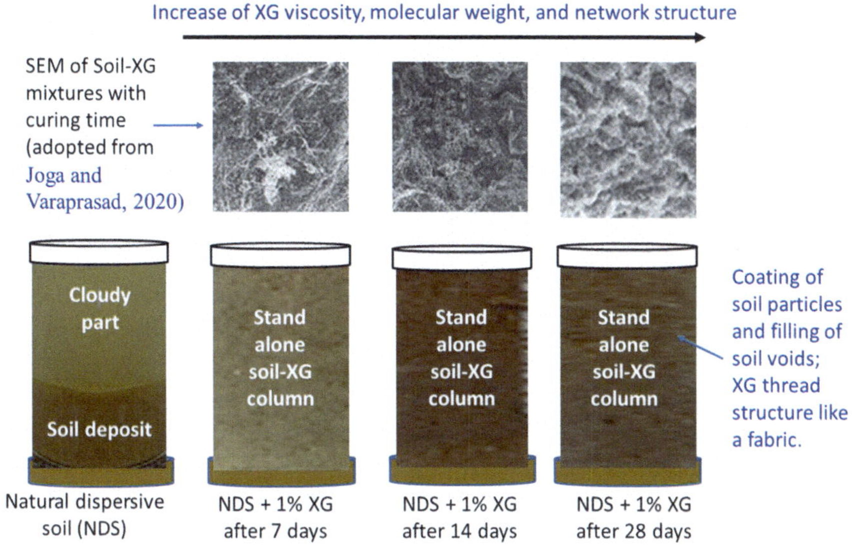

Fig. 5.12 Schematic representation for the cylindrical dispersion test, investigating natural and 1% XG-stabilized dispersive soil

(iii) specific adsorption of biopolymer molecules and cations present in the pore fluid on the soil particle surfaces; and (iv) formation of soil aggregates.

In the double hydrometer test (Decker and Dunnigan 1977), the percentage of particles that are less than 0.005 mm in size, as established by the standard hydrometer test, is determined, and the percent dispersion (i.e., dispersivity) is calculated as the ratio of percent finer at 0.005 mm without and with the XG additive. Then, based on the dispersivity criteria, the soil is classified as "*dispersive*" if the dispersivity is greater than 50%, "*intermediate dispersive*" if it ranges between 30 and 50%, and "*non-dispersive*" for values of less than 30% (Elges 1985). The results have indicated that upon the addition of 1% XG, the natural dispersive soil became non-dispersive in behavior (Fig. 5.13).

In another study, Kwon et al. (2022) investigated the dispersion of kaolinite–XG mixtures for different XG concentrations using the cylindrical dispersion test, assessing the time required for the kaolinite–XG mixtures to reach an equilibrium sediment volume. They used X-ray CT images to evaluate the final structure of the various stabilized mixtures. Figure 5.14 illustrates the sedimentation behavior of the kaolinite and kaolinite–XG mixtures. The untreated kaolinite showed uniform sedimentation of a dispersed system of soil particles, as noted from the X-ray CT images (Fig. 5.14c). However, the kaolinite–XG suspensions behaved in a completely different manner. As the XG concentration increases, a flocculated system is produced due to the electrical forces (attraction and cationic bridging) between the XG molecules and the kaolinite particles. The degree of flocculation (and hence soil aggregation) increased with increasing XG concentration in the

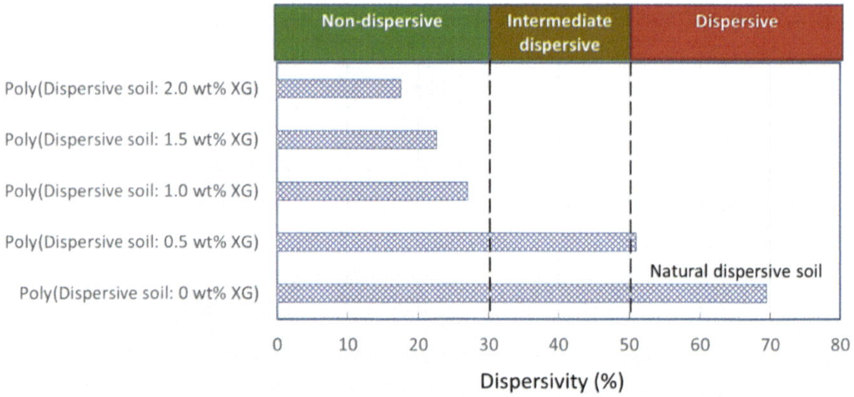

Fig. 5.13 Dispersivity of a natural dispersive soil stabilized by 1% XG (data from Joga and Varaprasad (2020))

solution, resulting in a denser sediment formation. The kaolinite blended with 2% XG resulted in a larger formation at the bottom of the column, and smaller sized aggregates were also observed for increasing height from the bottom. Whereas the untreated kaolinite showed an even formation distribution along the column height. The final stable heights (or volumes) were, in a decreasing order, untreated kaolinite, stabilized kaolinite with 0.5% XG, and equally for the stabilized kaolinite with 1 and 2% XG.

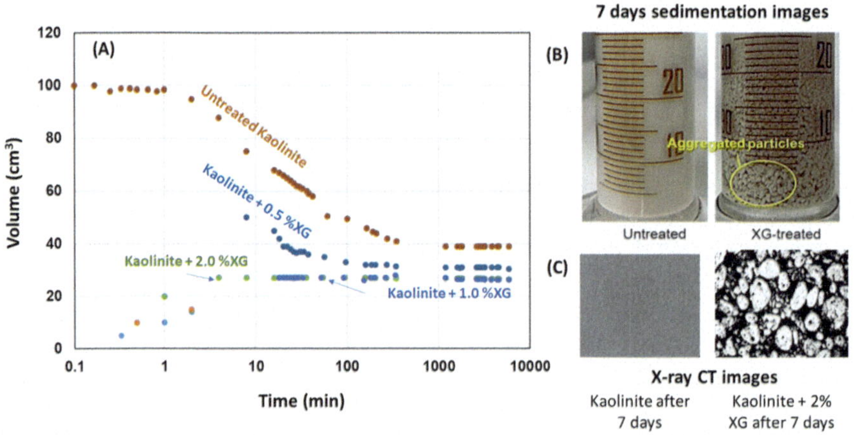

Fig. 5.14 Sedimentation behavior of kaolinite and kaolinite–XG mixtures: **a** sedimentation column results; **b** images of settled mixtures after 7 days; and **c** X-ray CT images of settled mixtures at the bottom of the columns (adopted from Kwon et al. (2022))

5.3.4 Plasticity Indices

The effect of XG content on the plasticity indices of XG-stabilized soil can be demonstrated using the results reported by Singh and Das (2020). They used XG ($C_{35}H_{49}O_{29}$) to stabilize high plasticity silt soil (LL, PL and shrinkage limit (SL) values of 86%, 25%, and 11%, respectively) that had a high swelling behavior, with a measured differential free-swelling (or free swell index) value of 217%. X-ray diffraction (XRD) analysis confirmed the presence of the montmorillonite clay mineral in the soil. In addition, the soil had an OMC for standard Proctor-compaction of 21%, achieving a γ_{dmax} of 16.1 kN/m^3. The XG was added to the soil using the dry-mixing method, investigating different XG contents of 0, 0.2, 0.5, 0.8 and 1.0 wt%.

The variation of the plasticity index (i.e., PI = LL – PL) with XG content for the stabilized high-plasticity silty soil, considering maturing periods of 1 and 29 days, is shown in Fig. 5.15. Referring to this figure, the PI increased as the XG content increased up to 0.5 wt%, which is explained as follows. At low XG content, XG forms disordered structures of coils that absorb more water, thereby increasing the LL of the stabilized soil. However, for XG contents in excess of 0.5 wt%, XG might form an ordered structure of single or double helices and tends to be aggregated; hence both the LL and the PI decreased. Therefore, the degree of aggregation depends on the XG content in the stabilized soil.

To further explain the experimental results in terms of the diffused double layer (DDL), the results are plotted in terms of the percent increase in LL against XG content (Fig. 5.16). When clay mineral particles are immersed in water, soluble cations in the pore fluid will interact with the clay particles' negative charged surfaces. Moreover, when XG interacts with water, the carboxylic group (–COOH) of the XG starts to deprotonate H^+ ions, thereby creating hydrogen bonding with the clay particles. On the other hand, due to this deprotonation, the –COO group starts to attract and share the positive ions present in the pore solution, thereby creating repulsive forces between neutralized XG and neutralized clay particles.

Returning to the high plasticity silt soil investigated in Singh and Das (2020), there was about a 20% increase in the LL on the addition of 0.2% XG to the soil, which could be attributed to: (i) the formation of XG coiled structure (of 50 Å long), with its negative charges on the surface attracting positive charges from the pore fluid; (ii) increase of the repulsive forces between neutralized XG with neutralized soil particles; (iii) increase in the viscosity of the diffused ion layer; and (iv) increase of the molecular mass of XG within the DDL. With the addition of 0.5% XG, the LL increases by 40%, which could be attributed to XG structural transformation (i.e., from coiled to coiled-disordered), with greater molecular sizes and viscosity, in addition to the increase of the repulsive forces. Further increase in XG (0.8 wt%) would lead to the formation of helix structures with larger sizes (350 Å long) that could be single or double helical; hence more aggregation that contributes to further LL increase (of 47%).

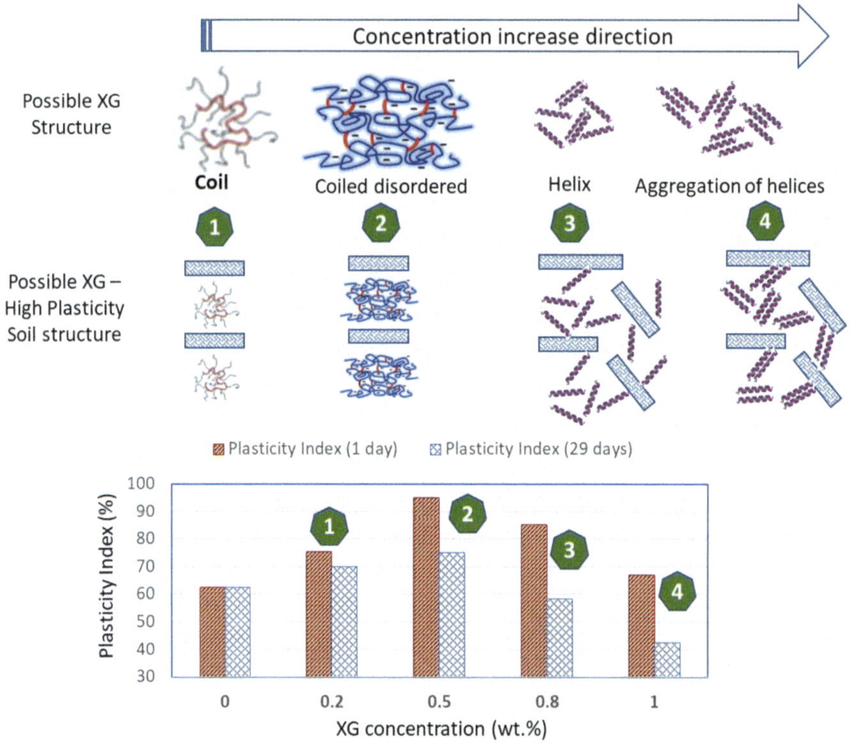

Fig. 5.15 Schematic representation of possible stabilized-soil structure formation with the addition of XG in view of the results of the plasticity index variation with XG content (data in chart from Singh and Das 2020)

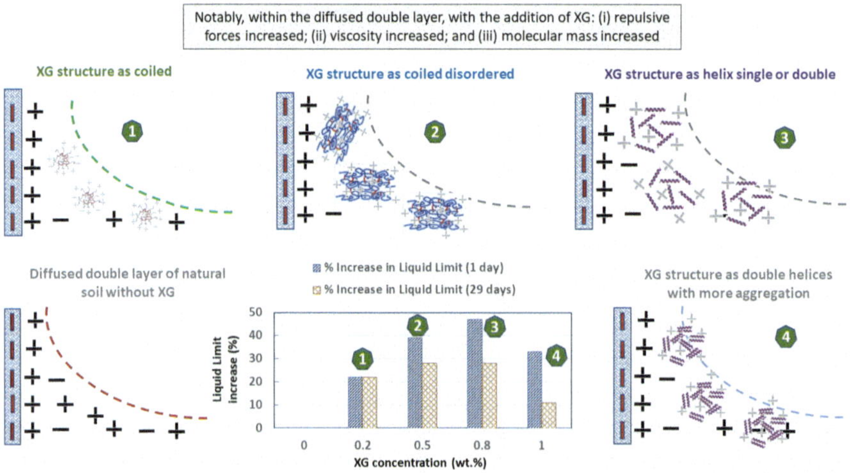

Fig. 5.16 Schematic representations of the Diffused Double Layer (DDL), with and without the addition of XG at different contents

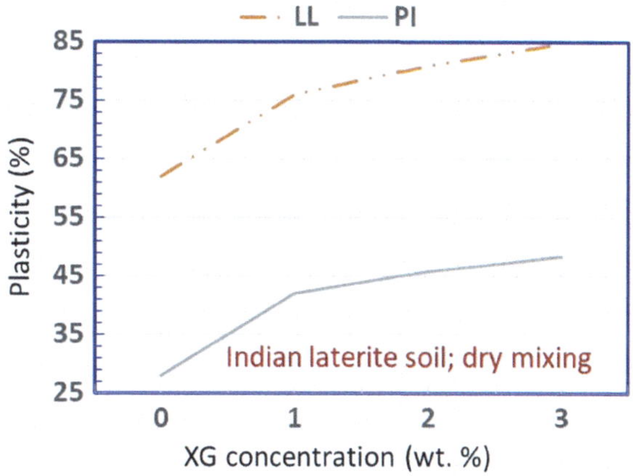

Fig. 5.17 Variations of plasticity parameters for XG-stabilized Indian laterite soil (data from Sawant et al. (2019))

However, for 1 wt% XG content, the structural formation will tend towards double helices, and more aggregations, that lead to a decrease in the LL, although the LL remains 33% higher than that of the untreated soil. Notably, further increase in XG content will lead to gelation status. It is interesting to note that the experimental results have indicated a decrease in LL upon maturing (for 29 days), which clearly indicates the capacity of the XG to aggregate and change its conformational structure over time. Therefore, it is of utmost importance to characterize the XG properly before its use to stabilize soils.

In another study of soils with high plasticity, Sawant et al. (2019) studied the plastic behavior of XG-treated laterite soil from India. Laterite soil is formed in hot and wet tropical areas, with leaching of the soil leaving only oxides of iron and aluminum. The plasticity properties of the XG-stabilized laterite soil are shown in Fig. 5.17. As the XG content increases, the LL and PI increased due to the increase of the XG viscosity (i.e., an increase in the water holding capacity).

Bagheri et al. (2023) tested an Australian silt soil of low plasticity, i.e., with its measured LL (of 38%) < 50%, and showed that as the XG content (% wt. of the dry soil) increased, the soil plasticity index increased (Fig. 5.18). There was a 57% increase in the LL on the addition of 0.5% XG to the soil, and the LL increased by about a further 4% from 0.5 to 1.0 wt% XG, and an additional 4% from 1 to 2 wt% XG.

In another study, Barani and Barfar (2020) evaluated the plastic behavior of XG-treated pure kaolinite clay. Figure 5.19a presents the variations of the OMC for compaction and the maximum dry density achieved, while Fig. 5.19b shows the variations of the LL and PI parameters for the kaolinite–XG mixtures. As the XG concentration increases, the maximum dry density marginally decreased, and the OMC slightly increased due to: (i) increased ability of the XG to adsorb water (i.e., resulting in increased plasticity); (ii) higher viscosity of the pore fluid (in the presence of XG); and (iii) formation of an interconnecting network of hydrogen bonds and cross-links that render the soil matrix stiffer and resist compaction.

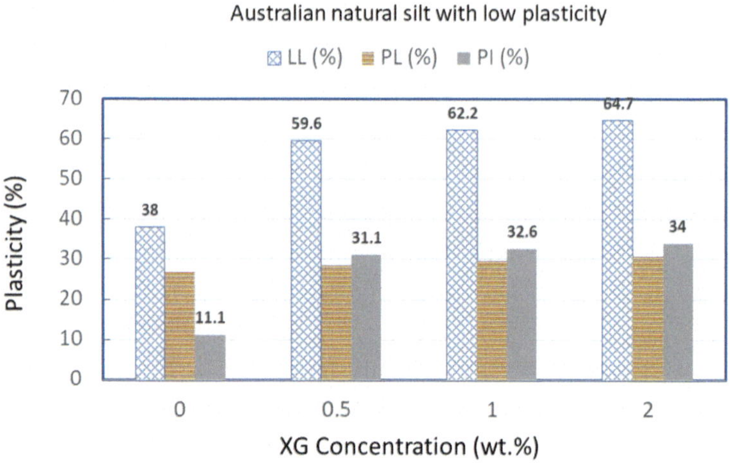

Fig. 5.18 Variation of plasticity parameters with XG content for an Australian silt with low plasticity (data from Bagheri et al. (2023))

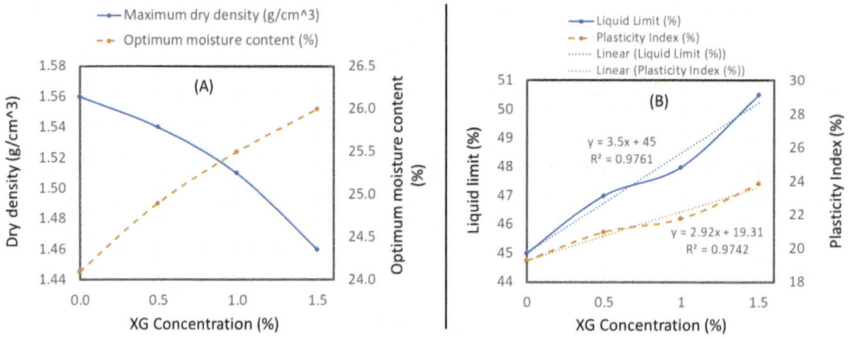

Fig. 5.19 Variations of compaction and plasticity characteristics with XG concentrations for kaolinite–XG mixtures (data from Barani and Barfar (2020))

Since the PI of the tested kaolinite was less than 20%, the soil was classified as soil with medium plasticity. However, with the addition of XG, the soil was classified as high plasticity (i.e., 20% < PI < 40%). In other words, the rise in the LL and PI of the kaolinite–XG mixtures causes a shift in the soil classification, according to the Unified Soil Classification System, in that for 0, 0.5, and 1% XG, the mixtures classify as "*low compressible clay*", whereas for 1.5% XG the mixture classifies as "*high compressible clay*".

Notably, the kaolinite crystal consists of repeating layers, each layer consisting of a silica sheet and an alumina sheet, sharing a layer of oxygen atoms between them (each one of these layers being three oxygen atoms in thickness). The layers are held together through hydrogen bonding between hydroxyls from the alumina sheet on one face, and oxygen atoms from the silica sheet on the opposite face. These forces are relatively strong, preventing hydration between layers, and allowing many layers to build up. The hydrogen bonding between the adjacent unit layers prevents expansion of the mineral beyond its basal spacing, such that kaolinite has a low colloidal activity (i.e., low plasticity and low swell–shrink volume change potential). Most of the CEC of kaolinite can be attributed to the dissociation of OH groups on clay platelet edges. The dissociation of H^+ leaves one non-neutralized negative charge in the octahedron. Such a dissociation reaction occurs at high pH and decreases at low pH levels. The magnitude of the negative charge also increases and decreases accordingly with the change in pH. When the pH is continually decreased from an alkaline to acidic condition, a point can be reached for the negative charges to be equal to zero. The pH at which this condition occurs is called the zero point of charge (ZPC). For kaolinite, the ZPC occurs at a pH of about 4.2. At the ZPC, the clay particles in soil–water system will not repel each other, but instead tend to aggregate and form larger particles (agglomerations), resulting in a decrease in soil plasticity. In contrast, negatively charged clay particles repel each other, resulting in dispersion and an increase in soil plasticity.

With the addition of XG to kaolinite clay, attraction occurs between the positive charges on the kaolinite edges and the anionic XG; whilst repulsion happens between the kaolinite clay surfaces and the XG, such that an aggregated structure will tend to occur. However, the increase in plasticity of the kaolinite–XG mixture with increasing XG content is due to: (i) the high ability of XG to adsorb water due to its high hydration energy (particularly when applied to the soil via dry-mixing); (ii) the viscous nature of XG, which increases with XG concentration; (iii) the high molecular weight of XG, which increases with viscosity; (iv) formation of hydrogel in the pore spaces of the soil matrix; (v) the mutual electrical interaction between the clay particles and the XG; (vi) the aggregation potential of the XG strands with the clay soil particles.

To further evaluate the effects of the mixing water quality (i.e., pore water chemistry) on the soil consistency of pure clay minerals, Fig. 5.20 shows the variation of LL with XG content for poly(kaolinite:XG) and poly(montmorillonite:XG) mixtures prepared using DW and a brine solution (2 mol/l NaCl). The chemical properties of the mixing solutions were: (i) for DW; dielectric constant = 80, electrical conductivity < 5 μS/m, and pH = 6.5; and (ii) for brine solution; dielectric constant = 55, electrical conductivity = 12 S/m, and pH = 6.5. For poly(kaolinite:XG) mixtures with DW, Chang et al. (2018) showed an increase in LL up to 0.5% of XG, followed by a decrease up to 1.0% XG, with the LL values remaining stable for concentrations greater than 1.0% XG. Such discrepancy may be attributed to the lack of standard methods for determining the consistency (Atterberg limits) properties of fine-grained soils mixed with XG biopolymer(s), specifically regarding the sample maturing time, i.e., after mixing the sample and prior to conducting the LL test, since the XG viscosity changes with maturing (curing) period.

For kaolinite mixed with brine solution, in the absence of XG, the LL is less than that for the DW because of the decrease in the thickness of the diffused double layer (DDL) (e.g., due to the lower dielectric constant of brine compared to DW), such that the amount of water being adsorbed is less. However, for the kaolinite–XG mixtures

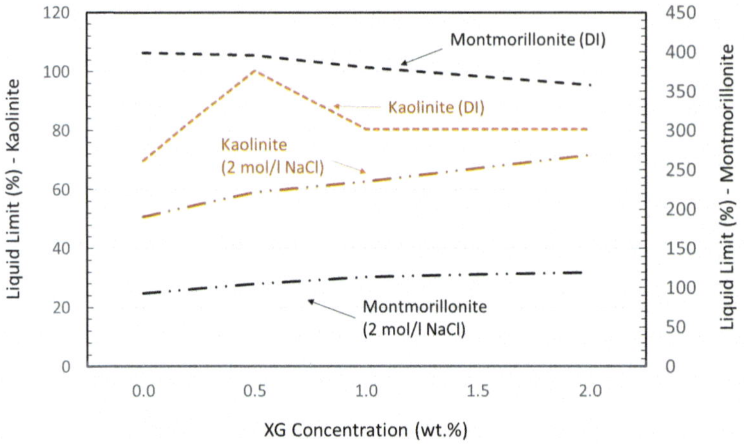

Fig. 5.20 Variations of liquid limit for kaolinite–XG and montmorillonite–XG mixtures with changes in mixing water quality (i.e., DW, deionized water, or brine solution (2 mol/l NaCl)) (data from Chang et al. (2018))

prepared using brine solution (Chang et al. 2018), the LL increases with increasing XG content because of the hydration of the XG and the increase of its viscosity.

For montmorillonite clay mixed with DW, the LL decreases with the addition of XG, probably due to the intercalation of XG between the layers of the montmorillonite crystal and reduction of the interlayer spacing; thereby reducing the amount of adsorbed water. Aggregation of the montmorillonite–XG structures could be another possibility. However, using the brine solution, and in the absence of XG, the LL decreased sharply because of the NaCl effect on the thickness of the DDL. Also, in the presence of NaCl and XG, the LL increases for increasing XG content due to the hydration of the XG.

5.3.5 Soil Water Potential

To elaborate further on the nature of the XG-treated soil's structure and the pore sizes, Zhang et al. (2022) evaluated the water potential characteristics of XG-treated natural silty soil sourced from Henan Province, China. The untreated soil had the following index properties: LL = 21.4%; PL = 12.6%; PI = 8.8%; $G_s = 2.70$; and a maximum dry density of 1.77 Mg/m^3 (obtained for the standard Proctor-compaction test).

In order to ensure the uniformity of the XG-silt mixture, dry soil was mixed evenly with XG. DW was then added to form the wet soil, which was then hermetically sealed at a constant temperature for 24 h until the water was fully infiltrated. The initial water content of the test-specimen was 15%, and 92% of the maximum dry density (measured as 1.63 g/cm^3) was selected as the initial dry density for the prepared

33-mm diameter by 6-mm high test specimens. For the water retention experiments, the saturated specimens were dried and wetted at constant humidity and temperature, with the suction intermittently measured using a WP4C dew point water potential meter.

The wetting/drying water retention behavior of the XG-treated soil is shown in Fig. 5.21. Overall, the shape of the experimental soil–water characteristic curves resembles those of remolded clay soils. As the XG content increases (from 0 to 2 wt%), the water retention capacity of the treated soil increases. At lower water potentials (or suctions), the volumetric moisture content of the treated soil significantly increases due to the addition of XG, which has a carboxylic acid (R–COOH) that disassociates into carboxylate ions (R–COO$^-$) and hydrogen (H^+). They could interact with hydrated cations and hydroxyl (–OH) ions present in the pore fluid, thereby generating more capillary forces that absorb and retain more water.

Also, with the addition of XG, the cumulative pore intrusion of mercury (in performing the mercury intrusion test) decreased by 4.95% and 6.5% for samples treated with 1 and 2% of XG, respectively, indicating a decrease in the soil pore-size distribution. The decrease was characterized as follows. When 1% XG was added to the natural silt (which had large pore void spaces measuring 10–80 μm in size), the pores were gradually connected by the XG hydrogels, and they reduced to medium-sized pores measuring 2–10 μm. For 2% XG addition, those pores were further decreased to smaller pore sizes measuring 0.1–2 μm, and their numbers were also decreased. In other words, the number of large and small pore sizes in the natural silt were decreased, and the number of the medium pore sizes increased (i.e., the

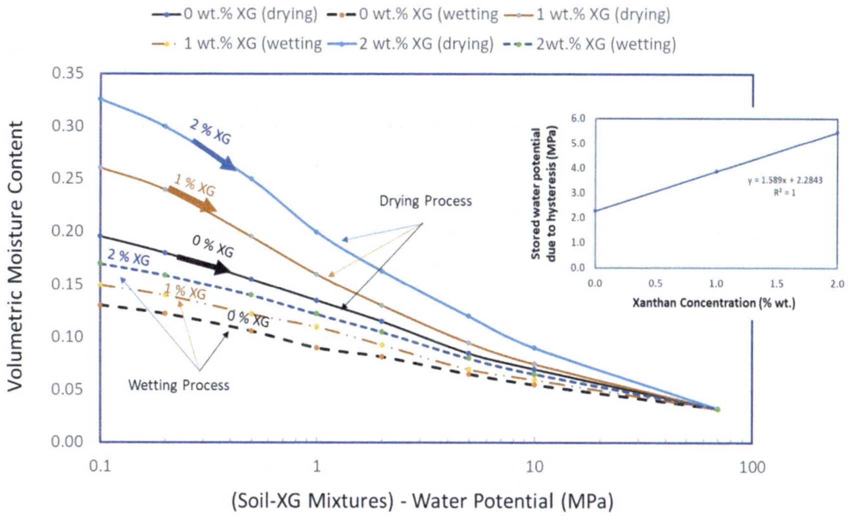

Fig. 5.21 Water potential characteristics of the XG-treated natural silty soil from Henan Province, China (data from Zhang et al. (2022))

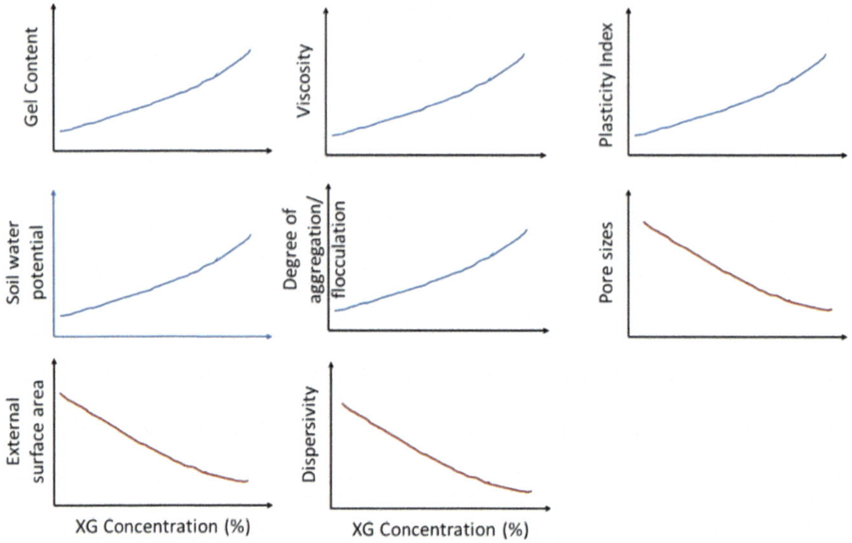

Fig. 5.22 Effects of XG content on some XG-stabilized fine-grained soil properties

pore-size distribution of the XG-treated silt transferred into a unimodal pore-size distribution).

It is also of importance to note the effects of XG content on the hysteresis behavior of the treated silt–XG mixtures. Referring to Fig. 5.21, as the XG content increased, the hysteresis effect (i.e., the area between the drying and wetting curves, with larger areas representing a stronger hysteresis effect) enlarged due to the nature of the XG pseudo-elastic behavior. The stored water potential due to the hysteresis effect was calculated by Zhang et al. (2022) to be 2.28, 3.88, and 5.46 MPa for 0, 1 and 2 wt% of XG, respectively; meaning that the stored soil–XG water potentials have increased by about 70% and 139% for additions of 1 and 2 wt% XG, respectively.

In summary, the effects of XG content on the basic geotechnical properties of XG-stabilized fine-grained soils are graphically represented in Fig. 5.22. It is worth noting that the presented exponential type of the curves has been used to highlight the expected behavior. However, actual experimental results can be flatter, as shown in the preceding discussions, depending on the soil and/or XG type.

5.3.6 *Compressive Strength*

In a study by Jan et al. (2019), natural soil, with PI of 27.45%, free swelling index of 60%, and standard Proctor-compaction properties of maximum dry density of 1.635 Mg/m^3 and optimum moisture content (OMC) of 19%, was mixed with XG at different concentrations of 0.5, 0.75, 1, and 1.25% by dry soil weight. At their OMC, the XG-treated soil showed an increase in the UCS with increasing XG concentration and curing time, with a maximum UCS at 1.25% XG (the highest XG content investigated in the study), due to bonding between (–COOH) and (–OH) of the XG molecules with the electrically charged fine soil particles.

Moreover, the XG-treated soil micrographs, at 1% XG concentration and curing of 7 days, have shown direct bonding associations between clay particles and the XG molecules (Laneuville et al. 2006; Marto et al. 2014; Chang et al. 2015; Latifi et al. 2015). These mechanisms are believed to contribute to the increase in UCS and stiffness of the XG-stabilized soil (Mitchell and Santamarina 2005; Ivanov and Chu 2008; Nugent et al. 2009; Latifi et al. 2015).

In another study by Bagheri et al. (2023), soil material from the Gold Coast area, Australia, with standard Proctor-compaction properties of maximum dry density = 1.72 Mg/m^3 and OMC = 21.7%, and consistency limits properties of LL = 38.0%, PL = 26.9% and PI = 11.1%, was treated with different XG concentrations (0.5, 1, and 2% of XG by dry soil weight). The dry XG powder was first uniformly mixed with the dry soil for about 15 min, and the required mass of water, corresponding to the OMC, was then sprayed into the mixture and thoroughly mixed. To avoid the formation of aggregations and ensure a homogenous mixture of the XG with soil particles, the resulting mixtures were sealed with double-layer plastic wrap and placed at a controlled room temperature for 24 h (Soldo et al. 2020). To make the UCS test-specimens, the prepared mixtures were placed into cylindrical metal molds and evenly compacted in five layers.

Regardless of the percentages of XG additive in the soil, the LL and PL for the XG-treated soil were higher than for the untreated soil. Significant increase in LLs (38, 59.6, 62.2, and 64.7%) and PIs (11.1, 31.1, 32.6, 34%) was observed for untreated soil and that treated with 0.5, 1.0, and 2.0% XG, respectively. A considerable increase (about three-fold) in the PI occurred when adding 2% XG, due to the gel type and hydrophilic nature of the XG, producing an increase in the water holding capacity of the soil and a potential decrease in its permeability (Mendonça et al. 2021a, b). After 10 days of curing, the 2% XG-treated soil showed almost 2.5 times higher UCS than the untreated soil due to soil bonding with XG additive. The scanning electron microscopy images of the XG-treated soil confirmed that the soil particles were coated and bonded with the XG.

In another study by Azimi et al. (2024), water–XG solutions with XG concentrations of 0.25, 0.5, 1.0, 1.5, and 2.0% (i.e., dry XG-to-water mass ratio) were used to stabilize an artificial soil composed of a mixture of montmorillonite and kaolinite at a dry mass ratio of 25% (mass of montmorillonite/mass of kaolinite). The resultant soil was characterized by its consistency limits properties of LL = 98.4%, PL = 35% and PI = 63.4%, and standard Proctor-compaction properties of $\gamma_{\text{dmax}} = 15.7$ kN/m^3 and OMC = 22%. It is of prime importance to note the difference in terminologies for the XG concentrations adopted in this particular study. The preceding XG concentrations were reported in reference to water mass ratio, and to express them in relation to the dry soil mass, such that one must consider the OMC used in compacting the soil. In other words, for an OMC of 22%, the XG concentrations (i.e., dry XG to dry soil mass ratio) would be 0.055, 0.11, 0.22, 0.33, and 0.44 wt%, respectively, which

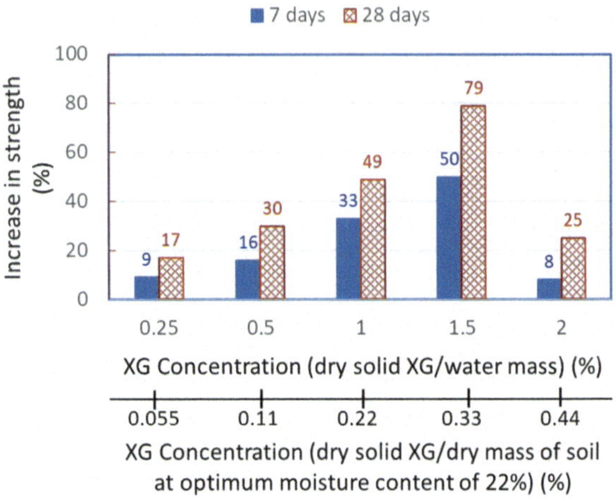

Fig. 5.23 Percent increase in unconfined compressive strength of XG-treated artificial soil (a mixture of montmorillonite and kaolinite at a dry mass ratio of 25%) with curing time (data from Azimi et al. (2024))

are below the reported XG concentrations used in literature for treatment of cohesive soils. It should be mentioned that a similar mixing/application method has also been reported for treating clay soils with sodium alginate biopolymer, polyacrylamide and sulfonated oil (Soltani et al. 2021, 2022a, b). The results (Fig. 5.23; data from Azimi et al. (2024)) showed an increase in the UCS by 9, 16, 33, 50, and 8% after 7 days of curing, and by 17, 30, 49, 79, and 25% after 28-day curing, for 0.25, 0.5, 1.0, 1.5, and 2.0 wt% XG concentrations (dry XG to water mass ratio), respectively, with maximum UCS values showing at 1.5 wt% XG concentration. The latter value corresponds to 0.33 wt% XG concentration when expressed in terms of the dry XG to dry soil mass ratio.

The UCS improvements can be attributed to:

(a) clay–XG interactions that induce flocculation of the soil fabric, with the flocculation mechanism being governed by the XG charge (in terms of type and density) (Seybold 1994). In such a system, electrostatic repulsion between negatively charged clay particles and the anionic XG molecules is dominant. Also, additional interaction may be developed through cationic bridging, whereby the exchangeable divalent cations present in the diffused double layer (DDL) serve as indirect connection/attraction bridges between the negatively charged carboxyl group (–COOH–) of the XG molecules and the negatively charged clay surfaces (Chang et al. 2015; Arab et al. 2019; Soltani et al. 2021; Bozyigit et al. 2023).

(b) formation of hydrogel (i.e. a network of cross-linked hydrophilic biopolymer chains retaining part of the pore-water), as well as hardening (due to dehydration) of the hydrogels over time, as a three-dimensional reinforcement network acting in favor of interlocking the clay flocs into a more coherent matrix of improved shear resistance (Farrés and Norton 2014; Chang et al. 2015; Soldo et al. 2020; Soltani et al. 2021; Mahmood et al. 2022).

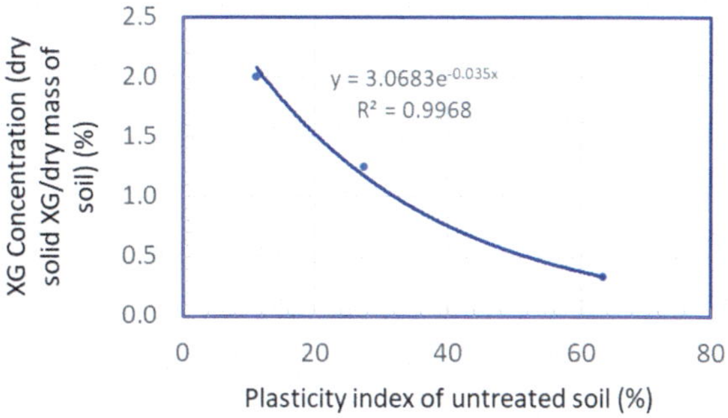

Fig. 5.24 XG concentration (dry XG to dry soil mass ratio) variation with plasticity index of untreated soils (limited to investigated PI domain of 11–63%)

From the preceding studies by Jan et al. (2019), Bagheri et al. (2023) and Azimi et al. (2024), we can simply demonstrate that as the PI of the untreated soil increases, the optimum amount of XG (in terms of the USC gain) required to treat the soils decreases (Fig. 5.24), with an initial guesstimate of the required XG amount needed (preferably using wet mixing to ensure full hydration of the XG) to treat soils obtained using the following equation:

$$XG = 3.0683e^{(-0.035\times PI)} \tag{5.4}$$

where XG is the concentration of xanthan gum (i.e., dry XG to dry soil mass ratio), and PI is the plasticity index of the untreated soil. The trend observed in Eq. 5.4 is somewhat expected, as higher PI is typically associated with the presence of more active clay minerals that undergo stabilization primarily through chemical DDL-based interactions rather than physical mechanisms. However, given the limited studies underlying Eq. 5.4, additional data is needed to validate and/or enhance its applicability.

However, the application of the free swell ratio parameter (defined as the ratio of the equilibrium sediment volumes of 10 g oven-dried soil (< 425 μm) when placed in distilled water (i.e., the polar liquid) to that in kerosene (i.e., the non-polar liquid)) can be a another viable option for estimating the XG concentration. The free swell ratio can be regarded as a somewhat direct representation of the clay mineralogy, with higher values alluding to higher activity levels (Soltani et al. 2024). Therefore, future research for the use of the free swell ratio to estimate the amount of XG that is required to treat cohesive soils is needed.

5.4 Chemical Bonding

5.4.1 Types of Inter-particle Forces

Chemical bonding is characterized as the bonding forces that hold the soil particles and the gelling agent together at their surfaces. At a microscopic level, the bonding depends mainly on the type of forces present at the interfaces between the soil particles and the biopolymer, which are ionic/electrostatic or covalent bonds (chemisorption), hydrogen bonding (strong polar attraction), and van der Waals forces (physical adsorption). Short-range bonds (ionic/electrostatic and covalent) have the highest bonding energies that form strong bonding. Long-range bonds, such as van der Waals forces that describe the interaction between dipoles in the bulk pore fluid within the soil aggregates, are the weakest bonds. In addition to these types of bonding, the increase in the concentration of the gelling polymer will lead to: (i) an increase in the thickness of the gelling material that extends over the soil particles and the soil pores; and (ii) an increase in both the viscosity and molecular weight of the gelling polymer that withstand the crystallization of the biopolymer macromolecule chain. This in turn will contribute physically to an increase in the strength of the ties that connect adjacent soil particles together, as well as over the soil pore void spaces.

Since XG is an anionic polysaccharide, it participates in hydrogen bonding and cationic bridging with the clay particles (Latifi et al. 2016). When XG is dissolved in water, the carboxylic acid (R–$COOH^-$) dissociates into carboxylate ion (R–COO^-) and hydrogen ion (H^+). The monovalent (H^+) ion acts as a bridge between COO^- on the XG tail and OH^- on the clay particle surfaces (cationic bridging). Also, the R–COO^- may link with cations (e.g., Ca^{2+} and Mg^{2+}) present in clay, resulting in the clay–XG biopolymer network. Since cohesionless soil particles carry no surface charges, these types of bonding (hydrogen and electrostatic attraction) are not present in such soils treated with XG (Mohamed and Paleologos 2018).

5.4.2 Energy of Interaction and Bonding Elasticity

The energy of interaction (adhesion energy) and the elasticity of the bonding between XG and kaolinite clay has been evaluated at different structural levels using different techniques. At the macroscale level, the adhesion energy was evaluated using data from tension tests, as well as fatigue-life experiments. At the nanoscale level, the elasticity of the kaolinite–XG bonding was evaluated by atomic force microscopy (AFM) imaging, employing a tip-scanning AFM system.

5.4.2.1 Adhesion Energy at the Macroscale Level

The adhesion energy of clay soils treated with XG can be indirectly evaluated at the macroscale level using the calculated fracture energy from tension tests. Fracture energy represents the energy that a material will absorb during failure (i.e., its ability for absorption of post-crack energy), and is expressed as the energy required to open a unit area of the crack surface. The experimental results reported by Barani and Barfar (2020) for 7-day cured beam specimens of kaolinite clay treated with XG, which were tested using the three-point bending of the single-edge notched beams with dimensions of $a = H/3$, $H/D = 1$ and $S = 3.3H$, where S indicates the span of the simple supports, D is the width of the beam, H is the depth of the beam, and a is the initial crack length of the sample. In this study, a support span of $S = 100$ mm was used to evaluate the adhesion energy.

Using an automatic rotator, the sieved oven-dried clay soil was mixed with 0.5, 1.0 and 1.5% of dry XG powder. Then the required amount of water was added and mixed until achieving a uniform mixture. After a homogenization period, beam specimens were prepared for testing at 95% of their γ_{dmax} (on the wet side of the OMC) according to their compaction characterizations. The fracture energy was analyzed using the size-independent fracture energy modelling approach, as per Elices et al. (1992) and Murthy et al. (2013); whilst fracture toughness modeling was performed based on the method presented in Amarasiri et al. (2011). The calculated results are shown in Fig. 5.25. For the same water content, as the XG content increases, the fracture (adhesion) energy of the poly(kaolinite:XG) mixtures increases because of the increase in the density of the crosslinking structures, and due to the adhesion between the XG–kaolinite clay structures. Conversely, for the same XG content, as the water content increases from initially 1.4% to within the range of 11–23%, the adhesion energy of the poly(kaolinite:XG) mixtures decreases due to the decrease in the density of the crosslinking structures.

The adhesion achieved between the XG and kaolinite clay involves two mechanisms: (i) interfacial bonding; and (ii) bulk energy dissipation within the poly(kaolinite:XG) mixture. The former is due to the interfacial bridging network,

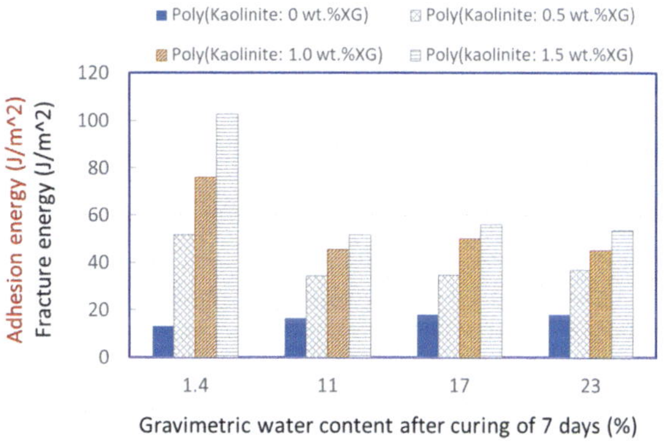

Fig. 5.25 Variations of fracture (adhesion) energy with gravimetric water content for 7-day cured poly(kaolinite:XG) mixtures (data from Barani and Barfar (2020))

between the XG and the kaolinite clay surface, via covalent bonding and physical interpenetration. To drive through an interfacial crack in performing the three-point beam test, either the XG hydrogel network or the bridging network needs to break. Then, the force required to break the bridging network is transmitted into the bulk poly(kaolinite:XG) mixture, leading to the breakage of sacrificial bonds (i.e., the ionically cross-linked XG–kaolinite mixture) and, hence, substantial bulk energy dissipation. As such, one could safely assume that the fracture energy is equal to that required to overcome the adhesive forces within the poly(kaolinite:XG) mixture.

In other words, for nanocomposite viscoelastic materials, such as XG–clay complexes, the energy is dissipated mainly at the XG–clay interfaces. Much like viscoelastic solids, including rubbers (Gent 1996), the energy required to advance a fracture plane (per unit area) consists of two parts: (i) the energy necessary to break the XG chains ahead of the crack tip; and (ii) the bulk viscoelastic energy dissipated around the crack tip. Such viscoelastic materials usually exhibit self-resilient and self-healing behaviors, as discussed in Chap. 3, due to the reversible nature of the XG physical bonds on drying and wetting cycles (Sun et al. 2015).

The reported data by Barani and Barfar (2020) for the fracture (adhesion) energy (Y) of the poly(kaolinite:XG) mixtures were further analyzed, using the response surface methodology and Minitab, in terms of the three independent variables (X_1 for XG content, X_2 for gravimetric water content, and X_3 for dry density). The following regression model was assumed for predicting the Y variable:

$$Y = b_0 + \sum b_i X_i + \sum b_{ii}(X_i)^2 + \sum b_{ij}(X_i)(X_j) \tag{5.5}$$

where b_0 is the constant coefficient; b_i the linear regression coefficient; b_{ii} the quadratic regression coefficient; b_{ij} the interaction regression coefficient; and X_i and X_j are the independent variables.

Using the data (reported by Barani and Barfar 2020), the adhesion energy (Y), as a function of (i) XG content (i.e., X_1 in wt%), (ii) water content after curing (i.e., X_2 in %), and (iii) dry density (i.e., X_3 in g/cm^3), can be expressed by the following equation:

$$\begin{aligned} Y = {} & 3083 + 185X_1 - 41.8X_2 - 3614X_3 - 23.1X_1^2 + 0.1645X_2^2 \\ & + 1066X_3^2 - 0.51X_1X_2 - 73X_1X_3 + 23.7X_2X_3 \end{aligned} \tag{5.6}$$

Response surfaces were generated using Eq. 5.6, and lack-of-fit tests were performed on the fitted model. For statistical tests, significance was defined at $p \leq 0.05$. The t-test was employed to compare the responses under optimized conditions with those predicted by models.

The regression analysis indicated that the XG content (X_1) had a significant role in improving the adhesion energy (Y) of the XG–kaolinite structure, followed by the water content (X_2) and the dry density (X_3). The regression model indicates: (i) for XG content (X_1), a positive effect for the linear term, and a negative effect for both the interacting and quadric terms; (ii) for water content (X_2), a negative effect for

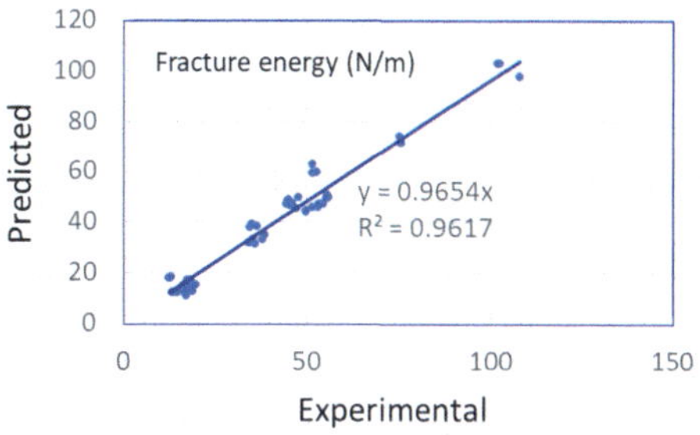

Fig. 5.26 Predicted versus experimental fracture (adhesion) energy values for 7-day cured poly(kaolinite:XG) mixtures

the linear term, positive for the quadric, negative for the interaction with XG content (X_1), and positive for the interaction with dry density (X_3); and (iii) for dry density (X_3), a negative effect for the linear term, positive for the quadric, negative for the interaction with XG content (X_1), and positive for the interaction with water content (X_2).

The determination of XG content, water content, and dry density regimes is an important aspect of the development of the adhesion energy.

The obtained relationship between the experimental and predicted values by Eq. 5.6 is shown in Fig. 5.26.

Typical surface plots showcasing adhesion energy variations with (i) water and XG contents (for a given dry density of 1.58 Mg/m^3), (ii) XG content and dry density (for a given water content of 12.5%), and (iii) water content and dry density (constant XG content of 0.75 wt%) are illustrated in Fig. 5.27.

Corresponding to Fig. 5.27; contour plots showing variations in adhesion energy with changes in (i) water and XG contents, (ii) XG content and dry density, and (iii) water content and dry density are shown in Fig. 5.28.

Considering the linear term of the model given by Eq. 5.6, the XG content (X_1) has a positive effect up to a content level of 1.2%, water content (X_2) has a negative effect up to 16% followed by a positive effect, and the dry density (X_3) has a negative effect up to 1.55 Mg/m^3 followed by a positive effect (Fig. 5.29). However, compared with X_1 and X_2, Y is notably less sensitive to the variations in X_3 (at least for the ranges examined).

Considering the interaction term of the model, the following points can be made:

(a) Irrespective of the XG content, the water content (X_2) has a negative effect, whereas the dry density (X_3) has a positive effect for lower XG contents and a negative effect for higher XG contents;
(b) For water content (X_2): (i) at low X_2 values, XG content interaction has a positive effect, while the dry density has a negative effect; and (ii) at high X_2 values, the dry density has a positive effect, whereas XG content has a positive effect up to 1.4 wt%, which is then followed by a negative effect; and
(c) For dry density (X_3): (i) at low densities, XG content interaction has a positive effect, and water content interaction has a negative effect up to 20%, which is then followed by a positive effect; and (ii) at high densities, XG interaction has

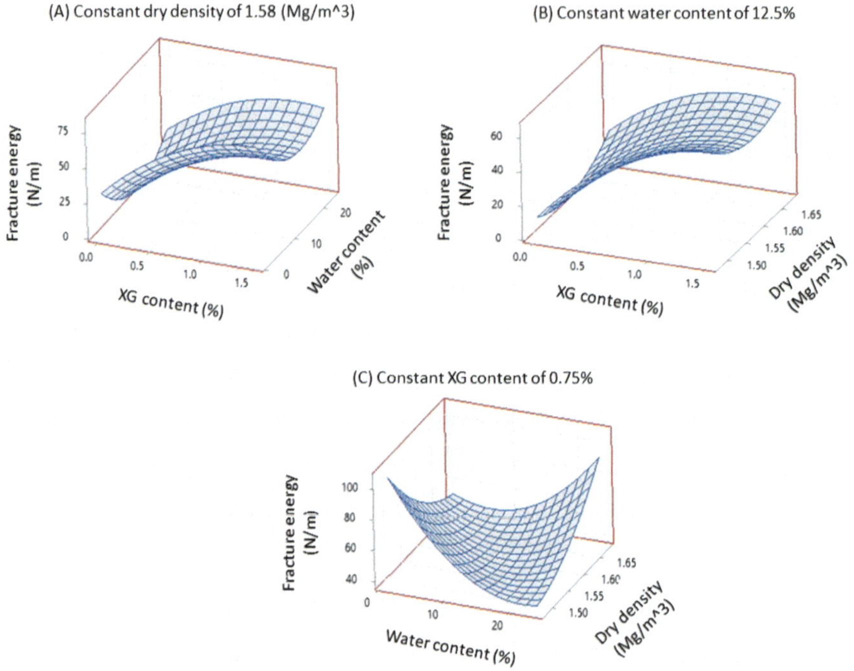

Fig. 5.27 Typical surface plots showcasing fracture (adhesion) energy variations with **a** water and XG contents, **b** XG content and dry density, and **c** water content and dry density for 7-day cured poly(kaolinite:XG) mixtures

no effect, whereas water content interaction has a negative effect up to 12%, which is then followed by a positive effect.

The adhesion energy increase due to the structural rigidity could be attributed to the increase of the XG content (X_1). This might lead to possible winding of the XG random coil into a stable ordered helix conformation, with an increase in intermolecular interactions, resulting in an increase in the hydrodynamic volume and expanding the chain configuration within the kaolinite–XG mixture. The net result would be an increasing polymer chain stiffness and hence overall structural rigidity.

5.4.2.2 Bonding Strength at the Macroscale Level

The strength of the interacting bonds (physical and chemical) of XG-stabilized clay soils can be indirectly evaluated using the results of fatigue life, which is the number of cycles at failure in constant-amplitude fatigue loading tests. Here, the experimental results reported by Ni et al. (2020) for XG-stabilized Shanghai clay sourced from the Yangtze River delta alluvial plain in China are considered. The tested soil had the following geotechnical characteristics: (i) sand, silt and clay fractions of 33.41%,

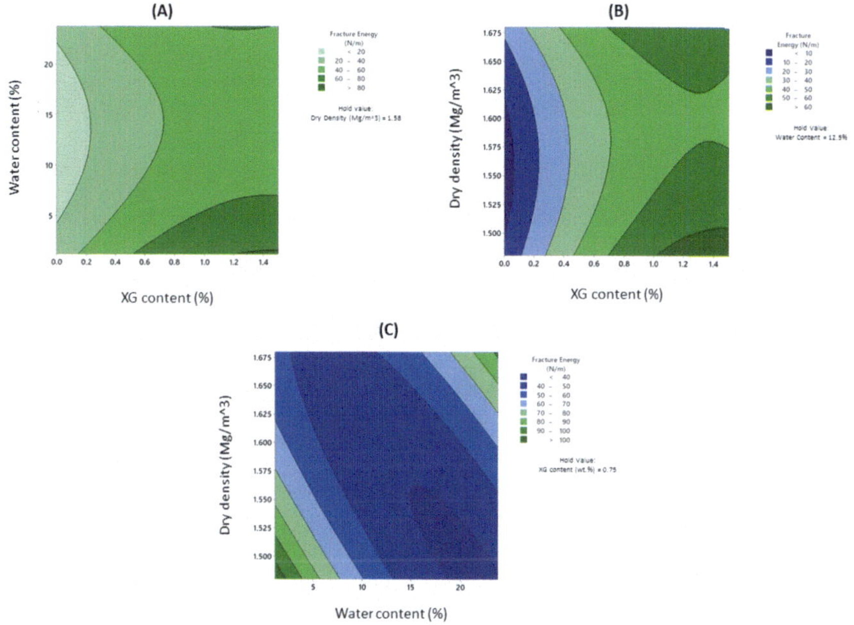

Fig. 5.28 Typical contour plots of fracture/adhesion energy (in N/m) with **a** water content and XG contents, **b** dry density and XG content, and **c** dry density and water content for 7-day cured poly(kaolinite:XG) mixtures

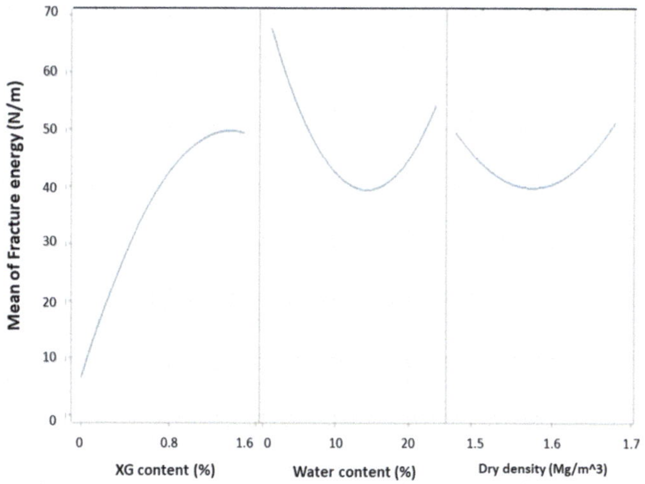

Fig. 5.29 Main effects plots for fracture (adhesion) energy of 7-day cured poly(kaolinite:XG) mixtures

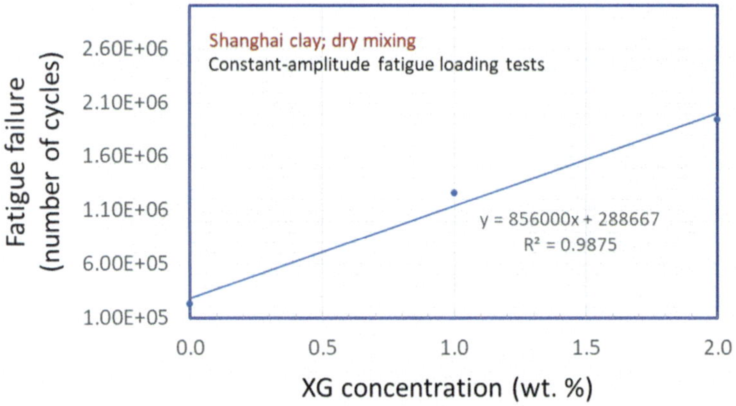

Fig. 5.30 Fatigue failure characteristics of XG-stabilized Shanghai clay soil (data from Ni et al. (2020))

54.86% and 11.73%, respectively; (ii) plasticity characteristics of LL = 37.9%, PL = 22.0% and PI = 15.9%; and (iii) standard Proctor-compaction characteristics of OMC = 23% and maximum dry density = 1.59 Mg/m^3. The XG was thoroughly dry mixed with the soil, investigating XG contents of 1 and 2 wt%. Then, the required amount of distilled water was added to reach an initial moisture content of 30% (i.e., on the wet side of OMC) for the soil–XG mixtures. To increase the XG solubility, the added water was first heated to 80 °C. The prepared samples were cured for 28 days in a controlled environment at 60% relative humidity and 20 °C. Then, constant-amplitude fatigue loading tests were conducted at the maximum stress to unconfined compressive strength (σ_{max}/UCS) ratio of 0.55, with the loading frequency set as a resonant frequency between the sample and the test apparatus, which occurred for about 40 to 45 Hz.

The results shown in Fig. 5.30 (data from Ni et al. (2020)) have indicated that the untreated soil sample failed after 233,750 loading cycles, whereas the 1 and 2 wt% XG-treated soil samples failed after 1.256 × 10^6 and 1.945 × 10^6 loading cycles, respectively. This implies that, compared to the untreated soil, the 1 and 2 wt% XG additions increased the fatigue life by 5.37 and 8.32 times, respectively. These improvements resulted from the increase in the bonding strength via water hydration of both soil particles and XG, water bridging between soil particles and XG, and XG tie formation between the soil particles.

5.4.2.3 Bonding Elasticity at the Nanoscale Level

The interaction of the kaolinite clay particles with XG at a 1-μm scale level was examined by Ramachandran et al. (2021) using AFM analysis. In their investigation, 0.5 wt% XG was thoroughly dry mixed with the kaolinite clay and left to cure in an oven maintained at 60°C for 7 days. Then, a sample of the dried kaolinite–XG mixture was prepared for nano-indentation and PeakForce Quantitative Nanomechanics (QNM) testing. It is worth noting that the PeakForce QNM test is generally used to map and distinguish between nanomechanical properties (i.e., modulus, adhesion, dissipation, and deformation) with up to atomic resolution in topography. The quantitative data produced can help researchers answer the critical question of what materials they are seeing in their topographic images.

The dried sample was first placed in a 25-mm diameter mold, and epoxy resin was poured under a vacuum in order to impregnate the resin into the sample. The thickness of the dry sample initially measured about 5 mm, with the addition of the resin increasing the thickness to 8 mm. The sample was then polished and prepared by focused ion-beam milling to produce a sample with minimum surface roughness. The nano-indentation tests were performed under the continuous stiffness measurement model that allows the measurement of contact stiffness at any point of the loading curve corresponding to any depth of penetration. The maximum displacement was restricted to 2 μm, and 25 indents were performed on the sample with a spacing of 20 μm.

In this study, the PeakForce-QNM test results were used to study the nanomechanical properties of the XG biopolymer at a high resolution. The indentation force applied to the sample was approximately 5 nN, and the depth of indentation was about 1 nm, which is significantly lower than that typically adopted in the nano-indentation method. Also, this technique was used to evaluate the degradation mechanism of the bonded kaolinite–XG mixture upon exposure to moisture. The force–indentation curves were analyzed at each pixel of the topographic image, and the local elastic modulus was determined at each contact point. The AFM imaging was performed using the Bruker Dimension Icon AFM system with PeakForce-QNM mode under ambient conditions. The imaging scan rate was maintained at 1.0 Hz, and a resolution of 256 $\times$ 256 pixels.

In the nano-indentation tests, the modulus of elasticity (E) and hardness (H) (i.e., maximum indentation force/contact area), were measured using the continuous stiffness measurement method developed by Huen et al. (2019). Both E and H were found to have a steady response at depths of 1000–2000 nm. The overall average E and H of the sample measured across its 25 indentation points were 5.02 $\pm$ 1.3 and 0.078 $\pm$ 0.03 GPa, respectively.

Figure 5.31 (data from Ramachandran et al. (2021)) shows the results of E, H, and the degree of anisotropy for the XG-stabilized kaolinite at nanoscale level. The variations of E within the sample were grouped into three regions: (i) "*low*", for the pore void spaces, with E = 2.3–3.0 GPa; (ii) "*medium*", for bonded kaolinite–XG, with E = 3.5–6.0 GPa; and (iii) "*high*", for the kaolinite clay particles, with E = 6.0–8.0 GPa (Fig. 5.31a). The calculated H values for the three regions were 0.02–0.04, 0.05–0.09 and 0.12–0.16 GPa for pore spaces, bonded kaolinite–XG, and kaolinite clay particles, respectively (Fig. 5.31b).

In analyzing the preceding outcomes reported by Ramachandran et al. (2021), the following set of reported results for E by other researchers must be considered: (i) Ulm and Abousleiman (2006); E = 40.3 $\pm$ 8.8 GPa for kaolin powder composed of 97% kaolinite; (ii) Zhang et al. (2009); E = 10–30 GPa for pure clays; (iii) Eliyahu

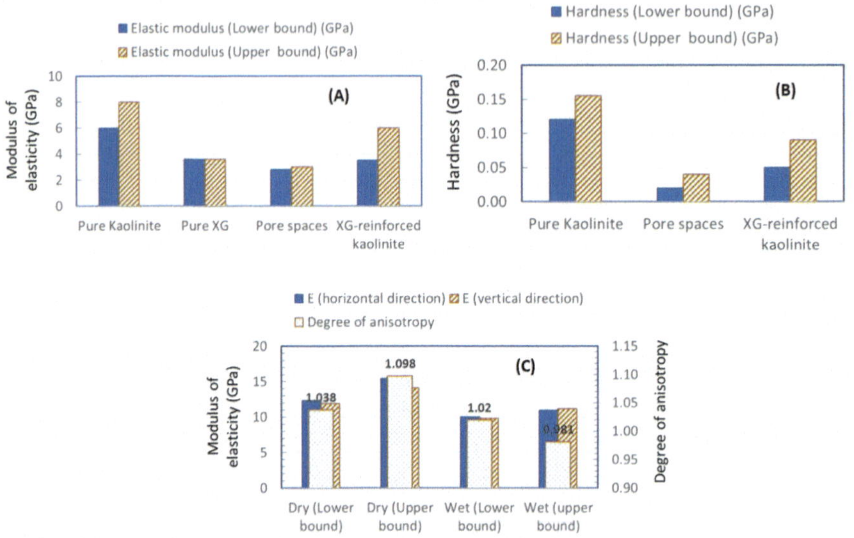

Fig. 5.31 Variations of modulus of elasticity and hardness characteristics at nanoscale level for XG-stabilized kaolinite: **a** modulus of elasticity; **b** hardness; and **c** degree of anisotropy; (data from Ramachandran et al. (2021))

et al. (2015); $E = 21–55$ GPa for clays; (iv) Ramachandran et al. (2021); an average E of 3.14 GPa for dry XG; and (v) Castrillo et al. (2015) reported on the effectiveness of kaolin clay in toughening epoxy polymer, even at small dosages.

Furthermore, the PeakForce-QNM results indicated that the average E depends on the measuring direction; i.e., either along the sample's horizontal axis (12.31 ± 3.19 GPa), or vertical axis (11.86 ± 2.25 GPa) directions, indicating that the dry bonded kaolinite–XG structure has a low degree of cross-anisotropy, ranging from 1.038 to 1.098 (Fig. 5.31c). After subjecting the sample to moisture uptake, the average E across the horizontal and vertical axes directions reduced to 9.99 ± 0.94 GPa and 9.79 ± 1.35 GPa, respectively; indicating again a low degree of cross-anisotropy, ranging from 1.02 to 0.981 (Fig. 5.31c). More significantly, this indicates that the bonding between the XG and the kaolinite clay particles has weakened upon moisture uptake, highlighting the effect of moisture on the XG. The disturbance of the XG gel via hydrophilic water absorption leads to the network of the XG breaking off from the kaolinite clay particles due to swelling of the XG. However, on drying, the XG will reattach to the kaolinite clay particles with a new bonding strength that is different from the original one. Therefore, intermittent drying and wetting impact the bonded kaolinite–XG at the microlevel, significantly affecting the mini- and macro-mechanical properties of XG-treated clay soils.

5.5 Summary and Concluding Remarks

XG is a biopolymer commonly used as a thickener, stabilizer and emulsifier in various industries, including food processing, pharmaceutical, environmental and geotechnical engineering, and petroleum industries. When applied to soil, XG can enhance the soil properties, including water retention and soil structure and fertility, making XG a promising soil stabilizer for agricultural, environmental and geotechnical engineering applications. The interaction between XG and soil is mainly driven by the development and propagation of a cohesive XG–soil bonding network, which can enhance the soil properties.

XG is a long-chain polysaccharide composed of repeating units of glucose, mannose and glucuronic acid, such that it can interact with soil particles through various types of bonds. Hydrogen bonds are one of the main types of bonds that can occur between XG and soil particles. These bonds are formed between the hydroxyl groups of XG and the oxygen atoms of the soil particles, such as clay minerals. Electrostatic interactions are another type of bond that can occur between XG and soil particles. These interactions are based on the attraction between the negatively charged XG molecules and the negatively charged soil particles via cationic bridging. Van der Waals forces are also important in the interaction between XG and soil particles. These forces are based on the attraction between the molecules of XG and the soil particles, and they can contribute to the stability of the cohesive network formed by XG.

The cohesive network formed by XG can have important implications for soil properties. For example, XG can improve soil structure by promoting the formation of aggregates, which can increase soil porosity and water infiltration. XG can also enhance the soil's water retention, which can improve plant growth, reduce the risk of soil erosion and stabilize expansive clays and sandy soils.

The interaction between XG and soil is complex and can be affected by several factors. For instance, the concentration and molecular weight of XG, the degree of branching of the XG molecule, and the biodegradability and environmental fate of XG can all affect the ability of XG to form a cohesive network and improve soil properties. The soil type and texture, the mineralogy of the clay particles and pore fluid chemistry can also influence the interaction between XG and the soil particles, with clay soils showing stronger effects compared to sandy soils. Furthermore, environmental factors, including pH and temperature, can alter the XG–soil interactions.

In conclusion, the formation of a cohesive network is a key mechanism underlying the interaction between XG and soil, and it can have important implications for soil properties, soil erosion control, land restoration, and the effectiveness of other soil treatments. Further research is needed to gain a full understanding of the basic mechanisms and factors that govern the interaction between XG and soils, thereby allowing to optimize XG's use as an effective soil-stabilizing agent.

References

Akpan EU, Enyi GC, Nasr GG (2020) Enhancing the performance of xanthan gum in water-based mud systems using an environmentally friendly biopolymer. J Petrol Explor Prod Technol 10:1933–1948. https://doi.org/10.1007/s13202-020-00837-0

Amarasiri AL, Costa S, Kodikara JK (2011) Determination of cohesive properties for mode I fracture from compacted clay beams. Can Geotech J 48(8):1163–1173. https://doi.org/10.1139/t11-031

Arab MG, Mousa RA, Gabr AR, Azam AM, El-Badawy SM, Hassan AF (2019) Resilient behavior of sodium alginate-treated cohesive soils for pavement applications. J Mater Civ Eng 31(1):04018361

Azimi M, Soltani A, Mirzababaei M, Jaksa MB, Ashwath N (2024) Biopolymer stabilization of clayey soil. J Rock Mech Geotech Eng (In Press). https://doi.org/10.1016/j.jrmge.2023.12.020

Bagheri P, Gratchev I, Rybachuk M. (2023) Effects of Xanthan gum biopolymer on soil mechanical properties. Appl Sci 13(2):887. https://doi.org/10.3390/app13020887

Barani OR, Barfar P (2020) Effect of Xanthan gum biopolymer on fracture properties of clay. J Mater Civ Eng 33(1):04020426

Bozyigit I, Zingil HO, Altun S (2023) Performance of eco-friendly polymers for soil stabilization and their resistance to freeze–thaw action. Constr Build Mater 379:131133

Castrillo PD, Olmos D, Sue HJ, González-Benito J (2015) Mechanical characterization and fractographic study of epoxy–kaolin polymer nanocomposites. Compos Struct 133:70–76. https://doi.org/10.1016/j.compstruct.2015.07.040

Chang I, Im J, Prasidhi AK, Cho GC (2015) Effects of Xanthan gum biopolymer on soil strengthening. Constr Build Mater 74:65–72. https://doi.org/10.1016/j.conbuildmat.2014.10.026

Chang I, Kwon YM, Im J, Cho GC (2018) Soil consistency and inter-particle characteristics of xanthan gum biopolymer containing soils with pore-fluid variation. Can Geotech J 56(2). https://doi.org/10.1139/cgj-2018-0254

Decker RS, Dunnigan LP (1977) Development and use of the soil conservation service dispersion test. In: Sherard JL, Decker RS (eds) ASTM special publication 623: dispersive clays, related piping, and erosion in geotechnical projects. ASTM International: West Conshohocken, PA, USA (pp. 94–109). https://doi.org/10.1520/STP26982S

Demirdal B, Miska SZ, Takach NE, Cunha JCS (2007) Drilling fluids rheological and volumetric characterization under downhole conditions [Society of Petroleum Engineers Latin American & Caribbean Petroleum Engineering Conference-Buenos Aires, Argentina (2007–04–15)] Latin American & Caribbean Petroleum Engineering Conference. https://doi.org/10.2118/108111-MS

Elges HFWK (1985) Dispersive soils. Civil Eng Siviele Ingenieurswese 27(7):347–354

Elices M, Guinea G, Planas J (1992) Measurement of the fracture energy using three-point bend tests: Part 3—Influence of cutting the P-δ tail. Mater Struct 25(6):327–334. https://doi.org/10.1007/BF02472591

Eliyahu M, Emmanuel S, Day-Stirrat RJ, Macaulay CI (2015) Mechanical properties of organic matter in shales mapped at the nanometer scale. Mar Pet Geol 59:294–304. https://doi.org/10.1016/j.marpetgeo.2014.09.007

Farrés IF, Norton IT (2014) Formation kinetics and rheology of alginate fluid gels produced by in-situ calcium release. Food Hydrocoll 40:76–84

Gent AN (1996) Adhesion and strength of viscoelastic solids. is there a relationship between adhesion and bulk properties? Langmuir 12(19):4492–4496

Hamza M, Nie Z, Aziz M, Ijaz N, Akram O, Fang C, Ghani MU, Ijaz Z, Noshin S, Madni MF (2023) Geotechnical behavior of high-plastic clays treated with biopolymer: macro–micro-study. Environ Earth Sci 82(91):1–18. https://doi.org/10.1007/s12665-023-10760-2

Huen WY, Lee H, Vimonsatit V, Mendis P (2019) Relationship of stiffness-based indentation properties using continuous-stiffness-measurement method. Materials (Basel) 13(1):97. https://doi.org/10.3390/ma13010097

Ivanov V, Chu J (2008) Applications of microorganisms to geotechnical engineering for bioclogging and bio-cementation of soil in situ. Rev Environ Sci Bio/technol 7(2):139–153

Jan R, Deepa D, Sharmila SMR (2019) Xanthan gum as sustainable biopolymer additive for soil treatment. Int J Recent Technol Eng (IJRTE) 8(4):10487–10492. https://doi.org/10.35940/ijrte.D4234.118419

Joga JR, Varaprasad BJS (2020) Effect of xanthan gum biopolymer on dispersive properties of soils. World J Eng 17(4):563–571. https://doi.org/10.1108/wje-05-2020-0152

Kasperski KL, Hepler CT, Hepler LC (2011) Viscosities of dilute aqueous suspensions of montmorillonite and kaolinite clays. Can J Chem 64(9):1919–1924. https://doi.org/10.1139/v86-316

Kumar A, Ramani S (2021) Assessing the potential of xanthan gum to modify in situ soil as baseliners for landfills. Int J Environ Sci Technology 19(1). https://doi.org/10.1007/s13762-021-03721-4

Kwon YM, Cho GC, Chang I (2022) Effect of anionic xanthan gum on the sedimentation of kaolinite. In: The 2022 world congress on advances in civil, environmental, & materials research (ACEM22) 16–19, August 2022, GECE, Seoul, Korea

Laneuville SI, Turgeon SL, Sanchez C, Paquin P (2006) Gelation of native betalactoglobulin induced by electrostatic attractive interaction with Xanthan gum. Langmuir 22(17):7351–7357

Latifi N, Marto A, Eisazadeh A (2015) Physicochemical behavior of tropical laterite soil treated with non-traditional additive. Acta Geotech 11(2):433–443

Latifi N, Horpibulsuk S, Meehan CL, Abd Majid MZ, Tahir MMd, Mohamad ET (2016) Improvement of problematic soils with biopolymer-an environmentally friendly soil stabilizer. J Mater Civ Eng 04016204. https://doi.org/10.1061/(ASCE)MT.1943-5533.0001706

Lee M, Im J, Cho GC, Ryu HH, Chang I (2021) Interfacial shearing behavior along xanthan gum biopolymer-treated sand and solid interfaces and its meaning in geotechnical engineering aspects. Appl Sci 11:139. https://doi.org/10.3390/app11010139

Mahmood A, Patel D, Hickson B, DesRochers J, Hu X (2022) Recent progress in biopolymer-based hydrogel materials for biomedical applications. Int J Mol Sci 23(3):1415

Marto A, Latifi N, Eisazadeh A (2014) Effect of non-traditional additives on engineering and microstructural characteristics of laterite soil. Arabian J Sci Eng 39(10):6949–6958

Mendonça A, Morais PV, Pires AC, Chung AP, Oliveira PJV (2021a) Reducing soil permeability using bacteria-produced biopolymer. Appl Sci 11:7278

Mendonça A, Morais PV, Pires AC, Chung AP, Oliveira PV (2021b) A review on the importance of microbial biopolymers such as xanthan gum to improve soil properties. Appl Sci 11:170

Mitchell JK, Santamarina JC (2005) Biological considerations in geotechnical engineering. J Geotech Geoenviron Eng 10(1222):1222–1233. https://doi.org/10.1061/(ASCE)1090-0241(2005)131

Mitchell RF, Miska S, Aadnøy BS (2011) Fundamentals of drilling engineering. Society of Petroleum Engineers, Richardson, 696p. ISBN: 9781555633387, 1555633382

Mohamed AMO, Antia H (1998) Geo-environmental engineering, Elsevier, Amsterdam, ISBN: 0-444-89847-6, 707p

Mohamed AMO, Paleologos EK (2018) Fundamentals of geo-environmental engineering: understanding soil, water, and pollutant interaction and transport. Elsevier, USA, imprint: Butterworth-Heinemann, eBook ISBN: 9780128051450; Paperback ISBN: 9780128048306; 708p

Murthy AR, Karihaloo B, Iyer NR, Prasad BR (2013). Determination of size-independent specific fracture energy of concrete mixes by two methods. Cem Concr Res 50(Aug):19–25. https://doi.org/10.1016/j.cemconres.2013.03.015

Ni J, Li SS, Ma L, Geng XY (2020) Performance of soils enhanced with eco-friendly biopolymers in unconfined compression strength tests and fatigue loading tests. Const Build Mater 263:120039. https://doi.org/10.1016/j.conbuildmat.2020.120039

Nugent RA, Zhang G, Gambrell RP (2009) Effect of exopolymers on the liquid limit of clays and its engineering implications. Transport Res Rec: J Transport Res Board 2101(1):34–43

Ramachandran AL, Dubey AA, Dhami NK, Mukherjee A. (2021) Multiscale study of soil stabilization using bacterial biopolymers. J Geotech Geoenviron Eng 147(8):04021074. https://doi.org/10.1061/(ASCE)GT.1943-5606.0002575

Sawant A, Patil A, Thonge R, Trankatwar V, Banne S (2019) Effect of xanthan gum on shear strength parameters of laterite soil in Konkan region. IOSR J Mech Civil Eng 16(3):24–29. https://doi.org/10.9790/1684-1603012429

Seybold CA (1994) Polyacrylamide review: soil conditioning and environmental fate. Commun Soil Sci Plant Anal 25(11–12):2171–2185

Singh SP, Das R (2020) Geo-engineering properties of expansive soil treated with xanthan gum biopolymer. Geomech Geoengin 15:107–122. https://doi.org/10.1080/17486025.2019.1632495

Soldo A, Miletić M, Auad ML (2020) Biopolymers as a sustainable solution for the enhancement of soil mechanical properties. Sci Rep 10(1):267

Soltani A, Raeesi R, Taheri A, Deng A, Mirzababaei M (2021) Improved shear strength performance of compacted rubberized clays treated with sodium alginate biopolymer. Polymers 13(5):764

Soltani A, Raeesi R, O'Kelly BC (2022a) Cyclic swell–shrink behaviour of an expansive soil treated with a sulfonated oil. Proc Inst Civil Eng Gr Improv 175(3):166–179. https://doi.org/10.1680/jgrim.19.00084

Soltani A, Deng A, Taheri A, O'Kelly BC (2022b) Intermittent swelling and shrinkage of a highly expansive soil treated with polyacrylamide. J Rock Mech Geotech Eng 14(1):252–261. https://doi.org/10.1016/j.jrmge.2021.04.009

Soltani A, Azimi M, O'Kelly BC (2024) Discussion: a fuzzy classification process for swelling soils [Transp Infrastruct Geotechnol 10(3):474–487]. Transp Infrastruct Geotech 11:1168–1180. https://doi.org/10.1007/s40515-023-00320-3

Sujatha ER, Atchaya S, Sivasaran A, Keerdthe RS (2020) Enhancing the geotechnical properties of soil using xanthan gumâan eco-friendly alternative to traditional stabilizers. Bulletin Eng Geol Envir. https://doi.org/10.1007/s10064-020-02010-7

Sun TL, Luo F, Kurokawa T, Karobi SN, Nakajima T, Gong JP (2015) Molecular structure of self-healing polyampholyte hydrogels analyzed from tensile behaviors. Soft Matter 11(48):9355–9366. https://doi.org/10.1039/C5SM01423A

Ulm FJ, Abousleiman Y (2006) The nanogranular nature of shale. Acta Geotech 1(2):77–88. https://doi.org/10.1007/s11440-006-0009-5

Zhang G, Wei Z, Ferrell RE (2009) Elastic modulus and hardness of muscovite and rectorite determined by nanoindentation. Appl Clay Sci 43(2):271–281. https://doi.org/10.1016/j.clay.2008.08.010

Zhang Z, Meng Z, Jiang T, Wang S, Zhao J, Zhao X (2022) Experimental study on the shear strength of silt treated by xanthan gum during the wetting process. Appl Sci 12(12):6053. https://doi.org/10.3390/app12126053

Chapter 6
Strength and Compressibility of Xanthan Gum-Treated Soils

6.1 Introduction

Xanthan gum (XG) is a high molecular weight anionic polysaccharide produced by the bacterium *Xanthomonas campestris*. Because of its unique rheological properties (as discussed in Chap. 3), XG is widely employed as a thickening, stabilizing, and gelling agent for a variety of industrial applications, including, food processing, cosmetics, pharmaceuticals, and oil and gas. Recently, XG has been explored as a chemical soil-stabilizing agent, owing to its ability to form a three-dimensional reinforcement-type network within soils for enhancing their geo-mechanical properties.

Understanding the strength, compressibility, and hydraulic conductivity properties of soil materials is crucial in a variety of settings, including civil engineering, geotechnical and geo-environmental engineering, and agricultural engineering. The strength characteristics of XG-treated soils have been evaluated in various studies, and the results indicate that the addition of XG increases the soil's shear strength, bearing capacity, and stiffness.

Recent studies highlighted the promising potential of XG treatment in enhancing the mechanical properties of different types of soils, improving the durability, lowering soil permeability through pore clogging, decreasing compressibility, and significantly reducing soil erosion (Chang et al. 2015). For example, Lee et al. (2019) evaluated the performance of XG-treated sandy soil in different proportions with bottom and fly ash mixes and observed considerable improvement in strength characteristics. Other studies on sand and collapsible soil reveal the development of biopolymer gel due to XG dehydration, which crystallizes, binds, and fills the soil pore voids, responsible for significant strength improvement and hydraulic conductivity reductions (Muguda et al. 2017). Similarly, with XG treatment, medium and high compressible silty clays exhibit reductions in compression index, coefficient of

A. M. O. Mohamed et al., *Sustainability in Ground Improvement: The Case of Xanthan Gum Biopolymer*, Green Energy and Technology,
https://doi.org/10.1007/978-3-031-75313-8_6

consolidation, and collapse potential (Sujatha and Saisree 2019). However, most of the studies documented in the literature were mainly focused on silty clays with low plasticity and residual and sandy soils, with limited work on problematic expansive clay soils.

Recently, a study conducted by Durga and Rasheed (2020) showed that the unconfined compressive strength (UCS) of a highly plastic clay soil (plasticity index of PI = 32.95%) treated with 2 wt% XG increased by 65% compared to the untreated soil. Notably, the study reported neither the XG mixing nor the compaction methods employed. Other studies are using XG in geotechnical engineering practice due to (i) the high soil strength achievable (e.g., 4.9 MPa of UCS with 1 wt% XG treatment on low plasticity soils) and (ii) formation of a network structure within the soil matrix that increased the inter-particle bonding and also reduced the soil compressibility (Khachatoorian et al. 2003; Bouazza et al. 2009; Chang et al. 2015, 2019, 2020; Ayeldeen et al. 2016; Latifi et al. 2016; Cabalar et al. 2017; Im et al. 2017; Lee et al. 2017, 2019; Qureshi et al. 2017; Kwon et al. 2020).

Furthermore, in a study by Hamza et al. (2023a), a problematic expansive subgrade soil with a PI of 31.7% was treated using XG with various concentrations (0, 0.5, 1.0, 1.5, 2.0, and 5.0%) and cured for extended aging periods (0, 4, 7, 14, 28, 60, and 365 days). The results showed that for soil treated with 1.5 wt% XG content, the UCS, modulus of elasticity, resilient modulus, and un-soaked and soaked California bearing ratio (CBR) values were notably increased by five to ten orders of magnitude from the untreated soil, which was attributed to the bridging of soil particles with stiff XG-hydrogel formation.

In summary, the application of XG as a soil stabilizer appears to be a promising technique for improving the mechanical properties of soils, although further research is needed to investigate the long-term durability and environmental impacts of XG-treated soils. Nevertheless, the findings of the above studies suggest that XG has the potential to be an effective and environmentally friendly soil stabilizer for a wide range of applications. The present chapter discusses the strength and compressibility characteristics of XG-stabilized soil materials.

6.2 Nature of the XG Gel

As discussed in Chap. 5, considering some geotechnical properties of XG-stabilized clay soils, the effect of increasing the XG content tends to: (i) increase gel content, pore fluid viscosity, and the soil plasticity index, water potential, and degree of aggregation and (ii) decrease the pore void sizes, external surface area of aggregates, swelling potential, and dispersivity. In view of these properties, the addition of XG to soils can increase their stiffness and reduce their compressibility.

The nature of the XG gel in stabilized soils can be evaluated by considering its dynamic properties, as discussed in Chap. 3. The frequency-dependent storage $[G'(\omega) = K'(\omega)^{n'}]$ and loss $[G''(\omega) = K''(\omega)^{n''}]$ moduli (with K', K'', n', and n'' being fitting coefficients) are used to characterize the nature of the XG gel, which in turn

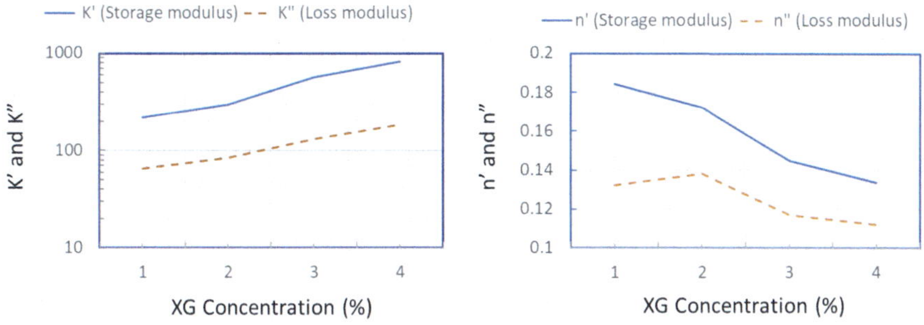

Fig. 6.1 Calculated power-law coefficients of storage and loss moduli for aqueous XG solutions with different XG concentrations (data from Song et al. (2006))

gives insight into the intensity of the XG interaction with the clay soil particles. The calculated power-law coefficients of storage and loss moduli for aqueous XG solutions with different XG concentrations are illustrated in Fig. 6.1.

Referring to Fig. 6.1, for all XG solutions (1–4 wt%), the K' values (for the storage modulus) are much greater than those of K'' (for the loss modulus), indicating the superiority of the elastic behavior (storage modulus) over the viscous behavior (loss modulus). Also, the values of n' (for G') and n'' (for G'') gradually decreased for increasing XG concentration, indicating that the frequency dependence of the linear viscoelastic functions becomes weaker as XG concentration increases. This implies that as XG concentration increases, the XG solutions become more elastic.

In addition, based on the n' values, the nature of the XG gel can be characterized as either a chemically cross-linked strong gel (for $n' > 0$) or a weak gel (for $n' = 0$). For the 1–4 wt% XG solutions shown in Fig. 6.1, the n' values approximately ranged from 0.13 to 0.18, indicating that the XG solutions do not form a stable (strong) gel, but instead exhibit a weak gel-like behavior. Furthermore, considering intermediate and semi-dilute XG concentration regimes (i.e., $0.05\% < C_{XG} < 0.5\%$), both conformational ordering of the XG backbone and intermolecular associations control the rheological properties of the XG dilute solution, whereas for the high concentrated regime ($C_{XG} > 1\%$), stress nonlinearities and liquid crystal type behavior are dominant.

As such, XG concentration plays a major role in respect of the stiffness and compressibility behaviors of XG-treated soils. Studies have indicated that the effects of XG content on the soil compressibility is not linear and that there is an optimum XG content beyond which the soil compressibility may start to increase again (Song et al. 2006). This is because the excess XG molecules would not participate in electrostatic interactions with the clay particles, as the available attraction sites for XG (e.g., available divalent cations in the Diffused Double Layer (DDL) for cationic bridging) have been exhausted (Soltani et al. 2021; Azimi et al. 2024). In this case, the excess XG molecules would tend to repel each other (rise in repulsive forces), or they may induce inter-particle lubrication effects by precipitating within the larger void spaces between the aggregated soil particles (Arab et al. 2019; Azimi et al. 2024). This would lead to the formation of local friable-to-stress/compression/deformation regions within the soil matrix, explaining the reductions in UCS and stiffness and the increase in compressibility beyond a certain optimum XG content.

6.3 Controlling Parameters

The strength, compressibility, and water flow properties of XG-treated soils depend on several factors, including:

(a) XG physicochemical properties (grain size, initial structural conformation, concentration, viscosity, molecular mass, quantity and distribution, shear rate and mixing temperature), as discussed in Chap. 3.
(b) Method of mixing XG with the soil (i.e., wet or dry mixing).
(c) Soil type (cohesionless or cohesive).
(d) Soil dry density and moisture content (i.e., compaction state parameters).
(e) Soil pore fluid chemistry.
(f) The environmental conditions that the XG-stabilized soil will be subjected to during and after XG application, in terms of pressure, temperature, pH variations from acidic to alkaline conditions, curing period/conditions, and cyclic wetting–drying and freezing–thawing actions.

6.3.1 XG Mixing Methods with Soils

The mixing method used to incorporate the XG into soil can affect the efficacy of XG as a soil stabilizer and its impact on the soil mechanical properties. Both wet and dry mixing methods, which relate to the status of the XG being in aqueous solution (wet) or powder form (dry), have been reported in the literature, and notably they have shown different effects on the resulting soil properties.

Wet mixing often involves adding XG aqueous solution (with known concentration) to soils based on a pre-defined moisture content (e.g., the optimum moisture content (OMC) for compaction) and then mixing the two components together. This method allows for better dispersion of the XG throughout the soil mass, resulting in more uniform soil stabilization. Wet mixing also helps in hydrating the XG, being a hydrocolloid that requires water to bind the soil particles together. The addition of XG using the wet mixing method can increase the water holding capacity of the soil, which can be beneficial in arid regions where water availability is limited. In addition to reducing the soil compressibility, wet mixing of XG has also been found to reduce soil erosion. Whereas in the dry mixing method, the XG powder is mixed with the dry soil before adding water and then thoroughly mixing them together.

The effects of wet and dry mixing methods on the resultant compressive strength properties of XG-stabilized Korean residual soil can be evaluated from the experimental results presented in Fig. 6.2 (data from Chang et al. (2015)), where both mixing methods were implemented at ambient room temperature. In the wet mixing

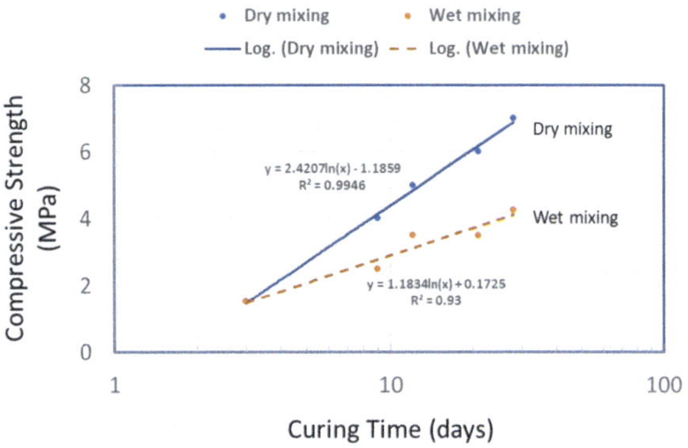

Fig. 6.2 Compressive strength variation with curing time for XG-treated Korean residual soil considering both dry- and wet-mixing sample preparation methods for the XG additions (data from Chang et al. (2015))

method, the dried soil was mixed with a water–XG solution for a soil moisture content of 60%. For proper mixing, the purified XG was first dissolved in distilled water, which was heated to 80 °C for enhancing the XG solubility, and the XG solution was mixed using a magnet stirrer to ensure uniform dissolution. The cooled XG solution with concentration of 1.67 wt% was then thoroughly mixed with the dry soil for the proposed stabilization (i.e., at room without heating). In the dry mixing method, the dried soil was thoroughly mixed with purified XG powder (w/w = 1%), and then the required amount of water (i.e., for achieving 60% water content) was added, followed by thoroughly mixing the components together. The soil–XG mixtures produced by the wet and dry mixing methods were placed into cubical molds (each measuring 40 × 40 × 40 mm) and then allowed to air dry at ambient room temperature of 20 °C for different curing periods.

It is worth noting that in these two types of XG mixing procedures, the XG-to-soil dry mass ratio was kept constant at approximately 1%. The experimental results presented in Fig. 6.2 have indicated that the dry mixing approach produced higher strength than the wet mixing method. To explain such an effect, one must investigate the solubility and viscosity of XG in water. Chang et al. (2015) conducted solubility tests on the XG and noted that, at room temperature, an XG content of 1.4 wt% (of water) is the maximum XG content that can be entirely dissolved in water; above 1.5 wt% XG content, it is extremely difficult to fully dissolve the XG due to the increased XG viscosity. Also, Chang et al. (2015) noted that a concentration of 1 wt% XG of dry soil leads to an approximate XG concentration of 1.7 wt% in water (i.e., for 60% water content in the Korean residual soil), which is well above the mentioned XG solubility concentration of 1.4 wt%. Therefore, full mixing of XG in water may not have been possible and it is unlikely an even distribution of XG occurred in the soil mass.

It is worth noting that (i) for the wet mixing procedure, Chang et al. (2015) increased the water temperature to 80 °C in order to enhance the solubility of XG in water at higher concentration; (ii) dry mixing is more effective only when the concentration of XG in water is above its solubility point; and (iii) when the soil is to be compacted at moisture content levels close the soil's OMC for compaction, the dry mixing method often becomes less effective, as the amount of water added is relatively low for achieving full hydration of the XG molecules, especially if they are in competition with active clay minerals (e.g., montmorillonite) present. This is consistent with the findings of Mendonça et al. (2021), who pointed out the issue of XG hydration with time for the dry mixing method, and the possibility of having partial hydration in the inner part of the hydrogel, which could be a problematic issue for high XG concentrations with the aggregated conformation structures. As such, more studies are required to determine which mixing method (i.e., wet or dry) should be used, and for which XG concentration levels and soil types.

Overall, the mixing method used to incorporate XG into soil can have a significant impact on the effectiveness of XG as a soil-stabilizing agent and its ability to enhance the soil strength and reduce the soil compressibility. Wet mixing is generally considered to be a more effective method for incorporating XG into soil for stabilization and erosion-control purposes. In other words, the wet mixing allows for better dispersion and hydration of XG throughout the soil mass, resulting in more uniform soil stabilization and potentially greater improvements in soil mechanical properties. However, the effectiveness of XG may also depend on other factors, including the soil type, and the XG concentration and soil pore fluid chemistry, as discussed in Chaps. 4 and 5.

It should be noted that the preceding examples for XG-treated soils are still inconclusive. As such, further research studies are required to establish specified procedures for preparing XG aqueous solutions to ensure full XG hydration, and to determine the basic rheological properties of the XG solution (e.g., its conformational structure, absolute and dynamic viscosities, molecular weight, and storage and loss moduli), as discussed in Chap. 4, prior to its application for treating soils.

6.3.2 Soil Type

Different soil types have different physical and chemical properties that can affect the binding ability of XG and, as a result, impact the compressibility of the soil matrix. The differences in the compressibility of dissimilar soil types treated with XG may be due to differences in their gradations and mineralogical compositions, and their densification or compaction states. These factors can influence the interaction between XG and the soil particles, and, therefore, the ability of the XG to bind the soil particles together and reduce the soil compressibility.

6.3.2.1 Pure Clay Minerals

In the study by Latifi et al. (2016), untreated and XG-stabilized kaolinite and Na-bentonite were compacted using standard Proctor effort to prepare the UCS test-specimens. The investigated clay soils were mixed with 0, 0.5, 1, 1.5, 2, and 2.5 wt% XG (by weight of dry soil) and distilled water at their OMC and then compressed in cylindrical molds to produce UCS test-specimens at the standard Proctor maximum dry unit weight. The compacted specimens were then wrapped in cling film and allowed to cure for the desired interval (0 days for untreated specimens and 7, 28, or 90 days for stabilized specimens) in a 20 °C temperature-controlled room.

The measured UCS values for the untreated and XG-stabilized kaolinite and Na-bentonite clays are presented in Fig. 6.3. The results indicate that the addition of XG increased the UCS of both kaolinite and Na-bentonite clays. For stabilized Na-bentonite with 1.0 wt% XG, a ten-fold strength increase was observed for 90-day curing, while for stabilized kaolinite with 1.0 and 1.5 wt% XG, five- and eight-fold strength increases, respectively, were reported for the same curing duration. In general, the strength gain over time for both XG-treated kaolinite and Na-bentonite could be attributed to the growth of XG amorphous structure, which covers the clay particles and fills the pore void spaces in the XG–clayey soil matrices. Since polysaccharides are known to have strong sorption and microstructural interactions with clay (Labille et al. 2004; Chang et al. 2015), XG monomers can directly bond clayey particles via cationic bridging and hydrogen bonding between the carboxyl group (–COOH) and the hydroxyl (–OH) groups of XG and the electrically charged clay particle surfaces (Gregory and Barany 2011; Awad et al. 2013).

To fully understand the mechanisms by which the strength increases with curing time in the study of Latifi et al. (2016), one should determine: (i) the rate of hydration of the interacting components (e.g., XG, kaolinite, and Na-montmorillonite) and (ii) the degree of XG coating to the kaolinite and the Na-montmorillonite particles.

In another study by Antonette et al. (2019), the XG itself was modified using bentonite prior to its use in stabilizing a sandy soil. First, XG gel with and without bentonite filler was prepared, as follows. A lab-scale magnetic stirrer was used to mix

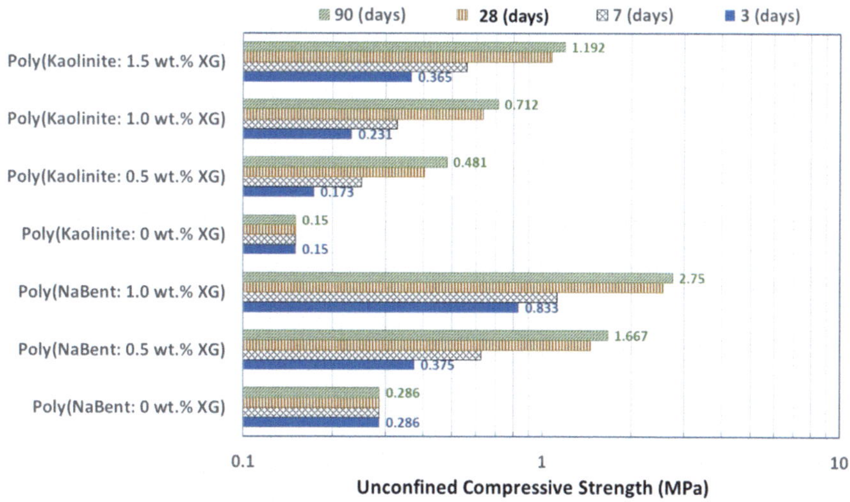

Fig. 6.3 UCS of untreated and XG-stabilized kaolinite and Na-bentonite clays (data from Latifi et al. (2016))

XG with deionized water at the desired percentages for 15 min. Each mixing batch consisted of 280 ml of deionized water, about 20 g of XG powder (corresponding to 1 wt% of dry weight of the soil used (2000 g)) and 10 g of dry bentonite when needed. Notably, the bentonite was initially mixed with the deionized water, and then the XG gel was added to the bentonite slurry. Then, the prepared XG gel (with and without bentonite) was added to the dry sand and thoroughly mixed slowly to prevent entrapping air voids. The XG-stabilized sandy soil mixtures investigated in the triaxial testing were prepared at the optimum Proctor moisture content of 14 ± 1.5%, with achieved maximum dry unit weights of between 16.2 and 16.8 kN/m^3.

The results shown in Table 6.1 for sand stabilized by 1 wt% XG (i.e., poly(sand: 1% XG)) and by 1% XG with 0.5% bentonite as a filler (i.e., poly(sand: 1% XG & 0.5% bentonite) have both shown an increase in the cohesion strength parameter, while a decrease in the friction angle was observed. Such behavior is expected due to the increase of soil plasticity with the addition of XG, both without and with the bentonite additive.

It is to be noted that water-saturated bentonite has a variable interlayer spacings (see Fig. 6.4); however, with the addition of XG gel, the interlayer spacings will be controlled, depending on the XG gel concentration. Therefore, the intercalation of XG gel into the interlayer spacings will increase the plasticity and effective cohesion and reduce the effective friction angle of the poly(sand: 1% XG & 0.5% bentonite).

Table 6.1 Combined effects of XG gel and bentonite on the effective friction angle (φ') and effective cohesion (c') of sand (data from Antonette et al. (2019))

Soil mixture	Effective friction angle, φ' (°)	Effective cohesion, c' (kPa)
Clean sand	32	0
Poly(sand: 1% XG)	23	215
Poly(sand: 1% XG & 0.5% bentonite)	15	749

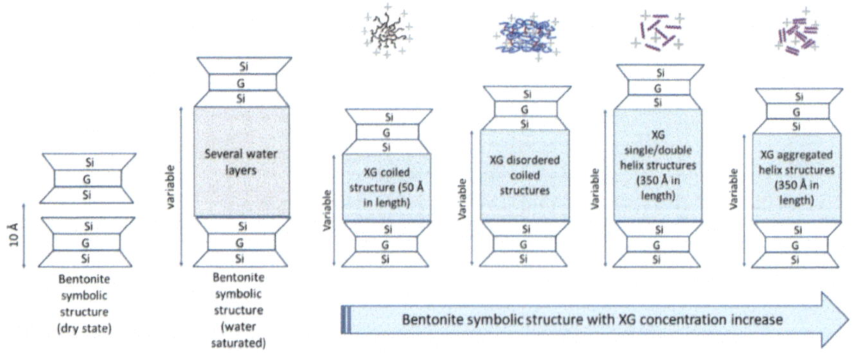

Fig. 6.4 Bentonite symbolic structures for dry state, water saturated, and water saturated with various XG hydrated gel concentrations

6.3.2.2 Cohesionless and Cohesive Soils

In a study by Chang et al. (2015), XG with different concentrations was used to stabilize the following soil types:

(a) Silica sand, which had a mean grain size of $D_{50} = 0.52$ mm, a coefficient of uniformity of $C_U = 1.35$ and a coefficient of curvature of $C_Z = 1.14$.
(b) Natural soil, classified as poorly graded sand with silt ($D_{50} = 0.32$ mm, $C_U = 3.02$ and C_Z of 0.92).
(c) Korean residual soil, originating from weathered granite, and which was composed mostly of kaolinite, halloysite, illite, vermiculite, and chlorite, with its main chemical composition being SiO_2, Al_2O_3, Fe, Mg, Na and K. It is classified as sandy lean clay, based on its geotechnical properties ($D_{50} = 0.058$ mm, liquid limit (LL) of 44% and plastic limit (PL) of 18%).
(d) Kaolinite clay, classified as clay with high plasticity, with geotechnical properties of $D_{50} = 0.044$ mm, LL = 56%, PL = 24%, and pH = 6.6.

Using the wet mixing method, the four dry disaggregated soils were separately mixed with XG at 1 wt% of the dry soil (i.e., w/w = 1%) to prepare soil–XG mixtures at moisture contents of 30% for the sand and natural soils and 60% for the Korean residual soil and kaolinite clay. For proper mixing, the purified XG was dissolved in distilled water, which was heated to 80 °C, and the XG solution was mixed using a magnet stirrer to form a uniform hydro-solution. Then, XG solutions, with concentrations of 3.33 wt% for the sand and natural soils, and 1.67 wt% for the Korean residual soil and kaolinite clay, were thoroughly mixed with representative samples from each of the soils. Additionally, for the sand, natural soil, and Korean residual soil, samples with 0.5 and 1.5 wt% XG were prepared to investigate the strengthening mechanism of soil composites comprising both coarse- and fine-grained particles. The wet mixing was performed under the same temperature condition (i.e., at room without heating). Control samples were prepared by mixing each soil with water (i.e., without XG) at the specified water content. Then, the soil and soil–XG mixtures were compacted into 40 × 40 × 40 mm cubical molds and allowed to air dry at ambient room temperature of 20 °C for a 28-day period.

The variations of the UCS and the elastic modulus of the four 1 wt% XG-stabilized soils with 28-day curing are shown in Fig. 6.5. Additionally, for comparison purposes, the same properties are shown for the Korean residual soil stabilized with 10 wt% cement. Addition of 1.0 wt% XG to the kaolinite soil resulted in significant increases in the UCS (from 0.44 to 2.5 MPa) and in the elastic modulus (from 25 to 125 MPa). The XG treatment of the sand produced an increase of its elastic property; hence, the cohesionless sand was transformed to a new material having a cohesive character, due to the dried XG coating the individual sand grains and forming ties between neighboring sand grains (crossing the void spaces). For cohesive soils, XG enhances the inter-particle bonding of clay soils, producing about a six-fold increase in UCS. This strength gain could be attributed to direct bonding of the clay particles thorough cationic bridging and hydrogen bonding between the carboxylic acid (–COOH) and

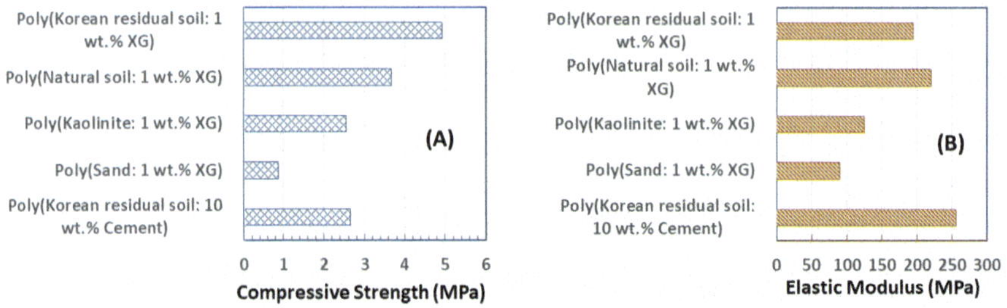

Fig. 6.5 Variations of **a** unconfined compressive strength (UCS) and **b** modulus of elasticity for four XG-treated soils and one cement-treated soil (data from Chang et al. (2015))

hydroxyl (–OH) groups of the XG and the electrically charged clay particle surfaces (Nugent et al. 2009).

6.3.2.3 Cohesionless Soil and Iron Mine Tailings Residue

In the following discussion of the results reported by Armistead et al. (2020), both dry and wet mixing methods were employed by these researchers. Using the dry mixing method, pure silica sand (SiO_2) was stabilized using 1 and 2 wt% XG concentrations (i.e., poly(100% SiO_2: 1% XG) and poly(100% SiO_2: 2% XG), respectively). Meanwhile, for using the wet mixing method, samples of pure sand and a mixture of 90% sand and 10% Fe_2O_3 were prepared using 0.2 M XG solution (i.e., poly(100% SiO_2: 0.2-M XG) and poly(90% SiO_2: 10% Fe_2O_3: 0.2-M XG), respectively).

For preparing the XG solution for poly(100% SiO_2: 0.2-M XG), the powdered XG was first added to a temperature-controlled (40°C) ultrapure water solution, while simultaneously agitating with a magnetic stirrer. Solutions were then incubated (10 min, 40°C) and subsequently sonicated (10 min). Once prepared, solutions were stored in darkness and used within 24 h of preparation.

However, for preparing the poly(90% SiO_2: 10% Fe_2O_3: 0.2-M XG) samples, the following procedure was adopted. The SiO_2 fraction (144 g, 90–150 μm) was first mixed with Fe_2O_3 (16 g, 224 ± 119 nm). Powdered XG was added to 44 ml (27.5%, M_{water}/M_{sand}) of ultrapure water to achieve a concentration of 0.2 ± 0.02 M. Solutions were then immediately mixed through 160 g of the tailings material to produce a homogenous mixture [(i.e., poly(90% SiO_2: 10% Fe_2O_3: 0.2-M XG))]. The resulting wet composite soil was compacted using a drop hammer into a 202-mm diameter by 42-mm deep sample mold. Samples were then extruded and left to cure (air dry) for 7 days at 20 °C.

Since Fe_2O_3 binding affinities with XG are pH dependent, it is worth noting that the binding affinity decreased as the solution pH increased over the investigated pH range of 4 to 8. As such, the poly(90% SiO_2: 10% Fe_2O_3: 0.2-M XG) showed a low binding affinity through pH levels of 5–8, indicating XG's unsuitability for use as

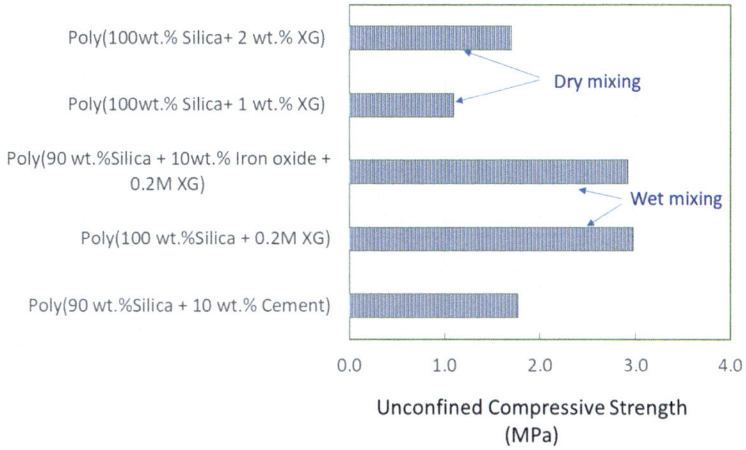

Fig. 6.6 7-day-cured unconfined compressive strength for silica sand and 90% sand–10% Fe_2O_3 mixture stabilized with XG for both dry- and wet-mixing sample preparation methods (data from Armistead et al. (2020))

a stabilization/solidification additive to fresh mine tailings composed of both SiO_2 and 10% Fe_2O_3. However, within acidic conditions (pH 4), the XG showed Fe_2O_3 binding affinity. The Fourier Transfer Infrared (FTIR) analyses indicated no changes in the surface spectra of XG-coated Fe_2O_3 particles compared to that of no-XG Fe_2O_3 particles.

From Fig. 6.6 (data from Armistead et al. (2020)), the UCS of the 7-day-cured wet-mixed poly(100% SiO_2: 0.2-M XG) and poly(90% SiO_2: 10% Fe_2O_3: 0.2-M XG) samples was 2.975 MPa and 2.925 MPa, respectively. Hence, the 10% Fe_2O_3 fraction did not affect the UCS properties of the XG-stabilized soil composite. Additionally, Fig. 6.6 shows the results of SiO_2 stabilized with 1% and 2% XG additions for the dry mixing method. The UCS of the 1% XG-treated SiO_2 is consistent with that shown in Fig. 6.5 (data from Chang et al. (2015)), but not with that reported by Armistead et al. (2020) for SiO_2 stabilized with 0.2-M XG (based on the wet mixing method). Such variations could be partly due to the ineffective hydration processes in the case of the dry mixing method.

6.3.2.4 Cohesionless Soil with Variable Moisture Contents

In a study by Lee et al. (2021), the shear strength properties of XG-stabilized sand samples prepared by the wet mixing method were evaluated. The initial moisture content of all the tested samples was 20%. The investigated ratios of XG powder to that of the dry sand were 0 (i.e., untreated/control), 0.5, 1.0, and 2.0 wt% (corresponding to 2.5, 5.0, and 10 ratios to the mass of water). The XG powder (obtained from Sigma-Aldrich: CAS number 11138-66-2) was dissolved in distilled water at ambient room temperature (21 °C), and the mixture was magnetically stirred to obtain a uniform XG aqueous solution (i.e., hydrogel). Then, the solution was mixed with dry sand, and specimens measuring 60-mm in diameter and 25-mm high were prepared for direct shear testing. The initial dry density of the specimens (1450 ± 10 kg/m^3), equating to a medium relative density (of 36–44%), was controlled by tapping to minimize the effects of variable particle packing density on the soil strengthening achieved. The specimens were prepared for three representative moisture states:

(a) Initial (i.e., at 20% moisture content), whereby the specimens were sheared immediately after preparation.
(b) Dried, whereby the specimens were sheared after oven-drying at 40 °C for a 21-day period (at this state the residual moisture content was less than 3%).
(c) Dried–submerged, whereby the dried specimens were fully immersed in deionized water at room temperature for a 24-h period (to achieve a full saturation condition) and then sheared.

The deduced cohesion and friction angle values of the XG-stabilized sand are presented in Fig. 6.7, considering the three different moisture states described above (i.e., "*initial*", "*dried*", and "*dried–submerged*"). The XG hydrogel in the pore voids of the stabilized sand highlights the effects of the moisture state on the cohesion and friction properties. In other words, the nature of the XG within the pore voids changed from its "*initial*" state of an XG-hydrogel to (i) a condensed XG-film in the "*dried*" state, thereby generating tension forces, and (ii) a disturbed and swollen XG structure, arising from exposure to water during submerging, which exhibits a more contractive behavior than that of the initial uniform XG-hydrogel.

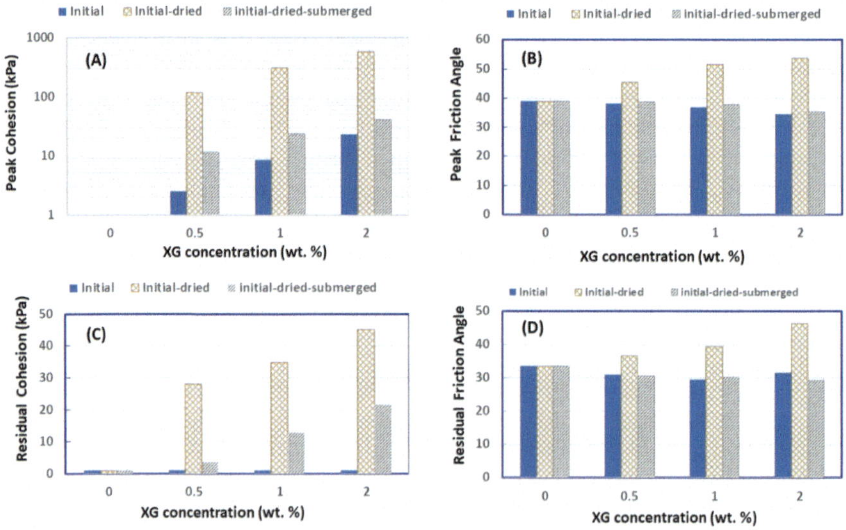

Fig. 6.7 Effects of XG concentration and moisture state on the shear strength properties (cohesion and friction angle) of XG-treated sand specimens tested in direct shear (data from Lee et al. (2021))

Referring to Fig. 6.7 (data from Lee et al. (2021)), in the "*dried*" condition, both the cohesion and friction angle properties progressively increased with increasing XG concentration. This increase is due to phase change of the initial XG-hydrogel to the XG-film state for the "*dried*" condition, thereby increasing the XG-tensile strength that led to increasing both the cohesion and friction angle. However, due to submerging, the cohesion decreased but remained higher than that of the "*initial*" state, whereas the friction angle exhibited a slight decrease from the "*initial*" state. Evidently, the effects of XG concentration on the mechanical properties of the treated sand are much greater in the "*dried*" state than that in the wet (i.e., "*initial*" and "*dried–submerged*") states. This highlights the remarkable improvements in the cohesion properties and strengthening of the XG-stabilized sand achieved upon drying.

To evaluate the consistency of the reported results from different researchers, let us compare the results shown in Table 6.1 (data from Antonette et al. (2019)) for poly(sand: 1% XG) with initial moisture content of 14% and maximum dry density of 16.2–16.8 kN/m^3, referred to as $S_{\text{Antonette}}$, and that reported in Fig. 6.7 (data from Lee et al. (2021)) for the same treated soil (i.e., "*initial*" condition) with a moisture content of 20% and dry density of 14.2 kN/m^3, referred to as S_{Lee}. For $S_{\text{Antonette}}$, $\varphi' = 23°$ and $c' = 215$ kPa, while for S_{Lee}, $\varphi' = 38°$ and $c' = 8$ kPa, highlighting the considerable difference between these studies, which could be attributed to the lack of standard soil–XG testing procedures.

6.3.2.5 Laterite Soil

Laterite soil is formed in hot and wet tropical areas, with leaching of the soil leaving only oxides of iron and aluminum. In a study by Sawant et al. (2019), Indian laterite soil (with LL = 62%, PL = 34%, and PI = 28%) was treated using dry mixing method with XG concentrations of 1, 2, and 3% (by the dry mass of the soil) and compacted at Proctor optimum (of the untreated soil). The direct shear test was conducted via the Indian Standard, where the unsaturated compacted sample was sheared quickly (1 mm/min), with the deduced shear strength properties shown in Fig. 6.8. As the XG concentration increases, the cohesion progressively increased, but the friction angle reduced slightly.

6.3.2.6 Loess Soils

Jiang et al. (2022) investigated XG stabilization of loess soil from Sanmenxia, Henan Province, China. The soil was classified as well-graded silty soil with low plasticity and had the following geotechnical properties: (i) composition of 77.2% silt, 14.1% clay, and 8.7% sand; (ii) $C_U = 2.57$ and $C_Z = 12.43$; (iii) LL = 26.9% and PI = 8.9%; and (iv) natural water content of 5.1%, natural dry density of 1.33 Mg/m^3, and OMC for compaction of 17.7%, producing a maximum dry density of 1.74 Mg/m^3. The XG powder used was commercial food grade, with a pyruvic acid content $\geq$

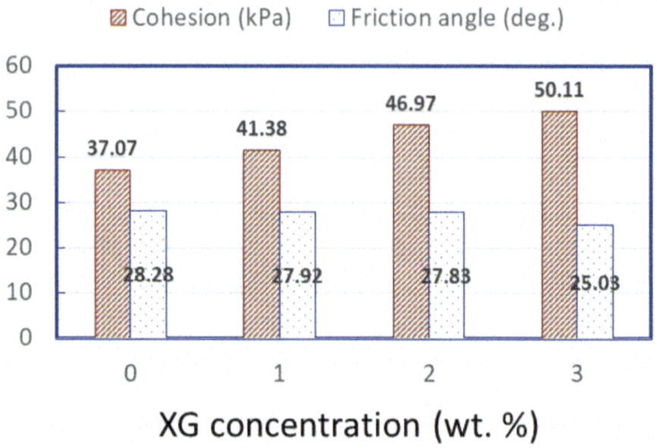

Fig. 6.8 Variations of cohesion and friction angle for XG-stabilized Indian laterite soil tested in direct shear according to the Indian Standard (data from Sawant et al. (2019))

1.5%, and a viscosity of 1475 cP (1 cP = 0.001 Pa.s). It is worth noting that the study did not indicate the conditions at which the viscosity test was performed (i.e., XG concentration in solution, ionic strength of the solution, and the testing temperature).

The initial moisture content of the specimens in all tests was designed to be 20% to realize dehydration. The XG concentrations investigated were 0.5, 1.0, and 2.0% (by mass of dry soil). The specimens were prepared by the dry mixing method, first evenly mixing the dry XG and loess powder materials, and then adding the required quantity of water. The XG–loess mixtures were placed in a sealed desiccator for maturing over 24 h. Specimens (63.5-mm diameter by 25.2-mm high) were prepared by compaction in cylindrical molds to achieve three different density states, after which they were allowed to cure at room temperature (25 °C), 50% relative humidity, with natural air drying.

Figure 6.9 presents the variations of splitting tensile strength with XG concentration and initial dry density of the cured, stabilized loess soil samples. Tensile strength showed an increase with both XG concentration and dry density; for instance, with increasing XG concentration from 0 to 2%, the tensile strength increased by factors of 6.58, 5.62, and 6.06 for dry densities of 1.55, 1.63, and 1.70 Mg/m^3, respectively. Additionally, the dense XG fiber networks led to the formation of crystal layers on the external surfaces of the specimens, hence reducing water loss occurring from the inner part of the specimens. The ability of the stabilized loess soil to resist the tension forces might be mainly due to the bonding strength between XG and the soil particles, which could be in the form of bridge connections between particles and the dense aggregated structure formed by the XG.

6.3.2.7 High Plastic Natural Soils

Singh and Das (2019) investigated the XG stabilization of highly plastic silty soil, with PI of 61%, OMC for compaction of 21% and maximum dry unit weight of 16.1 kN/m^3. This soil possesses high swelling behavior, and its differential free-swelling value is 217%. Dry pulverized soil samples were properly mixed with the required

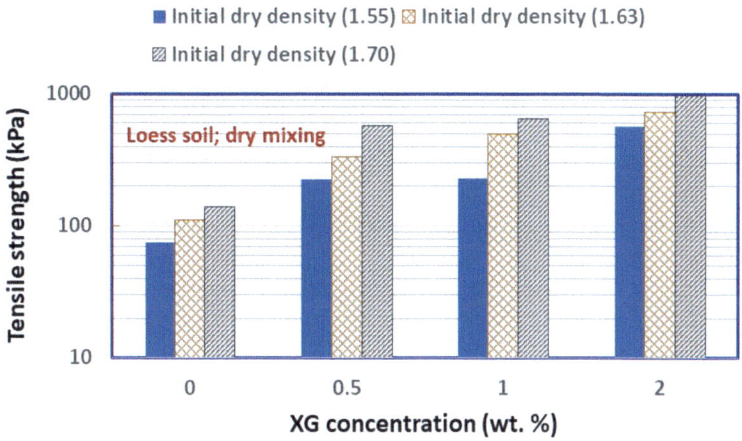

Fig. 6.9 Variations of splitting tensile strength with XG concentration and initial dry density of 7-day-cured, dehydrated XG-treated loess soil (data from Jiang et al. (2022))

amount of XG powder (i.e., 0, 0.2, 0.5, 0.8, and 1.0% by the weight of the dry soil), and distilled water was added corresponding to their respective OMCs. The compacted UCS samples were coated with paraffin wax (for better preservation of moisture) and were allowed to cure at room temperature (average temperature of 25 °C) for periods of 0, 3, 7, and 28 days.

The UCS values for the XG-stabilized soil are shown in Fig. 6.10. The results indicate that the UCS increases with increasing XG concentration, which is expected due to increased rigidity of the XG conformation structures and their compactness, increased gel content, and increased viscosity and molecular weight of the XG. Also, the external surface area of the XG-treated soil agglomerations tends to decrease with increasing XG concentration due to an increased degree of aggregation. In addition, the UCS substantially increased with increasing curing period. The scanning electron microscopy (SEM) analysis indicated that the untreated soil possessed flaky soil particles and large pore void spaces, whereas for the 1% XG-stabilized soil, the flaky soil particles were bonded together with a gel-type cementitious material, the nature of which varied with the curing period.

6.3.3 *Dry Density*

For cohesionless (sandy) soils, as the XG concentration increases, the dry density of XG-treated sandy soils increases slightly, attributed to the decrease in the void volume caused by dehydration of the XG (Chang et al. 2015). In this case, the XG forms amorphous matrices that coat the sand grains, forming bridges between sand grains and filling the pore voids. However, for cohesive soils, the XG molecules interact directly with the clay particles and the pore water. As the XG concentration

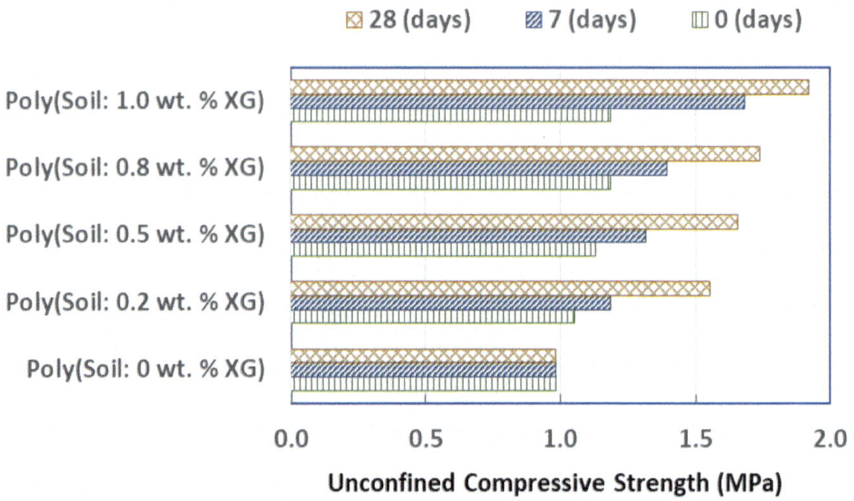

Fig. 6.10 UCS of XG-stabilized high plastic silty soil (data from Singh and Das (2019))

increases, XG monomers adsorb water and form a highly viscous suspension, which fills the void spaces between the clay particles; hence, the space between the clay particles increases, thereby contributing to an overall increase in the volume of voids within the treated mixture. Therefore, an increase in the XG content would lead to an increase in volume of voids and consequently a decrease in the dry density. For natural soil (with about 20% fines content), because of variations in the constituent particle sizes, it also showed a decrease in the dry density with increasing XG concentration.

6.3.4 Moisture Content

6.3.4.1 Drying Effect

The effect of the moisture content at the end of the sample curing period, and prior to testing, can be evaluated using the experimental results for kaolinite clay dry-mixed with 0.5, 1.0, and 1.5% wt% XG reported by Barani and Barfar (2021). The treated soil specimens were cured for 7 days at room temperature to achieve different residual moisture contents (due to drying) of 1.4% (i.e., dry state), 11%, 17%, and 23%, and then tested in tension using the three-point bending test.

Single-edge notched beam specimens, with a support span of $S = 100$ mm, and beam dimensions of $H = S/3$, $H/D = 1$, and $a = H/3$ were prepared, where S, D, H, and a are the span of the simple supports, the width of the beam, the depth of the beam, and the initial crack length of the sample, respectively. Loading of the beams was performed at a rate of 0.2 mm/min, so that they would fail in 5 to 10 min, as is common practice in fracture testing of quasi-brittle materials. The load–displacement

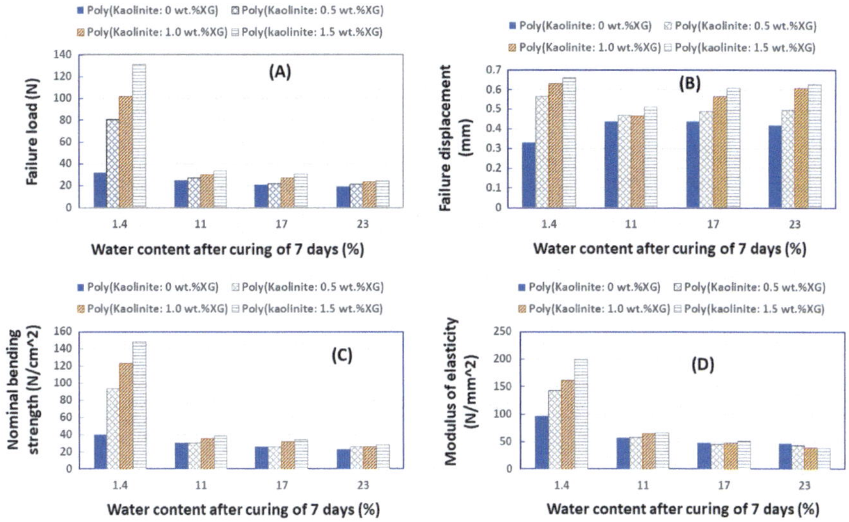

Fig. 6.11 Performance of poly(kaolinite: XG) mixtures in tension (three-point bending tests), as a function of the water content for 7-day curing: **a** failure load; **b** failure displacement; **c** nominal bending strength; and **d** modulus of elasticity (data from Barani and Barfar (2021))

curves from the three-point bending tests were used to calculate the failure load, the failure displacements, and nominal bending loads, as illustrated in Fig. 6.11.

The results indicated that the dry state (i.e., 1.4% moisture content) showed the highest failure loads, nominal bending strength, and failure displacements. Notably, in the dry state, some moisture still occurs within the hydrophilic groups, resulting in a residual moisture content of 1.4%. In addition, as the XG content increases for the 1.4% moisture content samples, the failure load, nominal bending strength, and elastic modulus all increased considerably, whereas, for the same XG content, as the moisture content increases, the failure load, nominal bending strength, and elastic modulus all decreased considerably.

For the kaolinite–XG system, the kaolinite and aqueous–XG macromolecules would interact with each other to form an interconnected network via cationic bridging and van der Waals forces within the hydroxyl (–OH) and carboxylic acid (–COOH) groups of the XG macromolecule and the electrically charged kaolinite clay particle surfaces and edges (Nugent et al. 2009; Mohamed and Paleologos 2018). However, during drying, water molecules tend to escape from the diffused ion layer that contains the XG chains, which results in the changing of the hydrogels from the rubbery to the glassy state, and overall shrinkage of the kaolinite–XG specimens (Muguda et al. 2017). It is to be noted that during drying, a crystallized surface layer is promoted by the microbial polysaccharide, which maintains moisture in the inner bulk soil.

6.3.4.2 Drying and Re-wetting Effect

In a study by Zhang et al. (2022), the dried natural silty clay soil (PI = 8.8%) was fully mixed with dry XG powder at 2% concentration (by the dry weight of soil). Deionized water was added, and the mixture was evenly stirred. The soil was then sealed at a constant temperature for 24 h until the water was fully infiltrated. The initial moisture content of the specimens was 15%, with compaction to 16.3 kN/m^3, i.e., 92% of the maximum dry unit weight. After their preparation, a vacuum pump was used to vacuum and saturate the specimens for testing.

Saturated untreated and 2 wt% XG-treated specimens were dried and re-wetted (Fig. 6.12). The initial moisture content of the saturated specimens was 22%. The specimens were uniformly dried to a state of 2% moisture content and then re-wetted to different moisture contents (of 4, 8, 16, and 22%) under constant humidity and temperature. Then, fast direct shear tests were performed under constant vertical stresses of 50, 100, and 200 kPa.

The effects of XG content on the shear strength properties of the treated silty soil are presented in Fig. 6.13. This figure shows that at low moisture content (of 2–4%), the cohesion and friction angle values of the XG-treated soil were substantially increased, with a 2.3-fold and 1.5-fold gains, respectively, compared to those of the untreated soil. However, at 8% moisture content, both cohesion and friction angle were only slightly increased. Note that because the XG is partially hydrated at lower moisture content, the friction angle of the XG-treated soil was higher than 45°, characteristic of (very) dense sandy soils.

The preceding experimental results (Barani and Barfar 2021; Zhang et al. 2022) for the two non-dispersive soils (kaolinite and silt) clearly highlight the simple fact

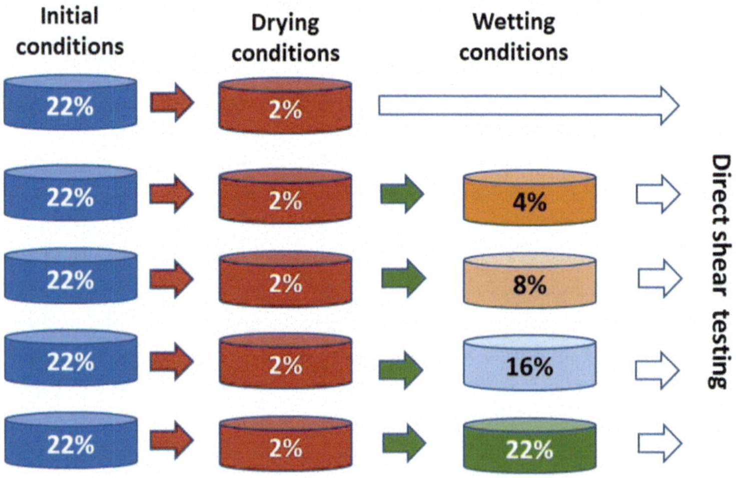

Fig. 6.12 Status of the tested specimens (by Zhang et al. (2022)) in direct shear testing with the drying and wetting paths

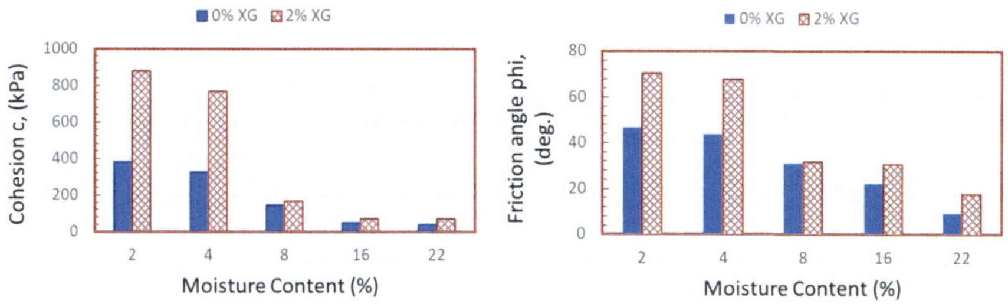

Fig. 6.13 Variations of cohesion and friction angle parameters for XG-treated natural silty soil from Henan Province, China (data from Zhang et al. (2022))

that in the dry state, XG-treated soils are highly stable (high strength, cohesion, and friction angle); however, once there is a moisture intake (e.g., due to rain infiltration or groundwater movement), the XG-treated soils became unstable with high collapsibility potential. Therefore, it is important to carefully consider the soil type and conditions under which XG-stabilization treatment can be applied in practice.

6.3.5 Curing Environment

Durability assessment is of vital concern since the XG biopolymers could be decomposed. Although XG is known to be a thermo-stable biopolymer below 250 °C (Zohuriaan and Shokrolahi 2004), it shows decomposition by bacteria or aerobic microorganisms (Hou et al. 1986). Also, polysaccharide is known to be stable against oxidation and is widely used as an antioxidation additive in oil and protein products (Sun and Gunasekaran 2009). Moreover, its stability under extreme conditions, including acidic and alkaline environments, as well as high salt concentrations, provides high potential for use in various industrial purposes (Katzbauer 1998). However, the mechanical stability of XG under cyclic wetting/drying and freezing/thawing actions should also be verified. Therefore, the long-term durability of XG-stabilized soils against biodecomposition must be verified to allow recommendation for its usage in geotechnical and geo-environmental engineering practice.

The effects of curing period and environmental conditions may be evaluated in view of the study reported by Joga and Varaprasad (2019). They treated a natural expansive clay soil (LL = 91%, PI = 52%, and a free swell index of 160%) with 1% XG content, employing the wet mixing method. Samples were prepared at their optimum standard Proctor compaction state. After 90 days of curing in an ambient environment, the UCS of the XG-treated sample increased from initially 0.42 to 2.21 MPa (5.26-fold increase) (see Fig. 6.14). Whereas when the XG-treated sample was maintained at a constant heating environment of 40 °C, the UCS increased from initially 0.58 to 3.27 MPa after 90 days curing (a 5.64-fold increase). However, when the XG-treated sample was immersed in water, the UCS increased from 0.42 MPa (for uncured sample) to 0.55 MPa after 90 days (1.31-fold increase). Such results are expected because of the nature of the amorphous bonding between XG and clay soils (see Fig. 6.15), which is completely different from that of cement-based stabilization, where inorganic cementitious products are formed. Such results are extremely important in the context of geo-engineering applications, in that the XG-stabilization approach might be suitable for arid lands (no rain and hot environment), but not suitable for regions with high humidity and rainfall.

In another study, Hamza et al. (2023b) evaluated the one-dimensional compressibility and swelling potential characteristics of a natural clay with high plasticity (PI = 30.5%) that was stabilized with 1.5 wt% XG, employing the dry mixing method.

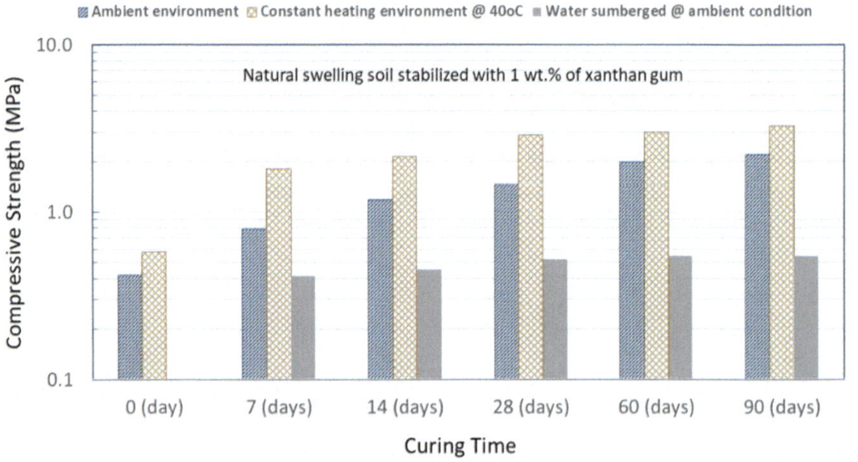

Fig. 6.14 Effects of curing conditions on the unconfined compressive strength of 1% XG-treated natural expansive clay soil (data from Joga and Varaprasad (2019))

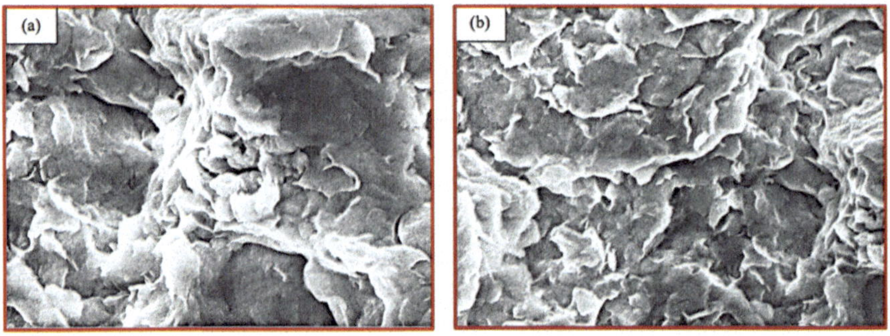

Fig. 6.15 Microstructure of 1% XG-treated natural expansive clay soil after **a** 7 days and **b** 28 days of curing (adopted from Joga and Varaprasad (2019)) (This article is an open access article distributed under the terms and conditions of the Creative Commons Attribution (CC-BY) license (http://creativecommons.org/licenses/by/4.0/))

The prepared mixtures were placed in sealed plastic bags for 24 h to achieve a homogeneous moisture distribution. Test specimens were prepared by standard Proctor compaction at the OMC to achieve the maximum dry density, and they were allowed to naturally air dry for periods of 0, 4, 7, 14, 28, and 60 days. The one-dimensional consolidation tests were conducted according to ASTM D2435M-11, investigating the applied vertical stress range of 10–800 kPa, each load increment maintained for 24 h.

From the one-dimensional consolidation test results, soil characteristics, including the compression and swelling (rebound) indices, swelling strain and swelling pressure, were deduced as a function of curing time, as shown in Fig. 6.16. With the increase of curing period, both compression and swelling indices decreased significantly (by an average of 78%) over the 60 days of curing, with the compression index reducing to about 0.06, which is well below the permissible upper limit (of 0.13) associated with low expansive soil (Holtz and Gibbs 1956). This indicates that the stiffness of the 1.5 wt% XG-treated natural high plastic clay increased with curing time due to XG interaction with the clay particles and formation of hydrogel networks, thereby providing an enhanced resistance to soil deformation in response to external loading. Referring to Fig. 6.16b, compared to the maximum swelling strain (of 5%) for the untreated natural expansive soil, arising from the presence of montmorillonite in the soil, the swelling strain for 1.5 wt% XG addition and 60-day curing had reduced to 0.9% (82% reduction), well below the permissible swelling potential limit of 1.5% (Hamidi and Hooresfand 2013). Similarly, the swelling pressure of the 1.5% XG-treated soil was reduced by 86% (from about 100 kPa initially to 14 kPa after 60 days of curing), which is also within the permissible range of < 50 kPa for the low swelling class defined by McCormack (1976).

The above results clearly indicate that XG-treated soil is strong due to curing (drying). However, in its wetting states, both compressibility and swelling are high, indicating that upon moisture intake, the XG treated soil became unstable. Such findings are extremely important in the context of geo-engineering applications; hence, XG stabilization might be suitable only for dry regions (i.e., no rain, with the groundwater level located below the stabilized zones).

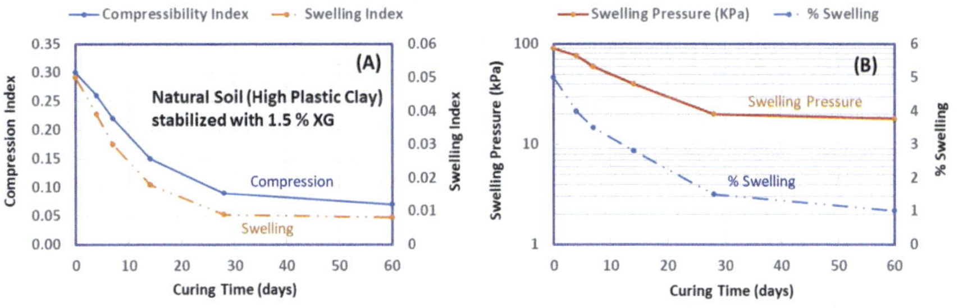

Fig. 6.16 Compressibility and swelling characteristics of a natural plastic clay stabilized with 1.5% XG (data from Hamza et al. (2023b))

6.3.6 Soil pH

XG is a negatively charged biopolymer that is sensitive to pH variations. At low pH levels (acidic conditions), XG can become partially or completely protonated, which reduces its negative charge and can lead to a reduction in its binding ability. As a result, the soil matrix may become less stiff, and its compressibility may increase. Conversely, at high pH levels (alkaline conditions), XG can become deprotonated, which increases its negative charge and can lead to an increase in its binding ability. This can result in a more rigid soil matrix, accompanied by a reduction in compressibility. However, the effects of pH on the compressibility of XG–soil materials may depend on other factors, such as the XG concentration and the type of soil.

6.4 Compressibility and Swelling

6.4.1 Effect of XG Content on Saturated Dispersive Soils

The effects of XG content on the compressibility and swelling of saturated dispersive soils for the unloading condition can be evaluated from the experimental results reported by Joga and Varaprasad (2020). They stabilized natural dispersive soil from India by XG addition at different concentrations (of 0 to 2.0 wt%), using the standard odometer mold (ring) testing procedure to determine both the compression and swelling indices (Fig. 6.17). The soil was characterized by its plasticity index (PI) of 39%, OMC for Proctor compaction of 19.2%, and sodium adsorption ratio of 34.4%. The XG-treated soil was compacted in the odometer mold (ring) at its OMC, and then saturated, following standard testing procedures. The results clearly show the impact of the XG addition on both the compressibility and swelling performance of the dispersive soil, with 1.0 wt% XG addition reducing the compression and swelling indices by 72% and 47%, respectively. For 2.0 wt% XG content, there were further minor decreases in both the compression and swelling indices.

The addition of XG induces biopolymer cross-linking, thereby forming an interconnected network of hydrogels in the voids of the soil matrix and causing preferential adsorption of the XG molecules and cations present in the soil pore fluid. These interaction mechanisms tend to reduce both the compressibility and swelling of dispersive soils. In addition, at lower XG concentrations (i.e., for $C_{XG} < 1$ wt%), XG exhibits a weak gel-like hydrogel, whereas at higher concentrations, ($C_{XG} >$ 1%), XG exhibits a liquid–crystal hydrogel, leading to filling and clogging of the soil pores.

It is worth noting that the dispersivity of this natural soil from India was measured using double hydrometer analysis (DHA) by the same authors, and the results indicated that the dispersivity for the untreated soil was 69.5%, while it decreased to 48, 30, 25, and 16.5% for additions of 0.5, 1.0, 1.5, and 2.0 wt% XG (by dry weight of the soil). In other words, the percent decrease in dispersivity ranged from 30% (for

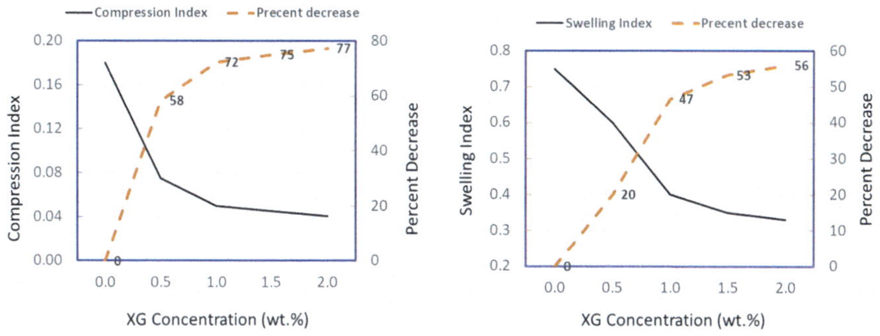

Fig. 6.17 Variations of the compression and swelling indices for XG-stabilized natural dispersive soil (data from Joga and Varaprasad (2020))

0.5% XG) to 76% (for 2 wt% XG), with the soil characterized as non-dispersive for the addition of $\geq$ 1 wt% XG.

6.4.2 Effect of Moisture State on XG–Kaolinite (Non-dispersive) Clay

In a study by Kwon et al. (2023), clay specimens were prepared by mixing dry kaolinite and XG powdered materials at different XG concentrations (0, 0.5, 1.0, and 2.0 wt% by dry mass of kaolinite). The kaolinite–XG mixtures were poured into the confining ring of the oedometer apparatus and compacted by tamping in three equal-height layers to achieve an initial void ratio in the range of 2.0 to 2.2. Before wetting using deionized water, the compacted test specimens were placed under a constant vertical stress of 5 kPa. Then, the volumetric expansion due to wetting under the maintained 5-kPa vertical stress was measured. Notably, these swelling experiments were conducted under the initially dry condition to maximize the variation of swelling potential with XG content due to biopolymer hydration.

Upon the uptake of water, the untreated specimen (0 wt% XG) experienced contraction, with a maximum compression strain of about 5%, which is expected because of the changes in kaolinite clay structure and pore sizes. However, with 0.5 wt% XG addition, the compression behavior also occurred during the first 10 min, but then swelling dominates, reaching a stable value of 5% swelling after 500 min. Specimens with 1.0 and 2.0 wt% XG exhibited swelling behavior from the start of the experiments, reaching stable swelling strains of 12% and 20%, respectively, after 500 min. The recorded swelling can be attributed to the hydration of the hydrophilic XG hydrogels.

6.4.3 Control of XG Swelling

The observed swelling due to XG hydration in the type of experiments performed by Kwon et al. (2023), and described in Sect. 6.4.2, can be controlled with the addition of a filler (e.g., bentonite) to the XG prior to the stabilization of non-dispersive soils. Such mitigation measure can be highlighted in view of a study reported by Antonette et al. (2019), whereby 0.5% bentonite was added to the 1.0 wt% XG prior to its use in stabilizing a sandy soil. The experimental results shown in Fig. 6.18a indicate that, by using bentonite clay as a filler (to the XG), once the shear rate increases beyond 0.1 1/s, there was no difference in viscosities of the 1.0 wt% XG + 0.5 wt% bentonite and 1.0 wt% XG aqueous solutions. Also, the results shown in Fig. 6.18b indicate that with the addition of 0.5% bentonite as a filler to the XG, after 10 min, about 70% of the free swelling that occurred with poly(sand: 1.0% XG) has been arrested, and by 100 and 2000 min, 84% and 90% of the free swelling were arrested. In this scenario, the use of bentonite leads to intercalation of XG between the bentonite layers and controls the swelling of both the XG and the bentonite itself.

6.5 Strength Dependence on XG Concentration and Soil's Physical Molding State

In this section, the results reported by Barani and Barfar (2021) for the three-point bending tests were used to evaluate the effect of 0.5, 1.0, and 1.5 wt% XG on the fracture behavior of kaolinite clay (PI = 19.3%, and standard Proctor OMC = 24.1%) during drying. The kaolinite–XG mixtures were prepared using the dry mixing method, with the required water added and mixed thoroughly until achieving a uniform mixture. Then, beam specimens of the kaolinite–XG mixtures were compacted in steel molds at 95% of their maximum dry density (i.e., on the wet

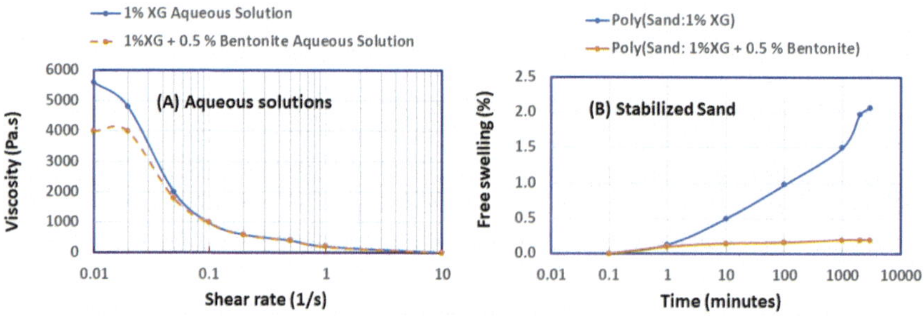

Fig. 6.18 **a** Variations of viscosities for 1 wt% XG and 1 wt% XG + 0.5 wt% bentonite filler aqueous solutions with shear rate at 25 °C; and **b** free swelling of XG-stabilized sand with and without bentonite filler (data from Antonette et al. (2019))

side of the OMC) using a standard compaction hammer. The compacted kaolinite–XG beams were allowed to cure for 7 days in a 20 °C temperature-controlled room. On reaching their desired (reduced) moisture contents, the beam specimens were firmly wrapped with a polyethylene cover for 48 h to equilibrate and maintain the moisture inside. The three-point bending test was performed on single-edge notched XG–kaolinite beams with dimensions of $a = H/3$, $H/D = 1$, and $S = 3.3H = 100$ mm, where S, D, H, and a are the span of the simple supports, the beam's width and depth, and the initial crack length of the sample, respectively. A coping saw, with a thickness of 0.5 mm, was used to cut a notch in the beam to a depth of 10 mm (= ~ $D/3$). Loading of the beams was performed at a rate of 0.2 mm/min, so that they would fail in 5 to 10 min, as per common practice for fracture testing of quasi-brittle materials. The beam samples were loaded until failure occurred, and both the failure load and the nominal bending strength were calculated.

The reported experimental data for these experiments are analyzed here, using response surface methodology (RSM) implemented in Minitab, for the failure load, Y_1, and the nominal bending strength, Y_2, in terms of the three independent variables, i.e., (i) XG content (denoted as X_1, and ranging 0–1.5 wt%), (ii) water content (X_2, ranging from 1.4% (dry state) to 23.7%), and (iii) dry density (X_3, ranging 1.55–1.64 Mg/m^3). Following RSM, a quadratic polynomial regression model was assumed for predicting the dependent (Y) variables, as follows:

$$Y = b_0 + \sum b_i X_i + \sum b_{ii} X_i^2 + \sum b_{ij} X_i X_j, \tag{6.1}$$

where b_0 is the constant coefficient, b_i is the linear regression coefficient, b_{ii} is the quadratic regression coefficient, b_{ij} is the interaction regression coefficient, and X_i and X_j are the independent variables.

The regression analysis was performed to fit second-order polynomial equations for the response variables Y_1 and Y_2. Response surfaces were generated using the fitted quadratic polynomial equations, and lack-of-fit tests were performed on the fitted models. For statistical tests, significance was defined at $p \leq 0.05$, with the t-test employed to compare the responses under optimized conditions with those predicted by the models.

The obtained equations relating the failure load, Y_1, and the nominal bending strength, Y_2, to the three independent variables described above are represented by second-order polynomial equations, as follows:

$$\begin{aligned} Y_1(\text{in Newtons}) = {} & 2243 + 373X_1 - 48.5X_2 - 2595X_3 - 24.2X_1^2 \\ & + 0.2953X_2^2 + 769X_3^2 - 1.73X_1X_2 - 190X_1X_3 \\ & + 25.4X_2X_3, \end{aligned} \tag{6.2}$$

$$\begin{aligned} Y_2\left(\text{in N/cm}^2\right) = {} & 305 + 661X_1 - 42.1X_2 - 374X_3 - 30.8X_1^2 \\ & + 0.3315X_2^2 + 137X_3^2 - 2.84X_1X_2 - 354X_1X_3 \\ & + 21.1X_2X_3. \end{aligned} \tag{6.3}$$

In the above equations, the negative sign represents the antagonistic effects, while the positive sign shows the synergistic effects among the variables.

Scatter plots showing the degree of correlation between the predicted and experimental (or measured) values for the failure load and nominal bending strength parameters are illustrated in Fig. 6.19.

Figures 6.20 and 6.21 illustrate typical surface plots for the failure load and the nominal bending strength, respectively, with midrange changes in (i) water content and XG content for a constant dry density, (ii) XG content and dry density for a constant water content, and (iii) water content and dry density for a constant XG content.

The contour plots corresponding to the surface plots shown above are presented in Figs. 6.22 and 6.23, respectively, for the failure load and the nominal bending strength, with (i) water content and XG content (constant dry density), (ii) XG content and dry density (constant water content), and (iii) water content and dry density (constant XG content).

The model equations indicate that, overall, the XG content, X_1, had a significant role in improving the mechanical rigidity of the XG–kaolinite structure, as indicated by a higher correlation, followed by the water content, X_2, and the dry density, X_3. For both failure load and nominal bending strength, the equations indicate the following: (i) for XG content, a positive effect for the linear term, and a negative effect for both the interacting and quadric terms; (ii) for water content, a negative effect for the linear term, positive for the quadric, negative for the interaction with XG content, and positive for the interaction with dry density; and (iii) for dry density, a negative effect for the linear term, positive for the quadric, negative for the interaction with XG content, and positive for the interaction with water content.

Among the three independent variables (and compared to the other two variables), the XG content, X_1, played a major role in increasing the failure load, Y_1, and the nominal bending strength, Y_2, due to the rigidity of the kaolinite–XG structure (see Figs. 6.24 and 6.25). The determination of XG content, water content, and dry density regimes is an important aspect of the development of the failure load and nominal bending strength. Considering the linear term of the models, the XG content has a positive effect up to 1.2 wt%, water content has a negative effect up to 16%,

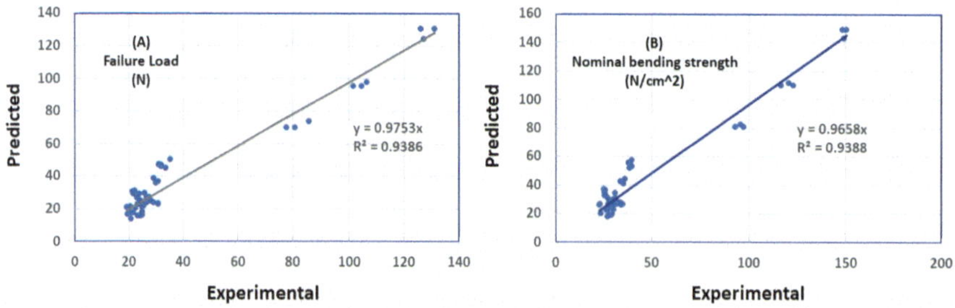

Fig. 6.19 Predicted versus experimental values of 7-day-cured poly(kaolinite: XG) samples in three-point bending tests: **a** failure load; and **b** nominal bending strength

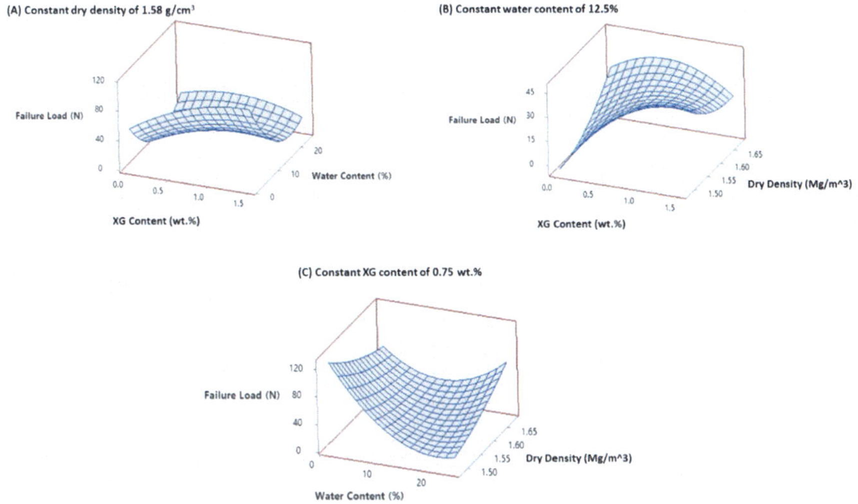

Fig. 6.20 Typical surface plots for the failure load with **a** water content and XG content at constant dry density of 1.58 Mg/m^3, **b** XG content and dry density at constant water content of 12.5%, and **c** water content and dry density at constant XG content of 0.75%

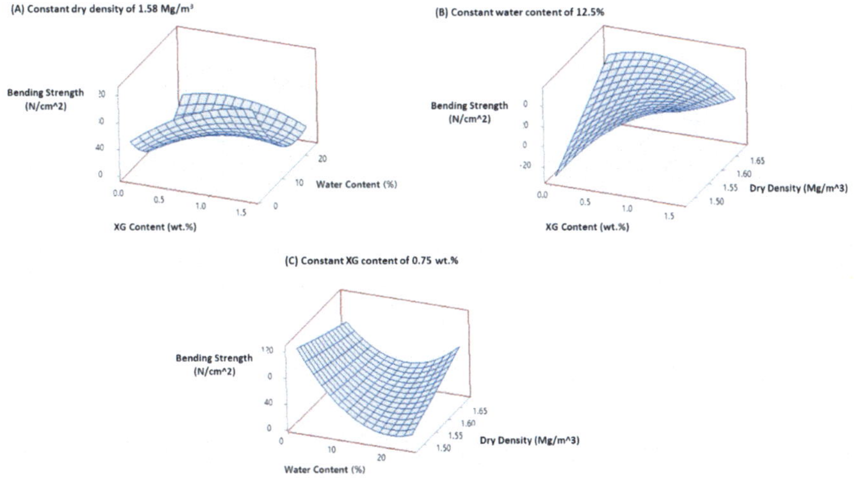

Fig. 6.21 Typical surface plots for the nominal bending strength with **a** water content and XG content at constant dry density of 1.58 Mg/m^3, **b** XG content and dry density at constant water content of 12.5%, and **c** water content and dry density at constant XG content of 0.75%

followed by a positive effect, and the dry density has a negative effect up to 1.55 Mg/m^3, followed by a positive effect.

Considering the interaction term of the model, the following observations are made: (i) irrespective of the XG content, the water content has a negative effect,

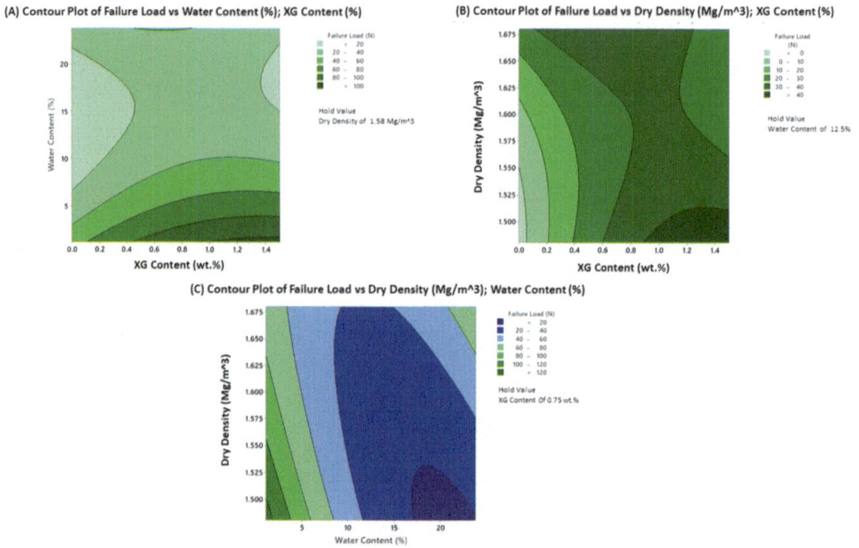

Fig. 6.22 Contour plots variations for the failure load with **a** water content and XG content (constant dry density of 1.58 Mg/m^3), **b** XG content and dry density (constant water content of 12.5%), and **c** water content and dry density (constant XG content of 0.75%)

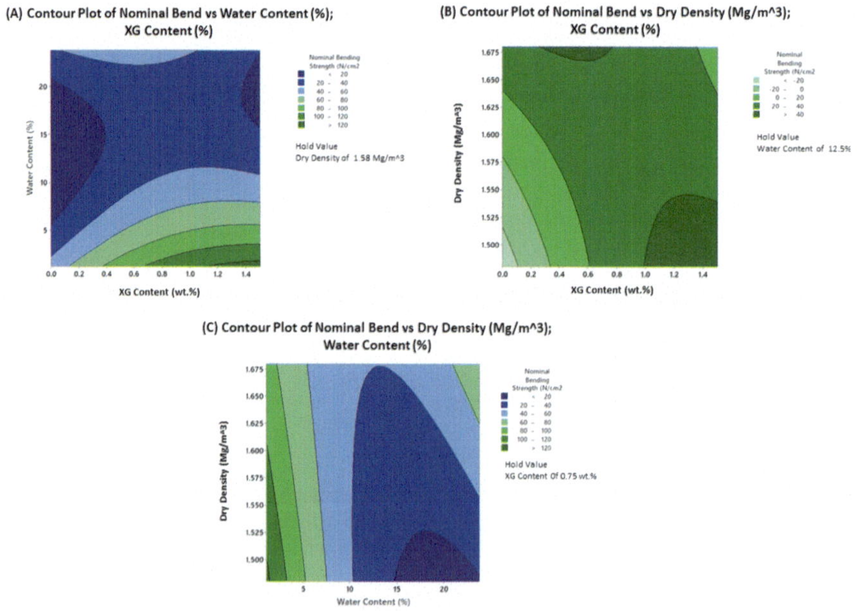

Fig. 6.23 Contour plots variations for the nominal bending strength with **a** water content and XG content (constant dry density of 1.58 Mg/m^3), **b** XG content and dry density (constant water content of 12.5%), and **c** water content and dry density (constant XG content of 0.75%)

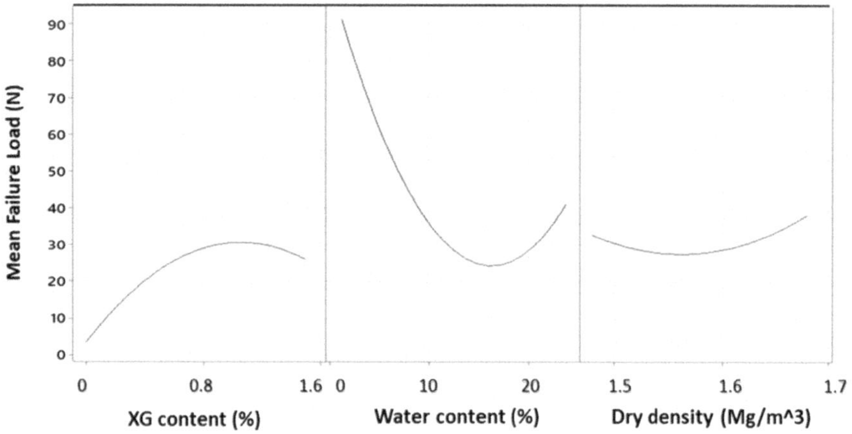

Fig. 6.24 Main effects plots for the failure load with respect to XG content, water content, and dry density, investigating 7-day-cured poly(kaolinite: XG) samples in three-point bending tests

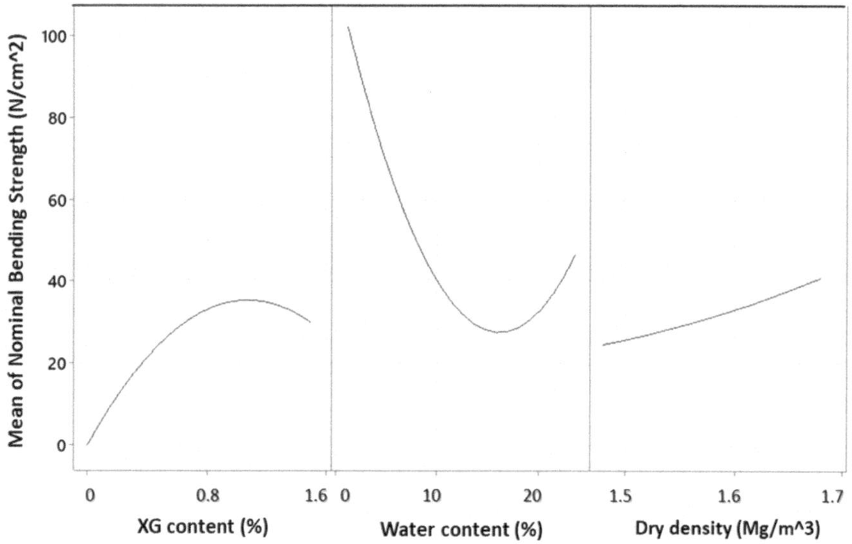

Fig. 6.25 Main effects plots for the nominal bending strength with respect to XG content, water content, and dry density, investigating 7-day-cured poly(kaolinite: XG) samples in three-point bending tests

whereas the dry density has a positive effect at lower XG contents and a negative effect for higher XG contents; (ii) for water content, (a) at low water contents, the XG content interaction has a positive effect, and the dry density has a negative effect, whereas (b) at high water contents, XG content has a positive effect up to 1.4 wt%, followed by a negative effect, and the dry density has a positive effect; and (iii) for dry density, (a) at low dry densities, XG content interaction has a positive effect, and water content interaction has a negative effect up to 20%, which is then followed by a positive effect, (b) at high densities, XG interaction has no effect, and water content has a negative effect up to 12%, followed by a positive effect.

It is to be noted that the failure load and nominal bending strength improvements due to the structural rigidity could be attributed to the increase of the XG content (X_1), which in turn might have led to possible winding of the XG random coil into a stable ordered helix conformation, and consequent increase in intermolecular interactions. This would produce an increase in the hydrodynamic volume, expanding the polymer chain configuration within the kaolinite–XG mixture, thereby increasing the polymer chain stiffness and overall structural rigidity.

6.6 Recommended Testing Protocol for XG-Treated Soils

6.6.1 XG Mixing Methods

XG hydrates in water, irrespective of its temperature (cold or hot). To achieve an optimum XG-treated soil stabilization performance, proper incorporation of XG into the water phase is of vital importance in overcoming the formation of XG lumps and for full dissolution of the XG. These lumps are formed when the XG is added to the water environment all at once, too fast, and using improper mixing methods that imped the complete hydration and the overall effective use of the XG additive. To overcome such difficulties, one would tend to achieve a uniform dispersion of the XG particles for a fast agitation of the hydration medium (e.g., by using a propeller stirrer). In this situation, the bulk XG will be at the upper part of the vortex where the mixing velocity is the greatest. This technique is especially applicable to XG with fine particle sizes, which are difficult to disperse with low shear rate. If, on the other hand, the XG particles are not uniformly dispersed, they might stick together and aggregate, and the fast hydration of the outer XG surface will lead to the formation of a gel layer, thereby preventing the entry of water into the inner XG surfaces that leads to partial hydration. The resulting swollen lumps can only be dissolved with high shear mixing rates and/or prolonged mixing times (Sandford et al. 1981).

To overcome these challenges, it is recommended to use one of the following approaches, depending on the XG form:

(a) *XG as a powder*: Depending on the nature of the added ingredients present in the mixture, dry mixing could be feasible. In this case, the particle sizes of the individual components (e.g., clay-size particles, salts) in the mixture should be similar to avoid their segregation (Sworn 2009), and the XG should have a fine particle size to ensure a homogeneous distribution and spatial separation of the single XG particles, thus facilitating XG hydration upon water addition.
(b) *XG as a slurry in non-aqueous liquids*: Non-aqueous (non-hydrating) liquids are used to enhance the homogeneous distribution of XG particles within the

soil mass. These liquids could be (partly) water-miscible liquids (e.g., glycerol, ethanol), as well as oils, with the following recommended mix proportions: (i) for oil, the XG to liquid ratio is 1:5 to 1:10; and (ii) for water-miscible liquids, the ratio is 1:2 to 1:10. Then, the slurry can be mixed with water, under agitation, for the full hydration of XG.

(c) *XG as a stock paste*: An aqueous stock solution of XG can be prepared with 6% XG concentration and stored long-term until later it can be diluted with water (Sworn 2009). With this method, however, a high shear rate must be applied during the preparation of the stock solution. Notably, for long-term storage of XG solutions, a preservative system is required by using, for instance, benzoates and sorbates.

(d) *XG as an easily dispersible commercial grade*: The most efficient way to ensure fast hydration with low shear rate, and no requirement to modify ingredients in the mixture, is the direct use of an easily dispersible, commercial grade XG, which is a specific manufactured grade of agglomerates with a porous structure, hence allowing for a fast entry of water and prevention of the swollen XG lump formation.

6.6.2 XG Concentration

XG is an effective agent for increasing water viscosity, even at low concentrations. For soil stabilization purposes, the recommended XG concentrations range from 0.2 to 1.0 wt% in relation to the aqueous liquid phase of soil pore fluid. However, if the XG is employed as a spraying solution, a concentration ranging from 0.08 to 0.10 wt% is recommended, depending on nozzle type and operational parameters.

6.6.3 Physicochemical Testing Evaluation

The following recommended tests are proposed in relation to gaining a complete picture for research studies:

(a) For the XG, the basic physical and chemical properties of the XG, including pH, viscosity, molecular mass, solubility, relative density, water absorption, gel rate, variations of moisture content with suction pressure, specific surface area, cation adsorption capacity and FTIR analysis, should be determined prior to its addition to the soil.

(b) For the soil to be treated, in addition to its basic physical and chemical properties, including gradation, Atterberg limits and compaction properties (maximum dry density and optimum moisture content), variation of moisture content with suction pressure, pH, pore fluid analysis, specific surface area, cation exchange capacity, and X-ray diffraction (XRD), FTIR and SEM analyses, should also be determined. Moreover, the soil strength properties (stress–strain response, cohesion, and friction angle), and pore-size distribution should be determined.

(c) For soil–XG mixtures, the same tests recommended for the untreated soil must be considered. In addition, durability in view of potentially exposed environmental conditions, such as cyclic wetting–drying and freezing–thawing actions and leachability with respect to pH variations, must also be evaluated. Moreover, erosion tests should be performed for stabilization of wind-blown sand.

6.7 Summary and Concluding Remarks

XG is a type of hydrocolloid that is commonly used in numerous industries, such as agriculture, food processing, pharmaceutical, cosmetic, and engineering. In recent years, there has been increasing interest in using XG as a soil additive to improve its geotechnical engineering properties.

While it could be hypothesized that dry mixing may facilitate a better initial distribution of the XG particles in the soil mass of uniform grain-size distribution, wet mixing would generally be preferred because it ensures full hydration of XG, with this being more relevant at low soil moisture contents. That is, hydration enables XG to achieve its gel-like conformation, which is critical for enhancing the bonding between soil particles and hence maximizing the strength performance of the stabilized soil.

In regard to the manner by which XG interacts with soils, the literature indicated that for cohesionless soils, XG does not have significant interaction; however, it coats the sand grains and aligns as threads and fibers, and the strength development is highly dependent on the concentration and the conformation of the XG within the coated layer to the sand grains. However, due to dehydration during the curing period, the coated layers would be contracted, become brittle, and act as a hard plastic. Meanwhile for cohesive soils, XG interacts with the soil particles, and, depending on the development of XG concentration-specific conformation, the pore void space arrangements and the overall strength development will be impacted. For example, in the case of montmorillonite-rich clays, XG intercalates between the crystal layers, as well as interacts with the montmorillonite external surfaces. Upon curing, volumetric shrinkage of the amorphous XG will occur; however, the XG remains ductile.

The strength properties of XG-treated soil materials are affected by several factors, including:

(a) *Mixing method*: The mixing method used for producing the soil–XG materials affects the strength properties of the resulting mixture. Overmixing can break down the XG molecules and cause a reduction in the strength properties, while under-mixing may result in non-uniform distribution of the XG, and hence cause weak spots to form in the treated soil mass. The optimal mixing method depends on the specific application and soil type and may involve a combination of mechanical mixing and manual blending. In addition, mixing XG on a dry or wet basis affects the XG hydration, uniformity, and structural development, all of which impact the soil strength development.

(b) *Concentration of XG*: The strength of XG-treated soil increases with increasing XG concentration up to a certain point. This is due to the formation of a more extensive network of XG molecules that bind the soil particles together. However, beyond a certain concentration, adding more XG may not increase the strength of the mixture any further, and may even have adverse effects.
(c) *Soil type*: Different types of soils have different physical and chemical properties that can affect the binding ability of XG and, as a result, affect the degree of compressibility of the soil matrix. Studies have confirmed that the compressibility behavior of XG-treated soils can vary depending on the type of soil. For example, some studies have reported that the compressibility of sandy soils treated with XG decreases significantly, while others have reported less significant effects on clayey soils. The difference in the compressibility of different soil types treated with XG may be due to differences in their gradation, mineralogy, and chemical composition. These factors can affect the interaction between XG and the soil particles, and therefore, the ability of XG to bind the soil particles together and reduce compressibility. Different types of soils may require different XG concentrations and/or application methods to achieve optimal stabilization outcomes (e.g., for minimizing compressibility). Accordingly, it is essential to consider the soil type when using XG for soil stabilization or erosion-control purposes.
(d) *Soil moisture content*: The moisture content is another important factor that influences the strength properties of soil–XG mixtures. XG requires water to form a gel-like network, so a higher moisture content typically results in a stronger material for the dried (cured) soil–XG mixture. However, excessive moisture can reduce the strength properties by promoting soil liquefaction (for sandy soils) or reducing the stability of the XG network. The optimum moisture content for achieving the best soil stabilization outcome(s) depends on the specific application and soil type.
(e) *Soil pore fluid chemistry*: The chemical composition of the fluid within the pore spaces of the soil can affect the interactions between XG and the soil particles, and, as a result, the compressibility of the soil matrix. Furthermore, the impact of pore fluid chemistry on the compressibility of XG-treated soils can depend on the type of soil and the XG concentration used. Therefore, it is important to consider the pore fluid chemistry when using XG for soil stabilization or erosion-control applications.
(f) *Soil pH*: The pH of the soil mixture is an important factor that affects the strength properties of soil–XG materials. XG is generally more effective at neutral to alkaline pH states because acidic conditions can break down the XG molecules and hence reduce the strength properties of the soil–XG mixture.
(g) *Temperature*: The temperatures at which soil–XG mixtures are prepared and cured also affect their strength properties. XG is less effective at lower temperatures because it forms a less stable gel-like network. On the other hand, higher temperatures can accelerate the curing process, but may also affect the stability of the XG network.

(h) *Curing time (e.g., drying state of the soil)*: The strength properties of soil–XG materials increase with increasing drying state of the soil. This is because XG forms a gel-like network that takes time to fully develop, which can range from a few hours to several days, depending on the specific application and soil type. In general, a longer curing (drying) time results in a stronger material, but it may also increase the cost and duration of the project.

(i) *Wetting state*: The strength properties of the soil–XG material decrease with increasing the moisture content of the soil mixture (from its dry state) due to water infiltration (e.g., rain or groundwater movement). Such behavior is extremely important in the context of geo-engineering applications; hence XG stabilization might be suitable only for dry regions (i.e., no rain, with the groundwater level located below the stabilized zones).

In summary, the strength properties of soil–XG materials are influenced by a combination of the above-mentioned factors. Therefore, it is essential to consider the specific application and soil type when optimizing the geo-mechanical properties of soil–XG materials.

References

Antonette J, Donza J, Jaradat KA, Darbari Z, Abdelaziz, SL (2019) Biopolymers. In: ASCE 8th international conference on case histories in geotechnical engineering—Philadelphia, Pennsylvania, Geo-Congress, pp 125–133. https://doi.org/10.1061/9780784482117.011

Arab MG, Mousa RA, Gabr AR, Azam AM, El-Badawy SM, Hassan AF (2019) Resilient behavior of sodium alginate–treated cohesive soils for pavement applications. J Mater Civil Eng 31(1):04018361. https://doi.org/10.1061/(ASCE)MT.1943-5533.0002565

Armistead SJ, Rawlings AE, Smith CC, Staniland SS (2020) Biopolymer stabilization/solidification of soils: a rapid, micro-macro, cross-disciplinary approach. Environ Sci Technol. https://doi.org/10.1021/acs.est.0c02001

Awad YM, Blagodatskaya E, Ok YS, Kuzyakov Y (2013) Effects of polyacrylamide, biopolymer and biochar on the decomposition of 14C-labelled maize residues and on their stabilization in soil aggregates. Eur J Soil Sci 64(4):488–499

Ayeldeen MK, Negm AM, El Sawwaf MA (2016) Evaluating the physical characteristics of biopolymer/soil mixtures. Arabian J Geosci 9:1–13. https://doi.org/10.1007/s12517-016-2366-1

Azimi M, Soltani A, Mirzababaei M, Jaksa MB, Ashwath N (2024) Biopolymer stabilization of clayey soil. J Rock Mech Geotech Eng 16(7):2801–2812. https://doi.org/10.1016/j.jrmge.2023.12.020

Barani OR, Barfar P (2021) Effect of xanthan gum biopolymer on fracture properties of clay. J Mater Civil Eng 33(1). https://doi.org/10.1061/(asce)mt.1943-5533.0003526

Bouazza A, Gates W, Ranjith P (2009) Hydraulic conductivity of biopolymer-treated silty sand. Géotechnique 59:71–72. https://doi.org/10.1680/geot.2007.00137

Cabalar A, Wiszniewski M, Skutnik Z (2017) Effects of xanthan gum biopolymer on the permeability, odometer, unconfined compressive and triaxial shear behavior of a sand. Soil Mech Found Eng 54:356–361. https://doi.org/10.1007/s11204-017-9481-1

Chang I, Im J, Prasidhi AK, Cho G-C (2015) Effects of xanthan gum biopolymer on soil strengthening. Constr Build Mater 74:65–72. https://doi.org/10.1016/j.conbuildmat.2014.10.026

Chang I, Kwon Y-M, Im J, Cho G-C (2019) Soil consistency and interparticle characteristics of xanthan gum biopolymer–containing soils with pore-fluid variation. Can Geotech J 56:1206–1213. https://doi.org/10.1139/cgj-2018-0254
Chang I, Lee M, Tran TPA, Lee S, Kwon Y-M, Im J, Cho G-C (2020) Review on biopolymer-based soil treatment (BPST) technology in geotechnical engineering practices. Transp Geotech 24:100385. https://doi.org/10.1016/j.trgeo.2020.100385
Durga S, Rasheed JA (2020) Effect of xanthan gum on unconfined compressive strength of clayey soil. Int J Res Appl Sci Eng Technol—IJRASET 8(1):753–754. https://doi.org/10.22214/ijraset.2020.1130
Gregory J, Barany S (2011) Adsorption and flocculation by polymers and polymer mixtures. Adv Colloid Interface Sci 169(1):1–12
Hamidi A, Hooresfand M (2013) Effect of fiber reinforcement on triaxial shear behavior of cement treated sand. Geotext Geomembr 36:1–9. https://doi.org/10.1016/j.geotexmem.2012.10.005
Hamza M, Nie Z, Aziz M, Ijaz N, Ameer MF, Ijaz Z (2023a) Geotechnical properties of problematic expansive subgrade stabilized with xanthan gum biopolymer. Road Mater Pavement Des 24(7):1869–1883. https://doi.org/10.1080/14680629.2022.2092027
Hamza M, Nie Z, Aziz M, Ijaz N, Akram O, Fang C, Ghani MU, Ijaz Z, Noshin S, Madni MF (2023b) Geotechnical behavior of high-plastic clays treated with biopolymer: macro–micro-study. Environ Earth Sci 82:91. https://doi.org/10.1007/s12665-023-10760-2
Holtz WG, Gibbs HJ (1956) Engineering properties of expansive clays. Trans Am Soc Civ Eng 121:641–663. https://doi.org/10.1061/TACEAT.0007325
Hou CT, Barnabe N, Greaney K (1986) Biodegradation of Xanthan by salt-tolerant aerobic microorganisms. J Ind Microbiol 1(1):31–37
Im J, Tran ATP, Chang I, Cho G-C (2017) Dynamic properties of gel-type biopolymer treated sands evaluated by Resonant Column (RC) tests. Geomech Eng 12:815–830. https://doi.org/10.12989/gae.2017.12.5.815
Jiang T, Zhao J-D, Zhang J-R (2022) Splitting tensile strength and microstructure of xanthan gum-treated loess. Sci Rep 12:9921. https://doi.org/10.1038/s41598-022-14058-4
Joga, JR, Varaprasad BJS (2019) Sustainable improvement of expansive clays using xanthan gum as a biopolymer. Civil Eng J 5(9):1893–1903. https://doi.org/10.28991/cej-2019-03091380
Joga JR, Varaprasad BJS (2019) Effect of xanthan gum biopolymer on dispersive properties of soils. World J Eng 17(4):563–571. https://doi.org/10.1108/WJE-05-2020-0152
Katzbauer B (1998) Properties and applications of xanthan gum. Polym Degrad Stab 59(1–3):81–84
Khachatoorian R, Petrisor IG, Kwan C-C, Yen TF (2003) Biopolymer plugging effect: laboratory-pressurized pumping flow studies. J Petrol Sci Eng 38:13–21. https://doi.org/10.1016/S0920-4105(03)00019-6
Kwon Y-M, Chang I, Cho GC (2023) Consolidation and swelling behavior of kaolinite clay containing xanthan gum biopolymer. Acta Geotech 18:3555–3571. https://doi.org/10.1007/s11440-023-01794-8;0123456789
Kwon YM, Ham SM, Kwon TH, Cho GC, Chang I (2020) Surface-erosion behavior of biopolymer-treated soils assessed by EFA. Géotech Lett 10:1–7. https://doi.org/10.1680/jgele.19.00106
Labille J, Thomas F, Milas M, Vanhaverbeke C (2004) Flocculation of colloidal clay by bacterial polysaccharides: effect of macromolecule charge and structure. J Colloid Interface Sci 284(1):149–156
Latifi N, Horpibulsuk S, Meehan CL, Majid MZA, Tahir MM, Mohamad ET (2016) Improvement of problematic soils with biopolymer—an environmentally friendly soil stabilizer. J Mater Civ Eng 29:04016204. https://doi.org/10.1061/(ASCE)MT.1943-5533.0001706
Lee S, Chang I, Chung MK, Kim Y, Kee J (2017) Geotechnical shear behavior of xanthan gum biopolymer treated sand from direct shear testing. Geomech Eng 12:831–847. https://doi.org/10.12989/gae.2017.12.5.831
Lee S, Im J, Cho G-C, Chang I (2019) Laboratory triaxial test behavior of xanthan gum biopolymer-treated sands. Geomech Eng 17:445–452. https://doi.org/10.12989/gae.2019.17.5.445

Lee M, Im J, Cho GC, Ryu HH, Chang I (2021) Interfacial shearing behavior along xanthan gum biopolymer-treated sand and solid interfaces and its meaning in Geotechnical engineering aspects. Appl Sci 11:139. https://doi.org/10.3390/app11010139

McCormack DE (1976) Foundations on expansive soils. Soil Sci Soc Am J 40:viii–viii. https://doi.org/10.2136/sssaj1976.03615995004000030011x

Mendonça A, Morais PV, Pires AC, Chung AP, Oliveira PV (2021) A review on the importance of microbial biopolymers such as xanthan gum to improve soil properties. Appl Sci 11(1):170. https://doi.org/10.3390/app11010170

Mohamed AMO, Paleologos EK (2018) Fundamentals of geo-environmental engineering: understanding soil, water, and pollutant interaction and transport. Elsevier, USA, imprint: Butterworth-Heinemann, eBook ISBN: 9780128051450; Paperback ISBN: 9780128048306; 708p

Muguda S, Booth SJ, Hughes PN, Augarde CE, Perlot C, Bruno AW, Gallipoli D (2017) Mechanical properties of biopolymer-stabilized soil-based construction materials. Géotech Lett 7(4):309–314. https://doi.org/10.1680/jgele.17.00081

Nugent RA, Zhang G, Gambrell RP (2009) Effect of exopolymers on the liquid limit of clays and its engineering implications. Transport Res Rec: J Transport Res Board 2101(1):34–43

Qureshi MU, Chang I, Al-Sadarani K (2017) Strength and durability characteristics of biopolymer-treated desert sand. Geomech Eng 12:785–801. https://doi.org/10.12989/gae.2017.12.5.785

Sandford PA, Baird J, Cottrell IW (1981) Xanthan gum with improved dispersibility. In Brant DA (ed) Solution properties of polysaccharides. ACS Symposium Series. American Chemical Society, Washington, pp 31–41

Sawant A, Patil A, Thonge R, Trankatwar V, Banne S (2019) Effect of xanthan gum on shear strength parameters of laterite soil in Konkan Region. IOSR J Mech Civil Eng 16(3):24–29. https://doi.org/10.9790/1684-1603012429

Singh SP, Das R (2019) Geo-engineering properties of expansive soil treated with xanthan gum biopolymer. Geomech Geoeng. https://doi.org/10.1080/17486025.2019.1632495

Soltani A, Raeesi R, Taheri A, Deng A, Mirzababaei M (2021) Improved shear strength performance of compacted rubberized clays treated with sodium alginate biopolymer. Polymers 13(5):764. https://doi.org/10.3390/polym13050764

Song KW, Kuk HY, Chang GS (2006) Rheology of concentrated xanthan gum solutions: oscillatory shear flow behavior. Korea-Australia Rheology J. 18(2):67–81

Sujatha ER, Saisree S (2019) Geotechnical behavior of guar gum-treated soil. Soils Found 59(6):2155–2166. https://doi.org/10.1016/j.sandf.2019.11.012

Sun C, Gunasekaran S (2009) Effects of protein concentration and oil-phase volume fraction on the stability and rheology of menhaden oil-in-water emulsions stabilized by whey protein isolate with Xanthan gum. Food Hydrocolloids 23(1):165–174

Sworn G (2009) Xanthan gum. In Phillips GO, Williams PA (eds) Handbook of hydrocolloids. Woodhead Publishing Limited, Cambridge, pp 186–203

Zhang J, Meng Z, Jiang T, Wang S, Zhao J, Zhao X (2022) Experimental study on the shear strength of silt treated by xanthan gum during the wetting process. Appl Sci 12:6053. https://doi.org/10.3390/app12126053

Zohuriaan MJ, Shokrolahi F (2004) Thermal studies on natural and modified gums. Polym Testing 23(5):575–579

Chapter 7
Xanthan Gum (XG) Sorption and Hydraulic Conductivity of XG-Treated Soils

7.1 Introduction

As discussed in Chaps. 4, 5, and 6, properties of xanthan gum (XG), soil, and soil solution influence adsorption and aggregative success of XG–soil structure. Molecular weight, size, conformation, and surface charge are all aspects of the XG that affect its sorption onto the soil particles. XG interacts more with the clay fraction of soils; as such, the particle surface area, type and amount of clay, soil structure, and pore-size distribution are soil attributes that influence the nature and level of XG sorption onto clay particle surfaces. The pH level, ionic strength, and electrolyte concentration of the soil solution also influence the sorption of XG onto clays.

In addition, as highlighted in Chap. 1, the reported hydraulic conductivities for XG-stabilized soils are inconclusive. For example, sand and silty soils stabilized/treated with 2% XG have shown a reduction of about 10% in the hydraulic conductivity (HC) from the untreated soils (Ayeldeen et al. 2016). However, the results of a mixture of XG (0.5 to 3%) with sand–bentonite have shown a slight increase in the HC with increasing XG content (Biju and Arnepalli 2020), highlighting the role of the fines content (i.e., silt and clay particles) in the soil composition, as well as the XG concentration, on the effectiveness/outcome of XG treatment.

Furthermore, in a study by Khachatoorian et al. (2003), a decrease in the HC of XG-treated Ottawa sand (0.43–0.6 mm particle-size range) over a curing period of 11 days was reported; however, the results reported in Cabalar et al. (2017) for XG-treated Narli sand (0.075–1.0 mm size range) did not show a clear tendency over the same curing duration. Furthermore, Mendonça et al. (2021) reported a slight decrease in the HC of 0.33% XG-treated silty sand during the first 14 days of curing, followed by an increase in HC for longer curing durations. Moreover, the gas permeability of a 3% XG-stabilized kaolin clay decreased by up to two orders of magnitude compared to that of the untreated clay (Ng et al. 2020).

A. M. O. Mohamed et al., *Sustainability in Ground Improvement: The Case of Xanthan Gum Biopolymer*, Green Energy and Technology,
https://doi.org/10.1007/978-3-031-75313-8_7

Therefore, quantifying the water flow characteristics and soil aggregation with XG treatment is of prime importance to determine the potential benefits of XG in increasing soil stability.

7.2 XG Sorption by Clay Minerals

7.2.1 Physicochemical Characteristics of Selected Clay Minerals

In a study conducted by Dontsova and Bigham (2005), sorption experiments were performed to assess the effects of polysaccharide concentration, pH, electrolyte concentration, and cation type on sorption of XG by different clay minerals. Figure 7.1 shows the changes of cation exchange capacity (CEC) and the total surface and external surface areas of the investigated clay minerals.

In these experiments, the samples were prepared as follow. Subsamples of the clays (0.15 g) were ultrasonically dispersed in water or a salt solution, adjusted to the desired pH level using 0.1 mol/l HNO_3 or saturated $Ca(OH)_2$ solutions, and mixed with the appropriate XG solution to form 40 ml of a final suspension with 3.75 g/l of clay. After 12-h equilibration on a reciprocal shaker (100 rpm) at room temperature (18 °C), the clay–XG suspensions were centrifuged for 30 min, and the total organic carbon (TOC) in the supernatant, at equilibrium, was measured.

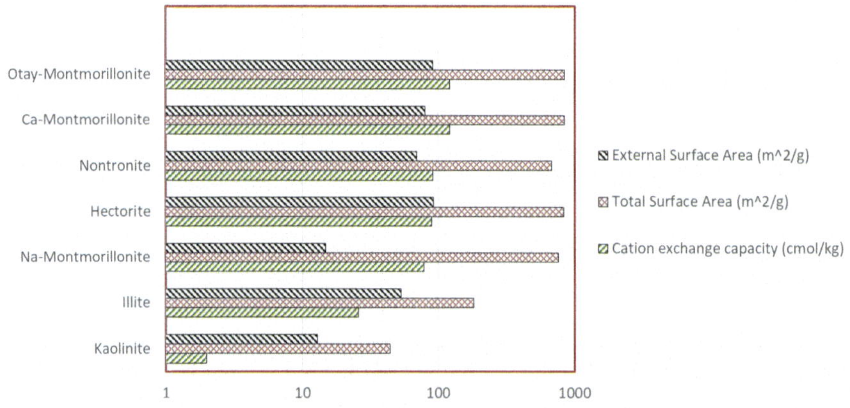

Fig. 7.1 Changes of cation exchange capacity (CEC) and the total surface and external surface areas of some selected clay minerals (data from Dontsova and Bigham (2005))

7.2.2 XG Sorption Equilibrium Time

The equilibration time (of 12 h) exceeded the actual time needed to reach equilibrium, which has been reported to be less than 1 h (Moavad et al. 1974; Guckert et al. 1975; Wen 2002). The quantity of XG sorbed was determined as the difference between the concentration of TOC in the supernatant liquid of the samples with and without clay. Control measurements (without clay) showed the TOC of XG in the supernatant to vary between 42% without electrolyte, and 31% in the presence of electrolyte.

7.2.3 XG Concentration Effect on Clay Minerals Sorption

Figure 7.2a illustrates the gradual increase in XG sorption behavior, suggesting the formation of multiple layers of polysaccharide on the clay particle surfaces, which results in the sorption of XG in quantities that exceed the available surface areas of the clay particles (Parfitt and Greenland 1970; Olness and Clapp 1973; Moavad et al. 1974). Whereas, for simple uncharged sugars that assume a flat configuration with sorption on montmorillonite (Greenland 1956), XG tends to retain its helical tertiary structure (Chenu et al. 1987). This structure results in an uncharged core ringed with charged groups that can enter into reactions with other compounds in solution, or with charged sites on the clay particle surfaces by bridging with multivalent cations.

The maximum XG sorption values by kaolinite, illite, and montmorillonite clay minerals with (i) 0 mmol/l $Ca(NO_3)_2$ were estimated as 3.5, 8, and 10 g/kg, respectively (considering 1 g/kg TOC is equivalent to 2.5 g/kg XG), and with (ii) 10 mmol/l $Ca(NO_3)_2$ as 9.75, 18.75, and 27.5 g/kg, respectively. These results agree with other studies, as discussed below:

(a) Sorption by Ca-exchanged sodium montmorillonite was about 2 g/kg TOC (5 g/kg XG) under neutral conditions from a 100 mg/l XG solution in 1 mmol/l $Ca(NO_3)_2$, and 9 g/kg TOC (22.5 g/kg XG) from a 200 mg/l XG solution in 10 mmol/l $CaCl_2$ (Dontsova and Bigham 2005).
(b) Sorption by calcium montmorillonite was about 4.8 g/kg of XG from a 50 mg/l solution and 27.8 g/kg from a 250 mg/l solution with a background of 1 mmol/l $CaCl_2$ (Clapp and Emerson 1972).
(c) Sorption of 20 and 30 g/kg from anionic polysaccharides by sodium montmorillonite (Finch et al. 1967; Labille et al. 2003).
(d) Sorption of 90 to 150 g/kg from anionic polysaccharides by montmorillonite (Finch et al. 1967).
(e) Sorption of up to 115 g/kg from other anionic polysaccharides by calcium montmorillonite (Labille et al. 2003).

Although the amounts of polysaccharide sorbed in the study by Dontsova and Bigham (2005) were relatively small, they would likely still contribute to the stabilization of soil structure. For instance, Clapp and Emerson (1972) showed that the

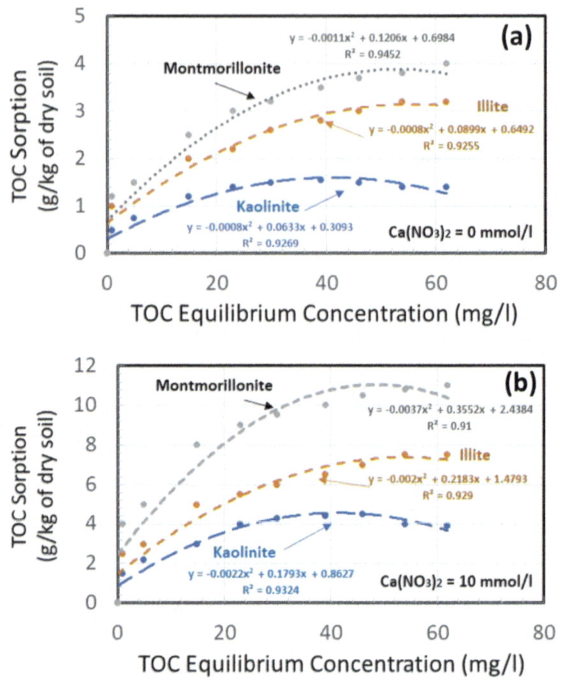

Fig. 7.2 Sorption of XG (represented as the total organic carbon, TOC) by kaolinite, illite, and montmorillonite clay minerals at room temperature (18 °C), as a function of XG concentration (0–200 mg/l) at: **a** pH 7 and zero $Ca(NO_3)_2$ concentration; and **b** pH 4 and 10 mmol/l of $Ca(NO_3)_2$ concentration (data from Dontsova and Bigham (2005))

sorption of as little as 1–2 g/kg of XG was enough to prevent dispersion of montmorillonite. In other studies, maximum aggregate stability was achieved at 10 g/kg for neutral polysaccharides (Chenu et al. 1987) and 20 g/kg for anionic polysaccharides (Labille et al. 2003). This means that 0.1–0.2 wt% XG would be enough to prevent montmorillonite dispersion, while 1.0 wt% XG would be required for achieving an optimal XG–montmorillonite aggregate stability.

7.2.4 Electrolyte Concentration Effect on XG Sorption by Clay Minerals

Referring to Fig. 7.2b (data from Dontsova and Bigham (2005)), as the $Ca(NO_3)_2$ concentration increased (from 0 to 10 mmol/l), the soil's sorption of XG was increased. At pH 4 (Fig. 7.2b), there was an approximate 2.3- to 6.5-fold increase in sorption with electrolyte. At pH 7 (Fig. 7.2a), the relative increase in sorption of XG in the presence of $Ca(NO_3)_2$ was also significant, and even greater than that obtained at pH 4 (i.e., approximately 4.5- to 141.5-fold) (Dontsova and Bigham 2005). The positive response to electrolyte agrees with the behavior described for anionic polymers by Theng (1982), who attributed this behavior to screening of the charge on the polyanion, with a resulting decrease in electrostatic repulsion between

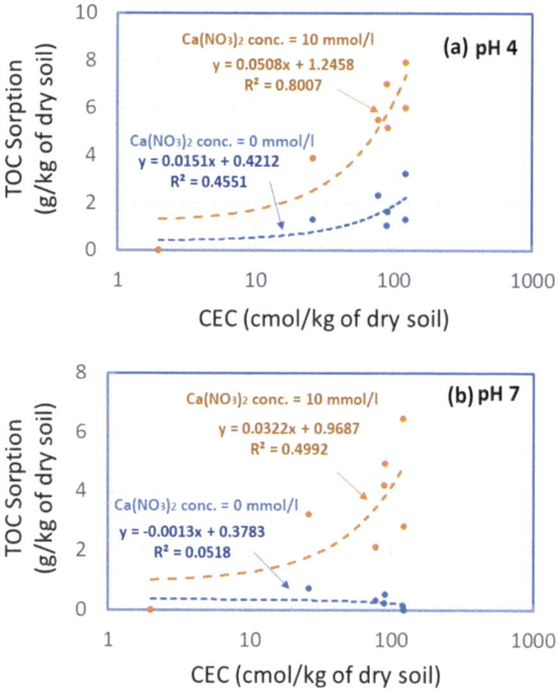

Fig. 7.3 Sorption of XG (represented as total organic carbon, TOC) by clay minerals (expressed in terms of cation exchange capacity, CEC), as functions of $Ca(NO_3)_2$ concentration and for solution pH of 4 and 7 (data from Dontsova and Bigham (2005))

XG surfaces. The results also indicated that sorption of XG by clay soils increased with the increase of CEC (Fig. 7.3), total surface area (Fig. 7.4) and external surface area (Fig. 7.5).

7.2.5 Electrolyte Type Effect on XG Sorption by Clay Minerals

Dontsova and Bigham (2005) reported on the sorption of XG by calcium montmorillonite from 100 mg/l solutions, using nitrate salts of various monovalent (K^+, Na^+, and Li^+) and divalent (Sr^{2+}, Ca^{2+}, and Mg^{2+}) cations as background electrolyte. They demonstrated that an increase in ionic strength from 0 to 30 mmol/l produced an increase in sorption, approaching a plateau in the upper concentration range. However, at the same ionic strength, divalent cations enhanced sorption to a greater degree than monovalent cations. The sorption of negatively charged XG by montmorillonite can be attributed to cationic bridging occurring due to the association between exchangeable cations and carboxylic groups of the polysaccharides (Parfitt 1972; Soltani et al. 2021; Azimi et al. 2024).

Previous research has shown a decrease in sorption of anionic polymers with an increase in cation size and electronegativity for monovalent cations (Mortensen 1962;

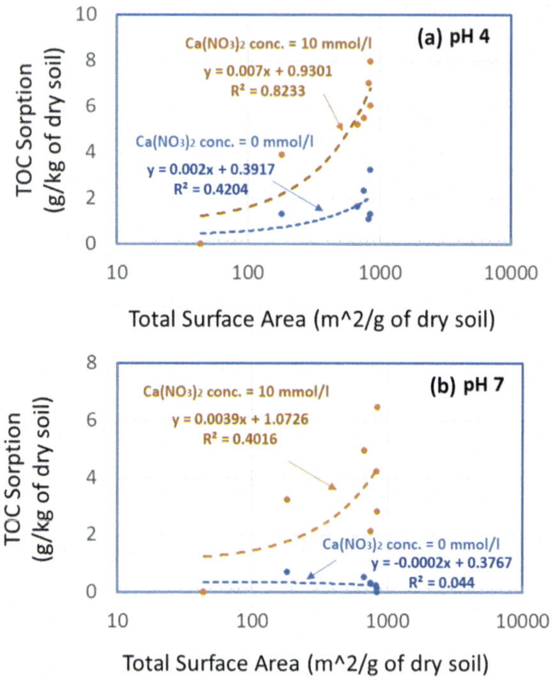

Fig. 7.4 Sorption of XG (represented as total organic carbon, TOC) by selected clay minerals (expressed in terms of total surface area), as functions of $Ca(NO_3)_2$ concentration and for solution pH of 4 and 7 (data from Dontsova and Bigham (2005))

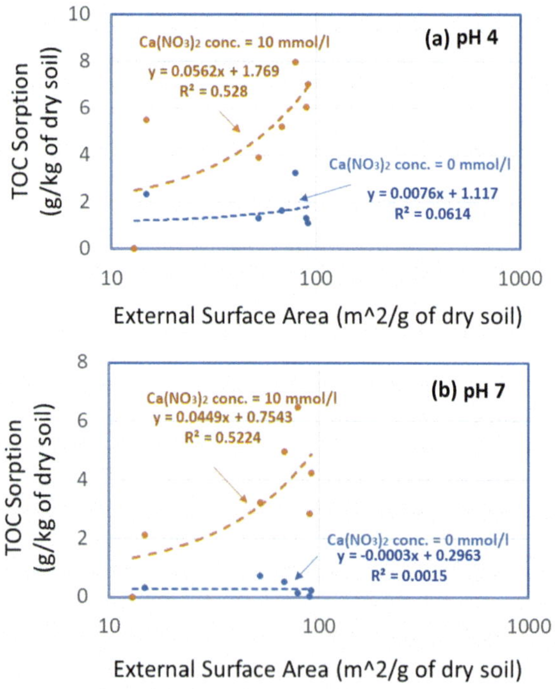

Fig. 7.5 Sorption of XG (represented as total organic carbon, TOC) by selected clay minerals (expressed in terms of external surface area), as functions of $Ca(NO_3)_2$ concentration and for solution pH of 4 and 7 (data from Dontsova and Bigham (2005))

Theng and Scharpenseel 1975), and an increase in sorption with a decrease in cation size for multivalent cations (Mortensen 1962; Theng and Scharpenseel 1975; Theng 1976). This behavior indicates that polyanions can remove water from monovalent cations and bind directly to the exchangeable cation, whereas with divalent cations, binding takes place through the water of hydration.

7.2.6 pH Effects on XG Sorption by Clay Minerals

Total sorption of XG was greater at pH 4 than at pH 7 for all clay minerals (see Figs. 7.3, 7.4, and 7.5). A 1.1- to 2.6-fold difference in sorption was observed between pH 4 and 7 in 10 mmol/l $Ca(NO_3)_2$. Such behavior was consistent with an increase in the negative charge of both the XG and clay particles as the pH increased and variable charge sites dissociated. Finch et al. (1967) showed that a decrease in the pH level from pH 7 to 4.4 reduced the electrophoretic mobility of soil–polysaccharide systems by a factor of about 10.

7.3 Influence of Polysaccharides on Flocculation/ Aggregation of Clay Minerals

7.3.1 Montmorillonite

Labille et al. (2005) evaluated the molecular properties of six different polysaccharides (Dextran, MWAP71, Rhamsan, XG, RMDP71 and YAS34) to determine their ability to flocculate diluted montmorillonite suspensions and compared them with succinoglycan in terms of aggregation rate and equilibrium aggregate size. The molecular structure and characteristics of these polysaccharides are presented in Fig. 7.6 and Table 7.1.

The experimentally determined flocculation rate and the final floc sizes at t = 10,000 s (i.e., 2.78 h) for various acidic polysaccharides are presented in Fig. 7.7. There are clear differences in both parameters for the seven investigated polysaccharides.

To explain these changes in terms of the polysaccharides' physicochemical parameters (Table 7.1), Fig. 7.8 shows the relationships between the final floc sizes and the following parameters: (i) molecular weight (M_w), molecular weight of repeat unit (M_u), mass per unit length (M_L) and mass per equivalent (M_q) (see Fig. 7.8a); (ii) contour length (L_c), persistence length (L_p) and radius of gyration (R_g) (see Fig. 7.8b), and (iii) surface coating per milligram (S), defined as $S = N\ (2 \times R_g)^2$, where N is the fiber number concentration per milligram (see Fig. 7.8c). However, no clear relationships are evident. Hence, as suggested by Labille et al. (2005), other structural parameters should be considered, such as the location of the charges in

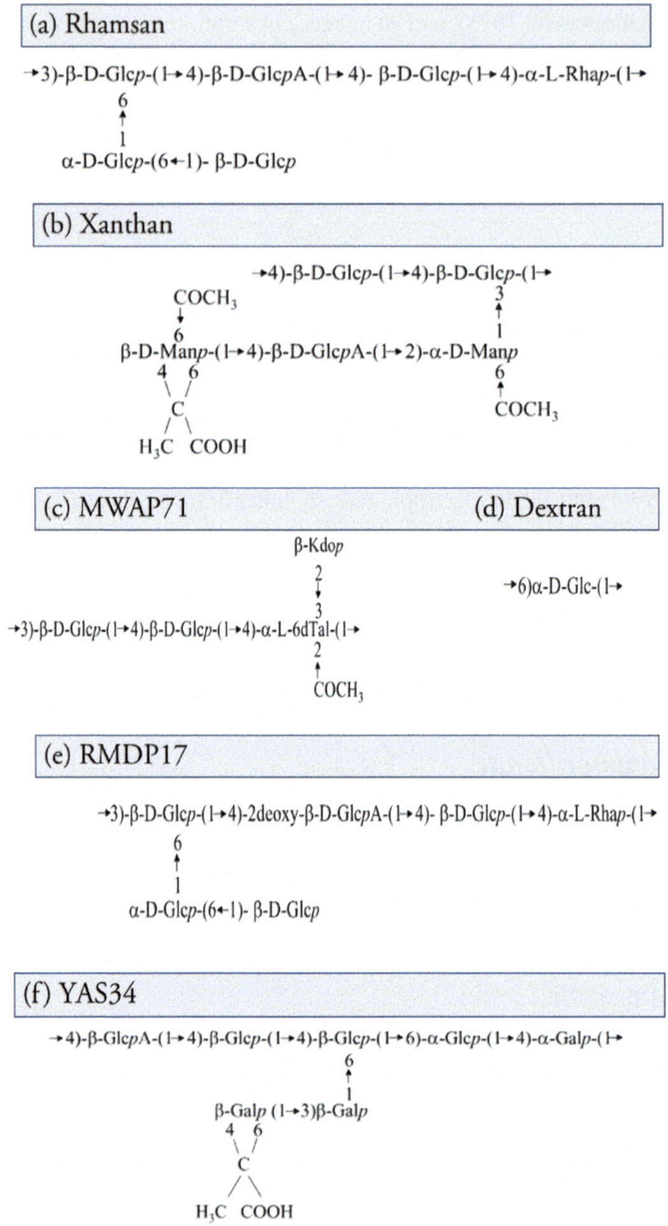

Fig. 7.6 Molecular structure of select polysaccharides: **a** Rhamsan; **b** Xanthan; **c** MWAP71; **d** Dextran; **e** RMD17; **f** YAS34; and **g** Succinoglycan

(g) Succinoglycan

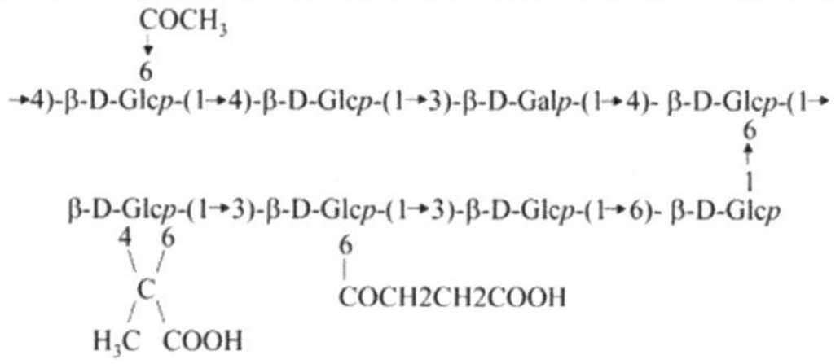

Fig. 7.6 (continued)

the macromolecule structure, whereby the acidic moieties located on the macromolecule sidechains play a major role in the ionic interactions with clay, as their presence significantly increases the kinetics of bridging flocculation.

In addition, acidic moieties located in the macromolecule backbone may also contribute, to a certain extent, to the attractive ionic interactions with coagulated clay. This effect of charge location is probably due to its different exposure in the macromolecule, as acidic sidechains are more accessible to clay particles than acidic moieties located in the innermost macromolecule backbones. Notably, since XG exhibits a very short di-saccharidic backbone repeat unit, and tri-saccharidic sidechains chains, it has a high linear mass of 2000 g/nm. Also, because the anionic sidechains are very close to each other, it has a high charge density ($\lambda = 2.7$) and strong electrostatic repulsions; hence, it is known to have a very rigid conformation ($L_p = 120$ nm) that does not favor bridging formation.

7.3.2 *Kaolinite*

A study by Nugent et al. (2009) investigated the effects of XG on the aggregation behavior of stabilized kaolinite–XG mixtures. Figure 7.9 shows the changes of the liquid limit (LL) as functions of XG concentration, XG viscosity and solution ionic concentration. In the case of deionized water, the LL exhibited a rise–fall response with increasing XG concentration, peaking at 0.5 wt% XG, such that beyond 2–3 wt% XG, the stabilized mixture produced lower LL values than the untreated soil.

The peak LL for 0.5 wt% XG is caused by an increase of the pore fluid viscosity and possible flocculation of kaolinite–XG structure. However, for concentrations greater than 0.5 wt% and less than 5.0 wt% XG, the decrease in the LL caused by the XG aggregating the kaolinite particles was stronger than the increase caused by viscosity, as shown from scanning electron microscopy (SEM) images obtained for the kaolinite at its LL state for (i) deionized water and (ii) 1.0 wt% XG solution (Fig. 7.10). The SEM image for the XG–treated kaolinite revealed significant aggregation compared with the untreated kaolinite. The strands of XG provide the mechanism for aggregation, binding multiple clay particles together.

Table 7.1 Physical properties of different polysaccharides and their experimentally determined flocculation parameters (data from Labille et al. (2005))

Polysaccharide	Dextran	MWAP71	Rhamsan	XG	RMDP71	YAS34	Succinoglycan
M_w	2.00E+06	5.70E+05	3.00E+06	9.00E+06	3.00E+06	2.00E+06	5.28E+06
M_u	1.80E+02	7.54E+02	1.09E+03	1.08E+03	1.08E+03	1.35E+03	1.68E+03
M_L	375	520	1094	2000	1080	1080	840
L_c	5.56E+03	1.10E+03	2.74E+03	4.10E+03	2.78E+03	1.85E+03	6.31E+03
λ	0.00E+00	4.90E–01	7.00E–01	2.70E+00	7.00E–01	1.10E+00	7.00E–01
M_q		7.54E+02	1.09E+03	5.40E+02	1.08E+03	6.74E+02	8.40E+02
L_P	1.00E+00	9.50E+00	8.00E+01	1.20E+02	8.00E+01	3.10E+01	3.30E+01
R_g	4.40E+01	5.49E+01	2.93E+02	5.07E+02	2.95E+02	1.44E+02	3.00E+02
N	3.01E+14	1.07E+15	2.01E+14	6.69E+13	3.01E+14	3.01E+14	1.14E+14
S	2.30E+00	1.29E+01	6.88E+01	6.88E+01	6.97E+01	2.51E+01	4.09E+01
Conformation	Flexible coil	Semi-rigid	Double helix	Double helix	Double helix	Double helix	Single helix

Note M_w, molecular weight (g/mol); M_u, molecular weight of repeat unit (g/mol); M_L, mass per unit length (g/nm); L_c, contour length (nm); λ, charge density; M_q, mass per equivalent (g/eq); L_p, persistence length (nm); R_g, radius of gyration (nm), defined as $R_g^2 \approx 0.08\, M_w \times 1.18\, L_p/(3\, M_L)$; N, fiber number concentration per milligram; S, surface coating per milligram (m^2), defined as $S = N\,(2 \times R_g)^2$; Conformation: spatial conformation in $CaCl_2$ (1.25×10^{-3} M) at 25 °C

Experimentally determined flocculation parameters

Flocculation rates, r (μm/s)	1.4	0.5	1.2	4.3	4.6	11	15.9
Final floc sizes, D_{50f} (μm) @ t = 10,000 s	140	Reversible	260	158	144	213	335

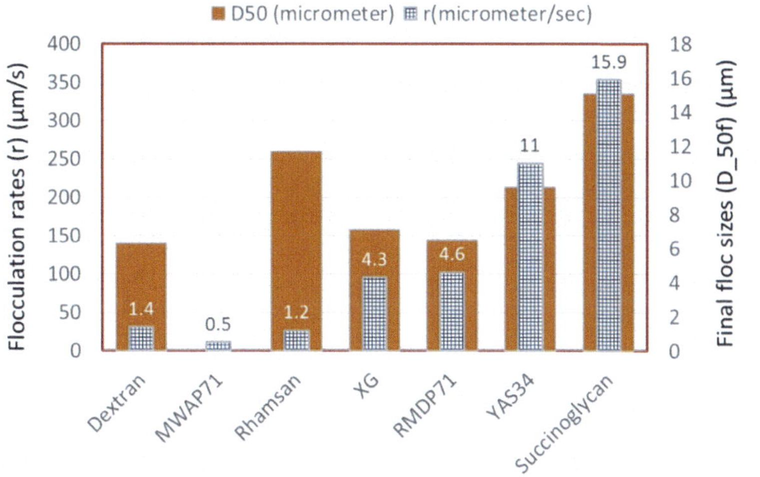

Fig. 7.7 Flocculation rates and final floc sizes at t = 10,000 s measured for seven different polysaccharides (data from Labille et al. (2005))

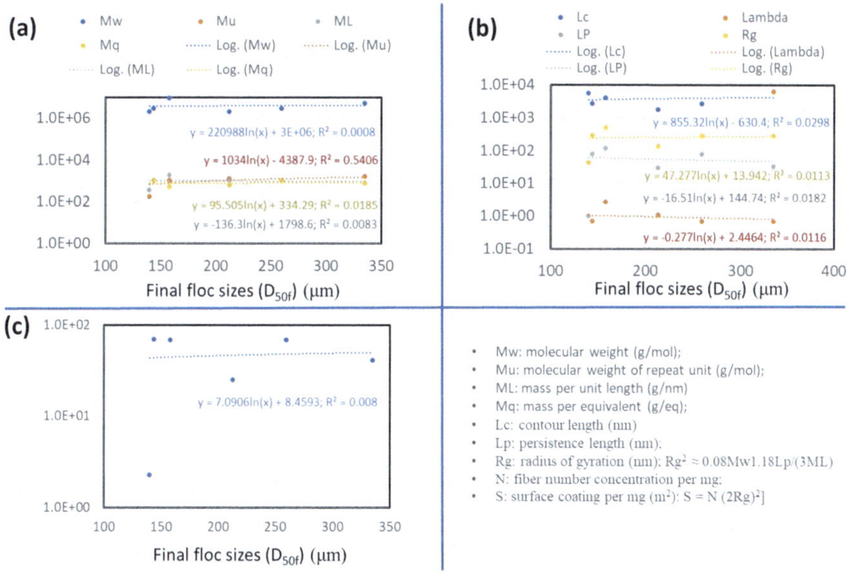

Fig. 7.8 Relationships between the final floc sizes with: **a** molecular weight (M_w), molecular weight of repeat unit (M_u), mass per unit length (M_L) and mass per equivalent (M_q); **b** contour length (L_c), persistence length (L_p), and radius of gyration (R_g); and **c** surface coating per milligram (S, in m^2), defined as $S = N\ (2 \times R_g)^2$, where N is the fiber number concentration per milligram

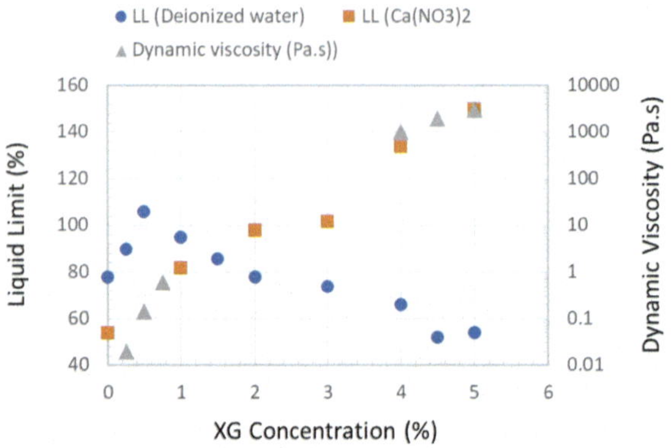

Fig. 7.9 Liquid limit variations with XG concentration and viscosity for XG-treated kaolinite clay (data from Nugent et al. (2009))

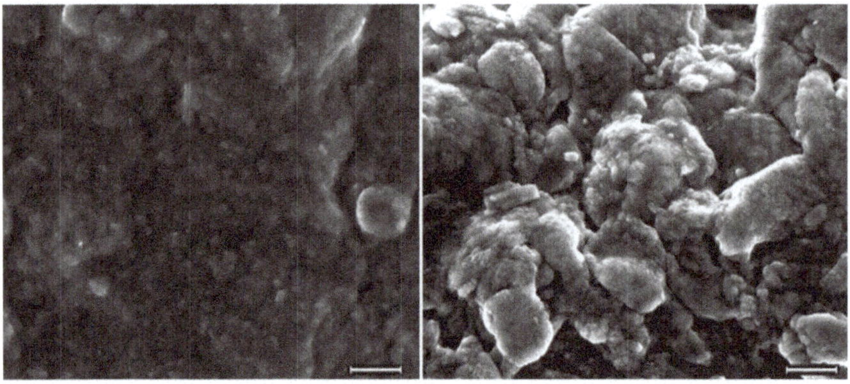

Fig. 7.10 SEM images of kaolinite at the liquid limit state for deionized water (left) and 1.0 wt% XG solution (right) (scale bars 20 μm) (images from Nugent et al. (2009))

When calcium nitrate solution (0.1 M $Ca(NO_3)_2$) was used instead of deionized water to prepare the XG solutions, which were added to the kaolinite for performing the LL testing, a dramatic increase in the LL occurred for XG concentrations of 2–3 wt% and greater (see Fig. 7.9). At 0 wt% XG (i.e., kaolinite clay alone), the addition of $Ca(NO_3)_2$ decreases the LL compared with that of deionized water due to the presence of Ca^{2+} ions in the pore fluid, which increases the solution ionic strength and reduces the thickness of the electrical double layer on kaolinite particle surfaces. At 1.0 wt% XG, the LL for the $Ca(NO_3)_2$ solution is still smaller than for deionized water because, even though XG polymer strand cross-linking occurs in the mixture, the degree of cross-linking was not significant enough to compensate for the reduction caused by the diffused double layer thickness. Increases in the LL produced through cross-linking suggest that clays containing natural polysaccharides can be further strengthened using common cross-linkers, such as calcium. Notably, this behavior, with the use of cross-linker and XG, is contrary to the known behavior of clays, i.e.,

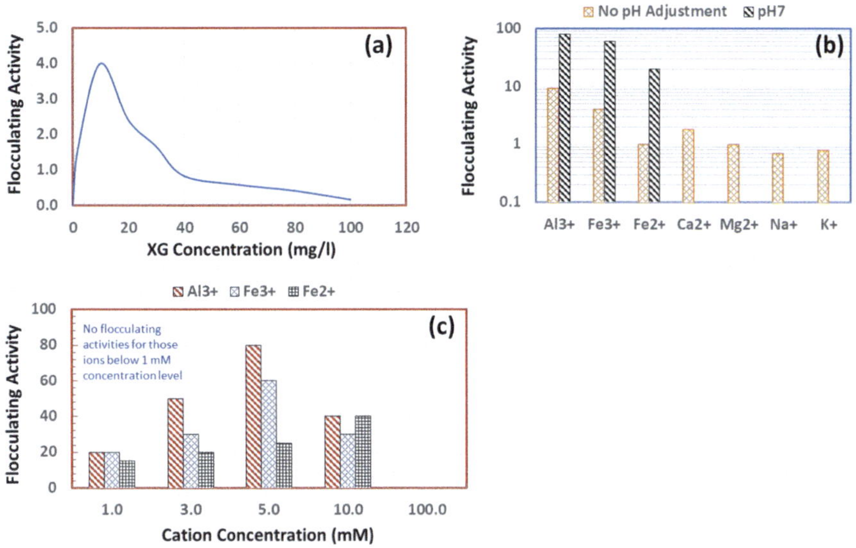

Fig. 7.11 Flocculating activities of XG-treated kaolinite suspension: effects of **a** XG concentration; **b** ionic species in the solution; and **c** cation concentrations at pH 7 (data from Yokoi et al. (1996))

as clay particles aggregate, the specific surface area of the clay decreases, and less liquid is required to fully hydrate the surfaces of the soil particles.

In another study by Yokoi et al. (1996), the flocculation properties of XG-treated kaolinite suspension were evaluated by calculating the mixture's optical density at 550 nm. Figure 7.11a shows the effects of XG concentration on the flocculating activity in a kaolinite suspension, without cationic species in the solution. An optimum XG concentration for its flocculating activity was recorded at 10.0 mg/l.

Figure 7.11b illustrates the effects of adding cationic species (5 mM) on the flocculating activity of XG. Solutions of $AlCl_3.6H_2O$, $FeCl_3.6H_2O$, $CaCl_2.2H_2O$, $MgCl_2.6H_2O$, $FeSO_4.7H_2O$, NaCl, and KCl were used as cation sources. The flocculating activity was enhanced by the addition of trivalent cations ($A1^{3+}$ or Fe^{3+}) and decreased by the additions of both divalent and monovalent cations under a condition without pH adjustment of the mixtures of the kaolinite suspension and cations (Fig. 7.11b). The trivalent cations enhanced flocculation activity by neutralizing and stabilizing the residual negative charges of the carboxyl groups of galacturonic acid and pyruvic acid in the XG molecules; thereby, forming bridges that bind the kaolinite particles to each other. However, at pH 7, which is the optimum pH for XG flocculating activity considering those metal ions, the flocculating activity was further increased by adding Al^{3+}, Fe^{3+} and Fe^{2+} (Fig. 7.11b). Based on metal ion speciation (i.e., metal ion species), at neutral pH, $A1^{3+}$, Fe^{3+} and Fe^{2+} are present as hydroxides, such as $Al(OH)_3$, $Fe(OH)_3$ and $Fe(OH)_2$, which might be effective in enhancing the flocculating activity of XG.

Figure 7.11c demonstrates the effects of cation concentrations on the flocculating activity of XG in a kaolinite suspension under a pH condition of 7.0. The optimum concentrations of Al^{3+}, Fe^{3+} and Fe^{2+} for the flocculating activity were 5, 5, and 10 mM, respectively. Notably, (i) there were no activities at 50 mM of these cations; (ii) the optimum reaction temperature was about 40–50 °C; and (iii) flocculating activity was stable at less than 80 °C (Yokoi et al. 1996).

7.3.3 Natural Sediments

Soil aggregate is defined as a "*naturally occurring cluster or group of soil particles in which the forces holding the particles together are much stronger than the forces between adjacent aggregates*" (Martin 1955). The forces that bond soil particles together are physical, chemical, and/or biological in nature. Aggregation affects soil physical properties, including, for instance, the soil density, stability, and structure, which in turn influence the movement of fluids, solutes, and heat through the soil medium (Warkentin 2008).

The three physical forces leading to aggregate formation are compressive, tensile, and shear forces. Compressive forces push soil particles together to create aggregates, whereas tensile forces pull soil particles apart to break aggregates. Shear forces act in a direction parallel to the plane of soil particles and consequently displace aggregates. Within aggregates, surface tension, microbial byproducts, and electrostatic interactions hold soil particles together (Frey 2005). Soil aggregates have an optimum size that produce maximum stability. When soil aggregates become larger that their optimum size, greater porosity causes destabilization, because the physical forces acting on the particle surfaces break the aggregates apart (Frey 2005).

When physical forces bring particles and aggregates together, biological activity causes rearrangement of the soil structure. Cycles of particle stabilization take place with bacteria (Feeney et al. 2006). An increase in bacterial activity leads to organic matter decomposition, which in turn weakens bonds between soil particles. Simultaneously, as bacteria metabolize organic substrates, they release extracellular polymeric substances, which bind soil particles together. Both actions of stabilization and destabilization lead to particle rearrangements.

On the biological level, three mechanisms lead to aggregate formation, namely adhesion, immobilization, and retention. Through the adhesion mechanism, electrostatic bonding flocculates clay particles. For particles smaller than two micrometers in nominal diameter, electrostatic interactions result from charges on microbes, soil particles, and electrolytes in soil solution. As the ionic strength of the soil solution increases, the physical distance between microbes and particles decreases (due to the lesser energy required for adhesion), which leads to clay flocculation (Gomez-Suarez et al. 2002).

Through the immobilization mechanism, vertical, horizontal, and oblique forces bind microbes to the soil particle surfaces. The combination of microbially produced polymers and clay particles enhances interactions on the soil surface, leading to aggregation of silt-sized particles ranging between 2 and 50 μm (Gomez-Suarez et al. 2002).

Through the retention mechanism, microbial cells, and byproducts bind the clay, silt, and sand particles together into aggregates larger than 50 μm. Larger aggregates have relatively higher porosity than small aggregates, leading to less dense/compact aggregates with larger microbial populations in the pore spaces (Frey 2005). Retention refers to the capacity of microbes to remain adhered to the soil particle surfaces after being subjected to an external force. Larger aggregates with more pore spaces would have greater microbial activity to increase retention through the formation of a biofilm (Gomez-Suarez et al. 2002).

A study by Land (2010) investigated the aggregation of three types of dredged sediments, without and with XG treatment. Sediments were collected from three sites in southern coastal Louisiana, USA: (i) Bayou Chevreuil, a freshwater site containing sediment with high clay content; (ii) Leeville, an intermediate salinity site containing sediment with moderate clay content; and (iii) Grand Isle, a marine site with predominantly sandy sediment. The sediments were prepared in the form of a slurry, similar in consistency to that produced from hydraulic dredging, and were stored at 4 °C for 7 months while preliminary tests and sediment characterization were conducted. The physicochemical properties of these sediments are summarized in Table 7.2.

The sediments were treated with XG at a ratio of 2:1 (sediment: polymer) resulting in final XG concentrations of 0.5 and 1 wt%. Two different saline solutions were applied to all samples for each sediment to simulate in-situ salinity ranges. Salinity treatments were 1.49 and 7.46 mS cm^{-1} for the freshwater sediment, 7.46 and 14.9 mS cm^{-1} for the intermediate sediment, and 22.4 and 37.3 mS cm^{-1} for the marine sediment.

Table 7.2 Physicochemical properties of the freshwater, intermediate, and marine sediments (data from Land (2010))

Soil properties	Freshwater	Intermediate	Marine
Moisture content (%, wet mass basis)	74.8	67.1	36.3
Redox potential (mV)	– 26	– 18	– 207
Soil pH	6.50	6.90	6.90
Cation exchange capacity (cmol charge/kg dry sediment)	125.0	84.6	27.7
Pore water salinity (ppt)	0.50	4.60	15.5
Sand content (%)	9.17	19.2	70.8
Silt content (%)	20.8	38.3	13.3
Clay content (%)	70.0	42.5	15.8
Organic matter content (%)	13.8	8.34	1.56
Soil textural class	Clay	Silty clay	Sandy loam

Samples were placed in polyethylene cups and set on the laboratory bench before being analyzed at predetermined time periods of 1, 8, 16, and 26 weeks. All samples remained uncovered for the designated time periods and were (i) kept at a steady temperature of 20 °C; and (ii) exposed to daily light regimes with all laboratory lights on for 12 h/day. For the duration of each time period, decanting of any remaining supernatant fluid and re-flooding with water of the appropriate salinity occurred every 8 to 10 days to maintain anaerobic flooded conditions. At the end of each designated time period, the redox potential was measured. Then, the samples were stored for a maximum of 3 days at 4 °C before further analysis.

Variations in the mean redox potential with respect to the designated times are illustrated in Fig. 7.12 (data from Land (2010)). Note that, due to their differences being marginal, the values obtained for XG concentrations of 0.5 and 1 wt% were averaged. For the freshwater sediment, the redox potential for XG-treated samples was only slightly different from their no-XG counterparts. However, for the intermediate and marine sediments, there was a large difference between XG-treated samples and their controls. For cases without XG, the higher redox potential of the freshwater sediment could be related to its lower pH, as compared to the other two sediment types. Irrespective of the time periods considered, the redox potential values for the intermediate and marine sediments were not significantly different from each other. By week 26, the redox potential for the freshwater sediment was significantly higher than that for the intermediate sediment.

Sediments with and without XG treatment tended to have lower redox potential values by week 8, followed by increasing redox potential values for the remaining

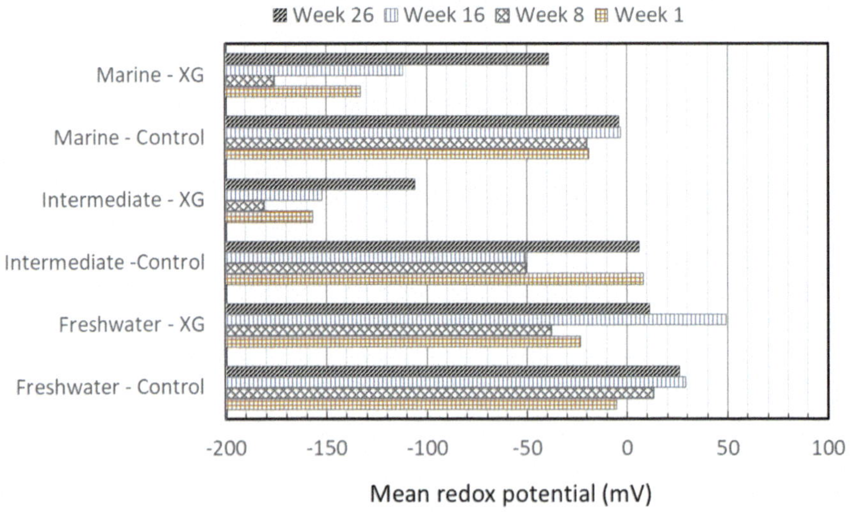

Fig. 7.12 Mean redox potential values for freshwater, intermediate and marine sediments treated with XG over time (data from Land (2010))

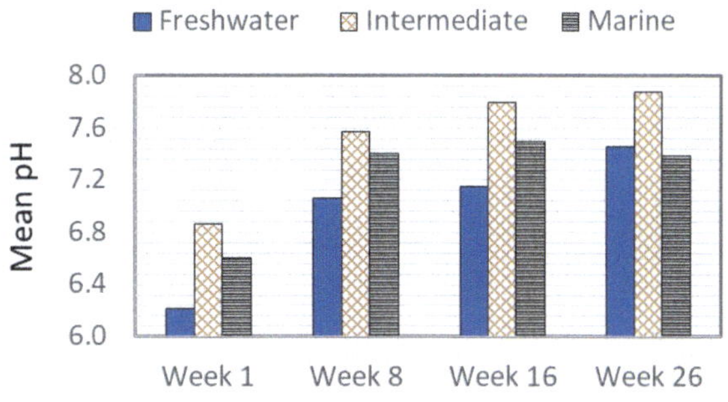

Fig. 7.13 Mean pH values for freshwater, intermediate, and marine sediments treated with XG over time (data from Land (2010))

times (see Fig. 7.12). For the intermediate and marine sediments, the redox potential values at week 8 were significantly lower than those at weeks 16 and 26. The lower clay content of the intermediate and marine sediments may have allowed the XG molecules to be more available for microbial activity, contributing to increased concentrations of electron donors and hence lower redox potential.

The mean pH values of the three XG-treated sediments with respect to time are shown in Fig. 7.13. The pH of the freshwater sediment was significantly lower than the pH of the marine sediment at weeks 1, 8 and 16. However, their pH values were not significantly different by week 26. Whereas the pH of the intermediate sediment was consistently greater than that of the marine sediment, although the difference was not significant for weeks 1, 8, and 16; however, the pH for the intermediate sediment was significantly higher than the pH for the marine sediment at week 26.

The moisture content values for the XG-treated freshwater, intermediate and marine sediments over time are illustrated in Fig. 7.14. Owing to their significantly different original water content values, after the sediment–XG mixtures had equilibrated for 24 h (i.e., the "Initial" reading plotted in Fig. 7.14), the moisture content for all three sediments was still significantly different—that is, water contents of approximately 82%, 76%, and 57% for the freshwater, intermediate, and marine sediments, respectively. With the passage of time, the moisture content for all three sediments was seen to progressively reduce, but remained significantly different from each other at weeks 1, 8, 16, and 26.

The freshwater sediment treated with 1% XG (Fig. 7.14a) exhibited the most reduction in the moisture content compared with other control or treated samples. The combination of XG and low ionic strength could have caused the XG to penetrate the expanding interlayers of clays, which would have pushed water molecules out of interstitial spaces. At each progressed time, however, the moisture content of the freshwater sediment samples with and without XG was not significantly different, suggesting that changes in the moisture content were from consolidation and dewatering processes. Moisture contents for all three sediments were always significantly different from one another.

Similarly, for the intermediate sediment (Fig. 7.14b), apart from their initial values, the moisture contents for the 0, 0.5, and 1% XG concentration levels were not significantly different from each other for any time period; however, as time progressed,

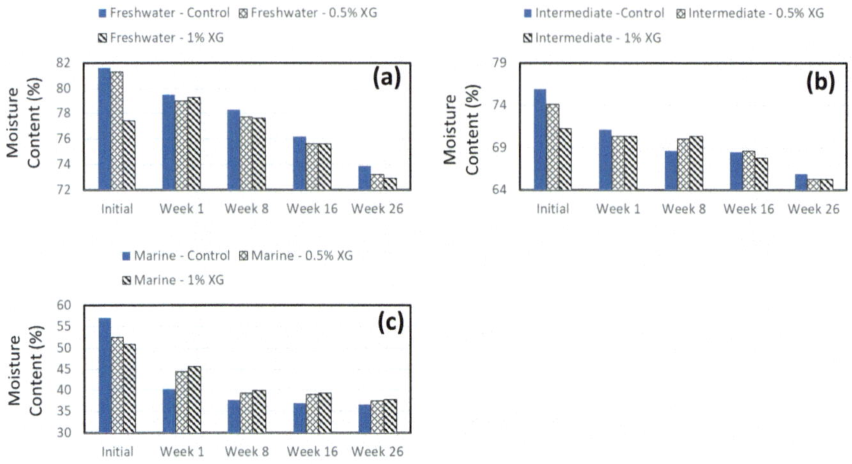

Fig. 7.14 Moisture content values for **a** freshwater, **b** intermediate, and **c** marine sediments treated with XG over time (data from Land (2010))

the moisture content values progressively decreased. Considering all the investigated cases, for the intermediate sediment, the moisture content decreased by about 7% over 26 weeks, with no significant differences between XG concentration levels.

For the marine sediment (Fig. 7.14c), moisture content values for the control samples were significantly different from the samples treated with 0.5 and 1% XG after 1 and 8 weeks. For the control samples, the moisture content decreased over time. By week 1, the moisture content significantly decreased from initially 57% to 40%. The decreasing trend continued for increasing time, with the overall moisture content reduction (from the initial state) obtained as approximately 20% after 26 weeks. For the marine sediment treated with XG, the moisture content values also decreased over time. By week 1, the moisture content decreased significantly from 52% to 45%; the overall moisture content reduction (from the initial state) was obtained as approximately 14% after 26 weeks.

In general, the presence of XG only had a significant effect on moisture content values for the marine sediment at week one. In this regard, XG-treated samples had moisture content values significantly higher than the control samples, possibly due to the amount of pore space in the marine sediment. High sand content led to greater and larger pore spaces, which increased the XG's adsorption of water. Whereas for the freshwater and intermediate sediments, the presence of XG was not significant even though the moisture content values significantly decreased over 26 weeks. After the XG degraded, sediments consolidated, thereby decreasing the moisture content.

Monitoring the amount of consolidation provides insight into whether XG prevents movement of sediment particles, thereby creating a more stable material. The consolidation values for the freshwater, intermediate, and marine sediments treated with XG over time are shown in Fig. 7.15. Note that the XG concentration did not significantly affect sediment consolidation. For the freshwater sediment, total

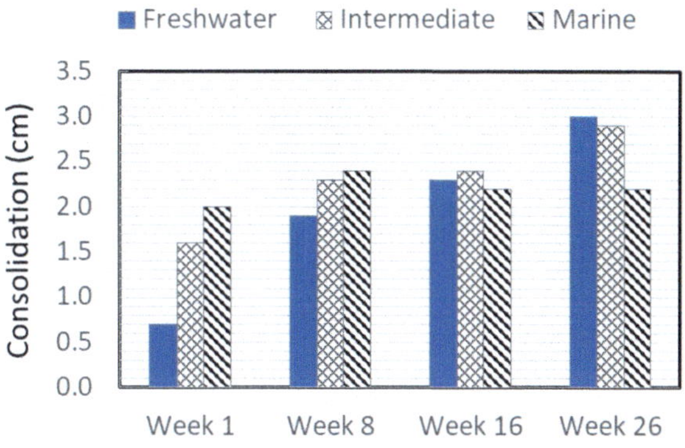

Fig. 7.15 Mean sediment consolidation values for the freshwater, intermediate and marine sediments treated with XG over time (data from Land (2010))

consolidation significantly increased over time, i.e., with all XG concentration levels averaged, the freshwater sediment exhibited an average consolidation of 0.7, 1.9, 2.3, and 3.0 cm after 1, 8, 16, and 26 weeks, respectively. For the intermediate sediment, the total consolidation after 26 weeks was significantly more than the total consolidation after 1 week, i.e., with all concentration levels averaged, the intermediate sediment consolidated 1.6, 2.3, 2.4, and 2.9 cm after 1, 8, 16, and 26 weeks, respectively. For the marine sediment, the total consolidation was not significantly different for different time periods, i.e., with all XG concentration levels averaged, the marine sediment consolidated 2.0, 2.4, 2.2, and 2.2 cm after 1, 8, 16, and 26 weeks, respectively.

To evaluate the effect of XG treatment on aggregate sizes, Land (2010) used a Beckman Coulter LS 13 320 Series multi-wavelength laser diffraction particle-size analyzer with an aqueous liquid module and sonicator system. This apparatus, which uses Mie scattering, Fraunhofer diffraction and polarization intensity differential scattering technology, can: (i) measure particles from 0.017 to 2000 μm; (ii) provide the percent volume of the sample falling into several size fractions: < 2 μm, > 2 μm, < 63 μm, > 63 μm and > 1000 μm; (iii) calculate mean aggregate size diameter for each sample; and (iv) assign classes of the soil fractions: 0–2 μm for clay, 2–63 μm for silt, and > 63 μm for sand.

Over time, the mean aggregate size was significantly different among all three sediments (Fig. 7.16; data from Land (2010)), but the presence of XG had no significant effect on the mean aggregate size. For the freshwater sediment, the average aggregate size increased over time with the following values: 90, 104, 121, and 173 μm for weeks 1, 8, 16, and 26, respectively. In other words, by week 26, the increase in aggregate size was significantly different from week 1, likely due to clay particle aggregation from consolidation and natural organic materials. For the intermediate sediment, the average aggregate size was not significantly different during different time periods, with the following values representing the average aggregate size over weeks 1, 8, 16, and 26: 73, 61, 64, and 67 μm. For the marine sediment, the average aggregate size decreased over the examined 26-week period, with the following values: 119, 100, 99, and 101 μm. By week 26, the decrease in aggregate

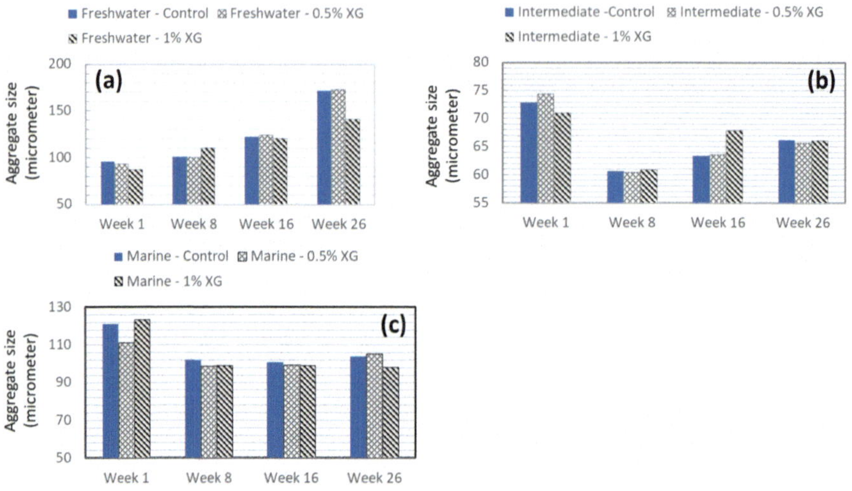

Fig. 7.16 Aggregate sizes for the **a** freshwater, **b** intermediate, and **c** marine sediments treated with XG over time (data from Land (2010))

size was significantly different from week 1, possibly due to less pore spaces from consolidation.

After averaging the percent volume values for all XG concentrations and time periods within each size class, the typical aggregate distribution of the three sediments was as follows (Land 2010): (i) the freshwater sediment had approximately 7%, 48%, and 45% of aggregates in the clay, silt and sand sized fractions, respectively; (ii) the intermediate sediment had approximately 5%, 60%, and 35% of aggregates in the clay, silt and sand sized fractions, respectively; and (iii) the marine sediment had approximately 4%, 32%, and 74% of aggregates in the clay, silt and sand sized fractions, respectively.

The freshwater and intermediate sediments had the highest percentage of aggregates in the silt fraction, with values of 48% and 60%, respectively. For the freshwater sediment, the percentage of aggregates in the sand fraction increased with increasing time, which was reflected by decreasing its silt fraction over time. The percentage of clay for the freshwater sediment did not significantly change over time. For both the intermediate and marine sediments, the percentage of aggregates in the silt fraction increased with increasing time, reflected by a decrease in the sand fraction over time. Additionally, the percentage of aggregates in the clay fraction of the intermediate and marine sediments increased over time.

In general, the presence of XG did not have any significant effect on increasing aggregation in the clay, silt or sand fractions of any of the three sediments. The XG addition was anticipated to increase particle aggregation as a process to enhance stability of newly deposited dredged sediment. Differences in size classes between the initially characterized sediments and those sediments for aggregate analysis arose from natural processes and naturally occurring soil organic matter.

In both the freshwater and intermediate sediments for aggregate analysis, the increase in sand fraction compared to the initially characterized sediments suggests aggregation of silt and clay particles from organic materials. In the marine sediments for aggregate analysis, the increase in silt fraction compared to the initially characterized sediments indicates that clay and silt particles filled the pore spaces between the sand grains and formed aggregates. Once again, differences between size classes of the initially characterized sediments and those sediments for aggregate analysis reinforce the importance of naturally occurring soil organic matter in natural aggregation processes. The XG had no notable effect on aggregation, likely because the particles in the natural sediments were already fully aggregated.

The preceding analysis with field sediments highlights the fact that XG treatment did not increase aggregation. This might be explained by considering the following characteristics of XG. Adsorption of anionic polymers onto clay particles depends on pH, ionic strength and type of cations present in the soil solution. Naturally, an anionic polymer and negatively charged clay particles tend to repel each other. Soils with low pH or high ionic strength experience greater adsorption of anionic polymers to clay particles (Wallace et al. 1986; Gu and Doner 1993). By lowering the pH or increasing the ionic strength of the soil solution, more cations become available to form cationic bridges between the polymer and clay particles. In addition, hydrogen ions in low pH soils may neutralize the anionic charge on the polymer, allowing polymer penetration into the spaces in between layers of expanding clay particles. Helical conformation allows anionic polymers a large "grappling" distance, which enables extension into the soil solution, promoting inter-particle bridging (Gu and Doner 1993). Adding a flocculating agent to the solution before adding an anionic polymer may enhance binding properties, as the polymer will interact with aggregates instead of particles.

To uncover the reasons behind this finding (i.e., no aggregation), the biodegradability potential of XG should be evaluated. Figure 7.17 (data from Land (2010)) illustrates the ratio (%) of the mean concentration of the soluble microbial biomass (C_{mic}) to the sediment organic matter concentration (C_{org}) for the freshwater, intermediate and marine sediments at the 0, 0.5, and 1% XG concentration levels after weeks 1, 8, and 16. The results indicated that the presence of XG had no significant effect on the C_{mic}/C_{org} ratio for each sediment type. Irrespective of time, the C_{mic}/C_{org} ratio for the marine sediment was significantly higher than that obtained for the freshwater and intermediate sediments. Moreover, the C_{mic}/C_{org} values for the freshwater and intermediate sediments were not significantly different from each other. Higher C_{mic}/C_{org} ratios for the marine sediment indicate that more carbon was available for microbial processing. The microbial community responded by assimilating carbon and increasing biomass.

To further evaluate the potential of biodegradability, Land (2010) evaluated the microbial respiration. Figure 7.18 shows the amount of carbon added from the XG, total amount of carbon respired, and percentage carbon increase after 125 days for the freshwater, intermediate and marine sediments treated with XG. The 1% XG treatment scenario did have a significant effect on the respiration rates, as reflected by the total carbon. In general, the freshwater sediment had the highest respiration

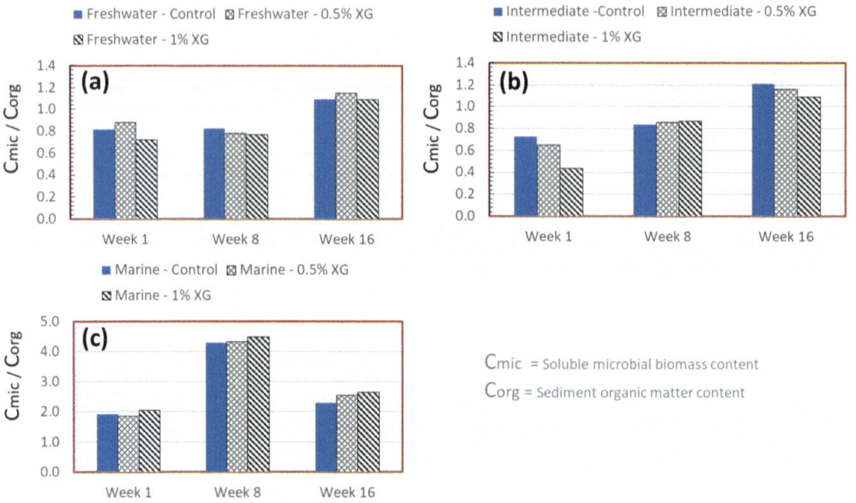

Fig. 7.17 C_{mic}/C_{org} ratios (in %) for **a** freshwater, **b** intermediate and **c** marine sediments treated with XG over time (data from Land (2010))

rate (25.8 g CO_2-C/kg of dry sediment), followed by the intermediate (12.0 g CO_2-C/kg of dry sediment) and the marine (4.33 g CO_2-C/kg of dry sediment) sediments. Note that respiration for the control samples resulted from organic carbon already present in the sediments and hence was not associated with XG treatment.

For all three sediments (fresh, intermediate, and marine), the microbial community in the samples treated with 1% XG respired significantly more than the control samples. Comparing the microbial respiration to the control samples; at 0.5% XG concentration, the percent increase was 233%, 271%, and 490% for the freshwater, intermediate and marine sediments, respectively. Meanwhile, at 1% XG, the percent increase was 446%, 612%, and 949% for the same sediments, respectively, which are approximately two times higher than those obtained at 0.5% XG concentration.

Comparing the cumulative microbial respiration contributed by the addition of 0.5 and 1% XG; at 0.5% XG concentration, microbes in all three sediments respired more carbon than needed for the added XG degradation, in the amounts of 3.6, 2.8, and 4.4 times more carbon for the freshwater, intermediate and marine sediments, respectively. Meanwhile, at the 1% XG concentration, more carbon was respired than needed for XG degradation in the amounts of 4.7, 4.2, and 4.9 times more carbon for the freshwater, intermediate, and marine sediments, respectively. These results are clear indication of the biodegradability of the added XG to the sediments, since the XG stimulates the microbes to metabolize carbon from microbial biomass turnover and soil organic carbon.

In conclusion, microbial respiration in these XG-treated sediments is above that of the control samples, in addition to being above the amount of respiration required to indicate complete mineralization of the XG due to (i) mineralization of soil organic carbon, and/or (ii) mineralization of additionally secreted microbial byproducts and

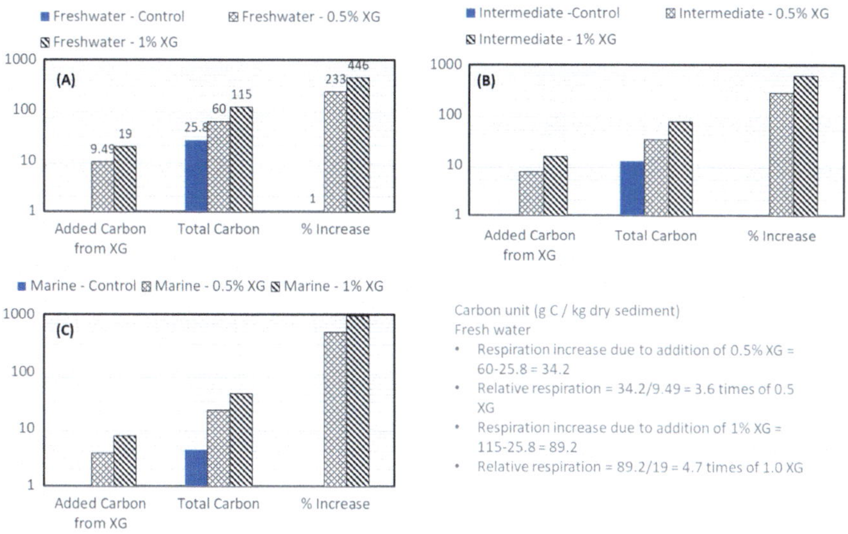

Fig. 7.18 Amount of carbon added from XG, total amount of carbon respired, and % carbon increase after 125 days for the XG-treated **a** freshwater, **b** intermediate, and **c** marine sediments (data from Land (2010))

biomass turnover. For either case (i or ii), metabolism of these materials removes organic material from the sediment that might otherwise lead to aggregation of soil particles. In other words, the application of XG to hydraulically stabilize dredged sediments may, in fact, be stimulating the growth of the microbial community, which inherently leads to increased mineralization of organic matter and release of carbon dioxide. While the addition of XG does not directly harm microbial communities in the sediments, they may be decomposed quickly and thus prevent any natural aggregation processes.

7.4 Effects of XG on Basic Geotechnical Properties

As discussed in Chap. 5, the effects of XG concentration on the basic geotechnical properties of XG-stabilized clay soils (Fig. 7.19) tends to: (i) increase gel content, viscosity, plasticity index, soil water potential and degree of aggregation; and (ii) decrease pore sizes, external surface area of aggregates, swelling potential, and dispersivity. In view of these properties, adding XG to soils can also affect the XG-treated soil's HC.

From the dynamic properties of the XG gel discussed in Chap. 3, the frequency-dependent storage (i.e., $G'(\omega) = K'(\omega)^{n'}$) and loss (i.e., $G''(\omega) = K''(\omega)^{n''}$) moduli (where K' is the storage modulus, K'' is the loss modulus, and n' and n'' are constants) are used to characterize the nature of the XG gel, which in turn gives insight into

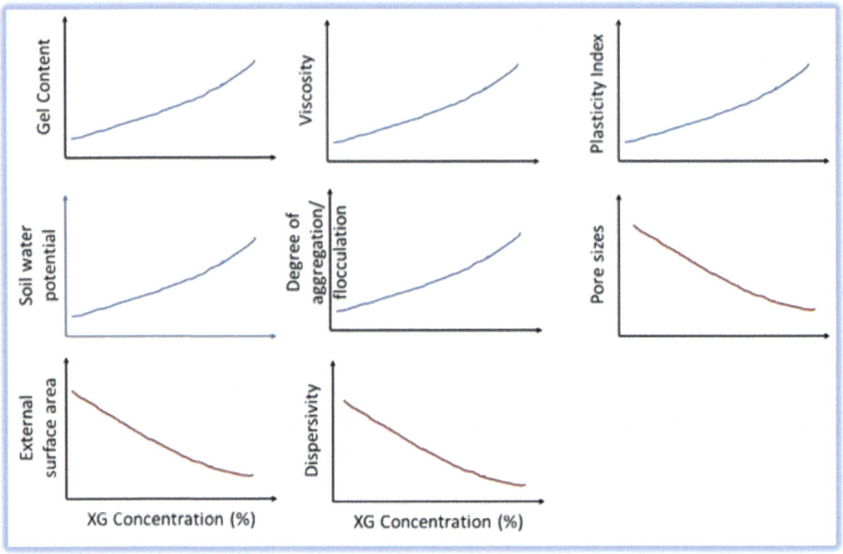

Fig. 7.19 Effects of XG concentration on basic geotechnical properties of XG-stabilized fine-grained soils

the strength of the XG interaction with clay soils, as well as the moisture flow mechanisms through soils. An example for the calculated power-law parameters of storage and loss moduli for aqueous XG solutions with different concentrations is shown in Fig. 7.20 (data from Song et al. (2006)). The K' values are much larger than those of K'' for all XG solutions, indicating the superiority of the elastic behavior (storage modulus) over the viscous behavior (loss modulus). Also, the values of n' (for the storage modulus) and n'' (for the loss modulus) are gradually decreased with increasing XG concentration, implying that as the XG concentration increases, the frequency dependence of the linear viscoelastic functions becomes weaker. In other words, as the XG concentration increases, the XG solution becomes more elastic in character.

Moreover, based on the n' values, the nature of the XG gel can be characterized as a chemically cross-linked strong gel or a weak gel. For $n' > 0$, the gel is known as physically weak, whereas for $n' = 0$, the gel is chemically strong. For the concentrated XG solutions (Fig. 7.20), the n' values ranged from 0.13 to 0.18; as such, the XG solution does not form a stable (strong) gel, but rather exhibits a weak gel-like behavior. Furthermore, in intermediate and semi-dilute XG concentration regimes ($0.05\% < C_{XG} < 0.5\%$), both conformational ordering of the XG backbone and intermolecular associations control the rheological properties of the XG dilute solution. However, in the high concentrated regime ($C_{XG} > 1\%$), stress nonlinearities and liquid crystal type behavior are dominant.

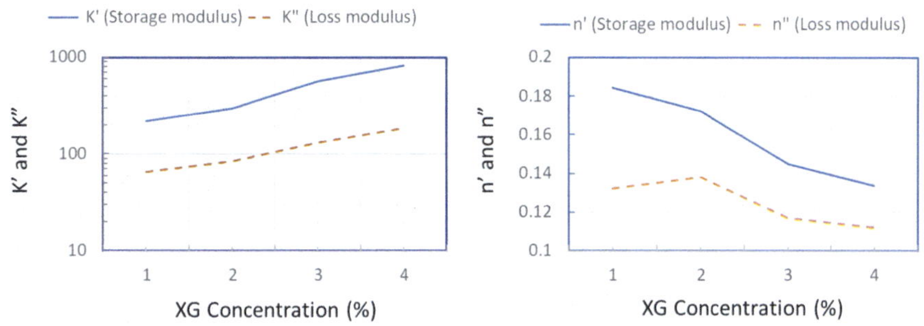

Fig. 7.20 Calculated power-law parameters of storage and loss moduli for aqueous XG solutions with different XG concentrations (data from Song et al. (2006))

7.5 Effect of XG on Soil HC

7.5.1 Questionable Reported Results

Generally, when XG is added to soil, the extended conformation of the XG molecules results in the formation of a network of polymer chains that fills the soil pore spaces. This network of polymer chains reduces the effective pore size of the soil, leading to a decrease in the HC that can be explained, as follows. XG is a polysaccharide that alters soil structure by binding the soil particles together. This binding process increases the stability of soil aggregates by forming a gel-like matrix that encapsulates the particles, reducing their tendency to disperse. As a result, the soil's HC is reduced because the stable aggregates create fewer and smaller pores, which restricts water flow through the soil. Additionally, XG increases soil cohesion and improves water retention, further contributing to the decrease in HC and enhancing the soil's resistance to erosion and compaction.

However, on first sight, the reported results in view of HC changes arising from XG addition to soil appear conflicting. Several studies reported a reduction in HC for soils treated with XG (Khachatoorian et al. 2003; Bouazza et al. 2009; Ayeldeen et al. 2016; Cabalar et al. 2017; Dehghan et al. 2019). For instance, in a study by Ayeldeen et al. (2016), the HC of sandy and silty soils with a 2% XG content was found to decrease by 10% of the untreated soils, which was attributed to the formation of a gel-like structure induced by XG. In addition, according to Czarnes and Hallett (2000), the reduction in HC is a result of the cross-linking elements that form within the soil matrices and fill in the pore voids, thereby obstructing the flow of water. However, a study by Biju and Arnepalli (2020) investigating the effects of XG on the HC of sand–bentonite mixtures observed an increase in the HC with increasing XG content. They reported that the increase in XG content could enhance the bonds between the clay particles (ionic/electrostatic, hydrogen, and van der Waals), thereby promoting the formation of more and bigger-sized aggregates of clay particles, leading to the widening of the water flow channels and a consequent increase in the HC (within

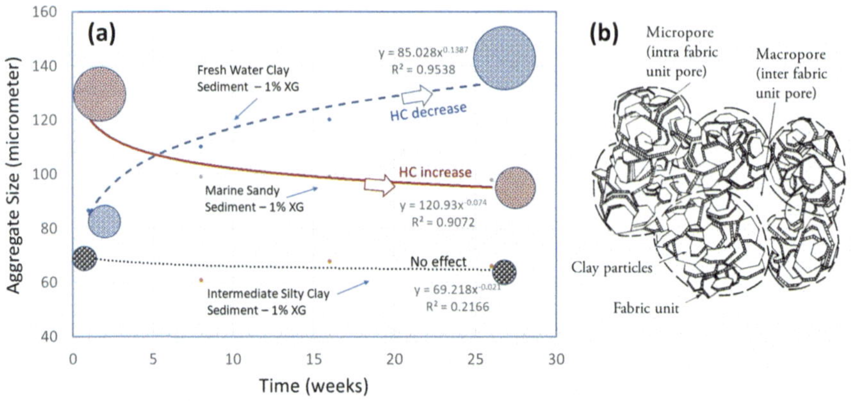

Fig. 7.21 **a** Aggregate size changes with time for sediments treated with 1% XG; and **b** schematic showing aggregates and micro- and macro-pores (data from Land (2010))

one order of magnitude). It is worth noting that the impact of aggregation on the HC is debatable and is discussed below in view of the results shown in Fig. 7.21 (data from Land (2010)).

In addition, the effects of curing time on the HC of XG-stabilized soils again show some contradictory results. For example, Khachatoorian et al. (2003) observed a decrease in the HC of treated Ottawa sand (0.43–0.6 mm) after 11 days of curing. Whereas Cabalar et al. (2017), who treated Narli sand (0.075–1.0 mm) with XG, did not observe a clear reduction tendency for HC over the same curing period. Furthermore, for longer curing durations, most of the experimental results showed an increase in the HC for mixtures of XG with sand and silt, which was attributed to the expected dehydration of XG over time (Ayeldeen et al. 2016; Cabalar et al. 2017). The increase in HC observed over time (in the long term) is more significant for XG-treated sands than XG-stabilized silts, since the finer silt particles delay water evaporation from the voids, as well as the dehydration of the XG hydrogels deposited within the pore spaces (Ayeldeen et al. 2016).

Let us evaluate the effect of the aggregate sizes on the expected changes in HC considering the previously discussed aggregate sizes results (i.e., data from Land (2010) presented in Fig. 7.16), and which are reproduced as shown in Fig. 7.21. However, let us first explain the aggregate formation and the pores within the aggregates. An aggregate could be one fabric unit or a group of fabric units (Fig. 7.21b). The pore spaces within the fabric units (i.e., macropores) are larger than the pore spaces between particles within the fabric units (i.e., micropores). The movement of water in the macro- and micro-pores is controlled by the amount of XG in soils. When XG is added to soils, the following conditions may well occur: (i) a decrease in the sizes of both the micro- and macro-pores; (ii) as XG concertation increases, the micropores may be completely blocked, and the size of the macropores becomes smaller; and (iii) the overall water movement through XG-treated soils will be through the macropores only. These processes will lead to a reduction in the water flow, as well as the HC of the XG-treated soils.

The results (Fig. 7.21a) indicate that over the 26-week time period: (i) for freshwater clay sediment, an increase of the aggregate size by about 1.6 times; hence a decrease in the HC would be expected due to aggregate formation as well as the changes of the sizes of the micro- and macro-pores, which contradicts the results reported by Biju and Arnepalli (2020) for sand–bentonite; (ii) for marine sandy sediment, a decrease in the aggregate size by about 1.3 times; hence an expected increase in the HC due to an increase in the macropore sizes as well as the overall soil porosity, which contradicts the results reported by Ayeldeen et al. (2016) for sand and silty soils and Khachatoorian et al. (2003) for Ottawa sand; and (iii) for intermediate silty clay sediment, a small decrease of the aggregate size by about 1.07 times; hence a marginal effect would be expected in the HC.

As highlighted above, some studies indicated XG addition produced a reduction in the HC, while others reported an increase in the HC. As such, it is essential to carefully consider the appropriate XG application rate and method for each specific situation. Therefore, understanding the effects of these factors is essential toward effectively using XG for soil stabilization applications. The following sections aim to discuss the effects of XG, mindful of the governing parameters described above, on the HC of XG-treated soils.

7.5.2 *Controlling Parameters*

It is known that HC is a measure of how easily water can flow through a porous medium (in this case soil). The HC depends on several factors, including the texture and structure of the soil, the amount, individual sizes, connectivity and distribution of the soil pore voids, the initial moisture content of the soil (degree of saturation), soil type, the XG application rate and method, the viscosity of the pore fluid, and the environmental conditions (e.g., curing time, temperature, humidity, and cyclic wetting–drying or freezing–thawing actions). Also, other factors such as XG concentration, soil compaction state, CEC and pH can affect the HC (Yong et al. 1992; Mohamed and Paleologos 2018). Understanding these parameters is essential toward effectively applying XG for soil stabilization applications.

7.5.2.1 Soil Type

The HC of soils treated with XG varies depending on the type of soil. For instance, soils with higher clay content tend to have lower HC due to their small pore sizes, whereas soils with lower clay content have higher HC due to their relatively larger pore sizes. Several studies have investigated the effects of XG on the HC of different soil types, some examples of which are included below.

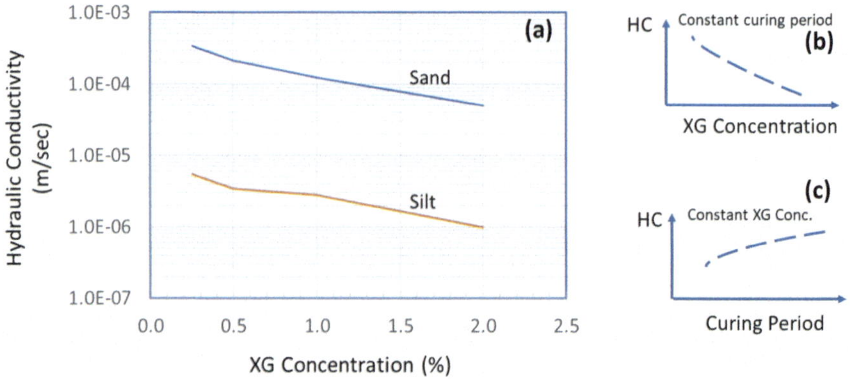

Fig. 7.22 Hydraulic conductivity variations with XG concentration and soil type after 5 weeks of curing in an oven at 30 °C (data from Ayeldeen et al. (2016))

Sand and Silt

This case can be evaluated with reference to the study reported by Ayeldeen et al. (2016), whereby the powered XG was mixed with deionized water (DW) to produce the XG gels. The XG solution concentration was calculated as the ratio of the weight of the used XG powder to the overall weight of the DW–XG solution (i.e., 0.25%, 0.5%, 1% and 2%). Then, samples of the dry disaggregated sand and silt soils investigated were mixed with the XG solution until homogeneity had been achieved in the mixture. This method is referred to as wet mixing. Specimens were prepared at their maximum density and were cured in an oven at 30 °C for predefined periods of 0, 1, 2, 3, 4, 5, and 10 weeks. The HC was measured using the falling-head permeability test in accordance with ASTM D5084–03.

The results shown in Fig. 7.22a clearly indicated that XG has a remarkable effect in reducing the HC for both sand and silt soils. After 5 weeks of curing in an oven at 30 °C (intended to simulate hot and dry environmental conditions), sand samples treated with 0.25% and 2% XG exhibited HC reductions of approximately 36% and 93% compared to the initial (no XG) value (of 3.4×10^{-4} m/s). The same trend was also observed for silt soil; however, the magnitude of HC reduction was slightly lower compared to similar sand–XG cases.

Furthermore, for XG-treated sand, the HC increased in the range of 150% to 220% between the first and the tenth week of oven curing at 30 °C; however, for the XG-treated silt, it was between 115% and 145% for the same curing period. For this case, the researchers stated that the HC increased more in the sand–XG mixture because the finer particles present in the silt–XG mixture delayed the dehydration process, and consequently the evaporation of moisture from the void spaces of the soil. Analysis of SEM images of XG-treated soil for different curing times (Ayeldeen et al. 2016) showed that shrinkage of the XG hydrogels within the matrices formed in the soil occurs over the curing time. This decrease in volume of XG within the pore void spaces consequently increases the sizes of the soil pore voids, as well as

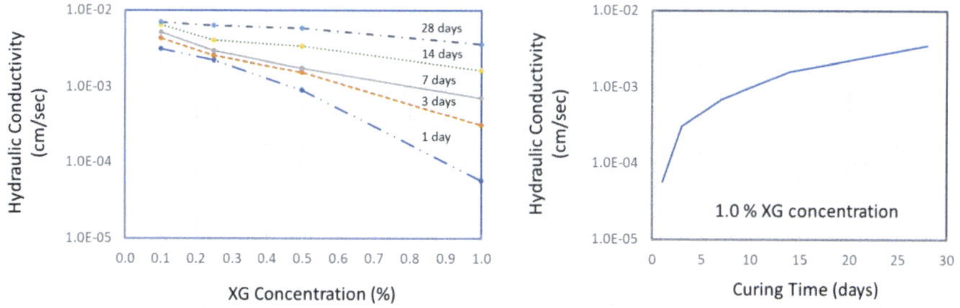

Fig. 7.23 Hydraulic conductivity variations with XG concentration and curing time for coarse sand (data from Wiszniewski and Cabalar (2014))

the HC. In general, it is considered that the final stage of shrinking in the XG is the transition phase between the rubber-like gel and the vitreous state. Through this transformation, the XG gains more strength, which was also verified in the shear strength tests.

Coarse and Medium Sands

A study by Wiszniewski and Cabalar (2014) investigated the changes in HC of XG-treated coarse and medium sands using: (i) constant-head permeameter tests performed according to ASTM D2434 for the coarse sand; and (ii) constant-head permeability tests performed in a triaxial apparatus for the medium sand (Figs. 7.23 and 7.24). For the coarse sand samples, the XG was added to the soil using the dry mixing method, and then placed without compaction into the testing mold and saturated with water. The specimens were allowed to cure for different periods of 1, 3, 7, 14, and 28 days (note that the nature of the curing condition was not specified). However, for the medium sand, the test samples were saturated by applying a pressure gradient from the bottom (30 kPa) to top (0 kPa) in the presence of a constant confining pressure of 50 kPa. Therefore, the HC for the two sands was determined using two different testing types (i.e., rigid boundary for the coarse sand, and flexible boundary for the medium sand).

The HC was found to decrease with increasing XG concentration for both sand types. Notably, the hydraulic gradient in the medium sand samples was steady (around 44) for all tests, whereas for the coarse sand, its value was maintained at between 3 and 10, depending on the XG concentration. Referring to Fig. 7.23a; for the same curing time, as the XG concentration increased, the HC decreased. For example, with increasing XG concentration to 1% and considering one-day curing, the HC decreased significantly from initially 7.16×10^{-3} to 5.75×10^{-5} m/s; however, after 28 days, the HC was found to have only decreased slightly from initially 7.16×10^{-3} to 3.55×10^{-3} m/s. Furthermore, for the same XG concentration, as curing time increased, the HC increased. For example, for 0.1% XG concentration, the HC increased slightly from 3.07×10^{-3} to 6.92×10^{-3} m/s for 1- and 28-day curing,

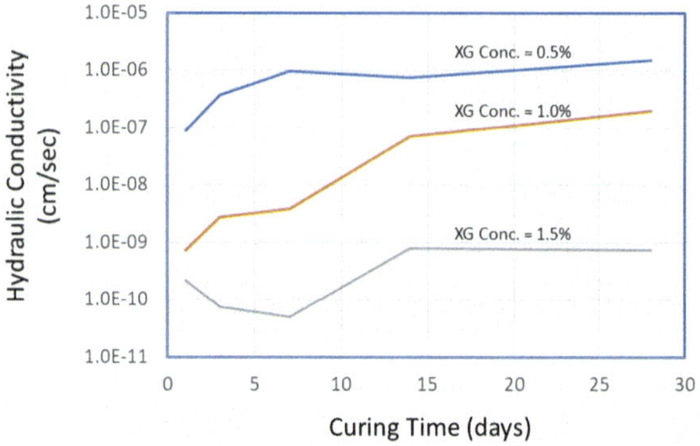

Fig. 7.24 Hydraulic conductivity variations with XG concentration and curing time for medium sand (data from Wiszniewski and Cabalar (2014))

respectively; however, for 1% XG concentration, the HC increased significantly from 5.75×10^{-5} to 3.55×10^{-3} m/s.

For medium sand (see Fig. 7.24), and considering the same curing time, as XG concentration increased, the HC decreased. For example, after 1 day of curing, the HC decreased substantially from its initial of 8.46×10^{-5} to 2.20×10^{-10} m/s with the addition of 1.5% XG. A similar change in the HC was noted after 28 days. For the same XG concentration, as curing time increased, the HC increased. For example, with 0.5% XG, the HC increased from 8.71×10^{-8} to 1.53×10^{-6} m/s. However, with 1.5% XG, the HC increased slightly from 2.20×10^{-10} to 7.27×10^{-10} m/s.

The effect of curing time on the HC of the XG-treated coarse and medium sands is of great significance. For instance, it was observed that the HC increases with curing time; e.g., for 0.5% XG-treated medium sand, it increased from 3.76×10^{-7} m/s at 3 days to 1.53×10^{-6} m/s at 28 days (Fig. 7.24). The aging influence appears slightly more pronounced when the XG concentration increased; e.g., for 1.5% XG-treated medium sand, the HC increased by a factor of 9.7 times (from 7.53×10^{-11} cm/s at 3 days, to 7.27×10^{-10} cm/s at 28 days). For the XG-treated coarse sand, however, the effects of curing time were less (a factor of 4.1), as all 28-day cured samples showed HC values similar to that of the untreated (i.e., no XG) sand (Fig. 7.23).

Therefore, noticeable differences were observed for the two sand types due to their different particle-size distributions. XG may have been more prone to leaching in the coarse sand; whereas, due to its finer gradation, the medium sand may have allowed a better adherence of sand particles to each other, thereby creating a more uniform/stable XG–sand linked network. Another reason could be due to the method of testing, whereby a triaxial cell was used in testing the XG-treated medium sand, and a confinement pressure was applied (flexible boundaries); thereby preventing water flow between the enclosing flexible membrane and the sample sidewall boundary. However, for testing coarse sand, where a standard permeameter apparatus was used (rigid boundaries), water flow between the sidewall of the permeameter cell and the soil sample was possible; as such, the HC values obtained for the medium sand are possibly more reliable.

Silty Sand

A study by Mendonça et al. (2021) investigated the HC changes of XG-treated inorganic non-plastic silty sand, composed of 72.7% sand and 27.3% silt-sized particles, and classified as SM (silty sand), as per the Unified Soil Classification System (USCS). The soil had an optimum moisture content for compaction of 14.3%, a maximum dry unit weight of 16.2 kN/m^3, and a pH of 7.1. Two types of XG were examined; the first variant was an XG-like biopolymer manufactured in the laboratory using *Stenotrophomonas maltophilia strain* Faro439, and the second variant was a commercially available XG produced by Sigma–Aldrich, supplied in powder form, and is obtained from the *X. campestris* bacterium through the fermentation of glucose/sucrose. A total of six sample groups were tested for HC: (i) natural/untreated soil; (ii) a mixture of soil with the XG-like biopolymer (Lab XG) (0.33%); (iii) a mixture of soil with the commercial XG (0.16% and 0.33%); and (iv) a mixture of soil with Portland cement (0.33% and 1.0%). Each mixture was tested at different hydraulic gradients (i.e., 6.9, 8.2, 9.4, and 10.6), and after four different curing times of 3, 7, 14, and 28 days. The grouting materials (Lab XG, commercial XG, and Portland cement) were mixed with the soil on a dry mass basis, and then compacted at the optimum moisture content to achieve the maximum dry density. Compacted specimens were cured in a humidity chamber, controlled at a temperature of 20 ± 2 °C and a relative humidity of 95 ± 5%. The HC was measured by the constant-head permeameter testing procedure.

Figure 7.25 shows that the XG-treated silty sand induced a notable decrease in the HC, which could be attributed to the development of some links and viscous hydrogels due to the contact of the XG with the water (i.e., hydration of the hydrogels) that fill part of the soil's voids (Bouazza et al. 2009). The results also reveal that the commercial XG is more effective than the Lab XG in reducing the soil's HC.

Silty Soils

In their investigation, Singh and Das (2019) examined the HC of an Indian high plastic silt treated with 0.5% and 1.0% XG under different consolidation pressures. They observed a gradual decrease (ranging between 20% and 50%) in the HC with increasing XG content and consolidation pressures. In another study by Bouazza et al. (2009), three different biopolymers, namely guar gum, sodium alginate, and XG, were found to reduce the HC of a silty sand (18% kaolinite and 82% naturally occurring sand). The decrease in the initial HC of the silty sand due to biopolymer-induced pore clogging exceeded four orders of magnitude with 0.5% XG, highlighting its superior pore plugging effect.

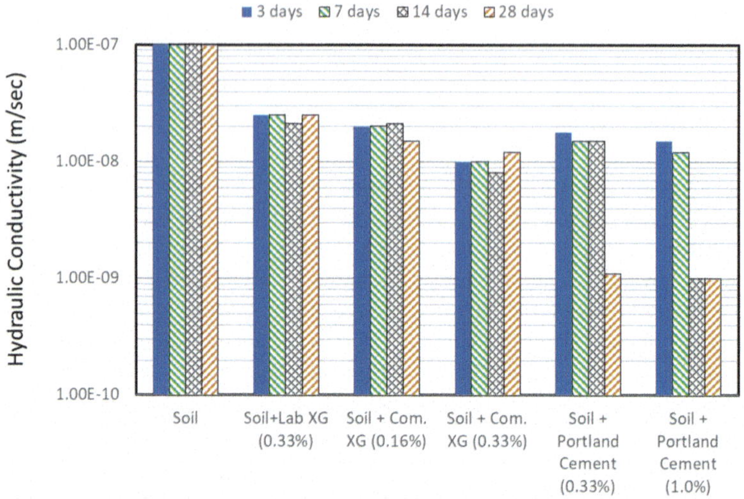

Fig. 7.25 Variations of hydraulic conductivity for silty sand treated with lab-produced XG (Lab XG), commercial XG (Com. XG), and Portland cement at various concentrations and curing times for a hydraulic gradient of 6.9 (data from Mendonça et al. (2021))

Natural Soil

The better performance achieved using XG is probably due to its very high affinity for water, as well as its ability to interact with clay minerals. In addition, XG polymer chains self-associate to form large, interconnected networks that can lead to bridging of clay particles, as indicated by Chen (1993). It is also interesting to note that a reduction in HC of at least two orders of magnitude has been reported by Martin et al. (1996) for (i) XG mixed with a fine-grained soil (clayey silt), or (ii) a flooding process through sand (Khachatoorian et al. 2003). In addition, XG's adsorption to sand can increase the silty sand–XG composite strength, as observed by Bouazza et al. (2001).

In a study by Kumar and Sujatha (2021), a natural soil (mixture of poorly graded sand and clay) was treated with XG and tested for HC using a rigid-wall permeameter test method (ASTM D5856 2015). The natural soil had a HC of 1.03×10^{-3} cm/s. Treating the soil with XG reduced the HC of the soil appreciably, making it suitable as a base liner material. A minimum XG dosage of 0.25% was found to produce a marked decrease (about three orders of magnitude) in the HC value to 2.8×10^{-6} cm/s after 24 h of curing. The reduction in HC can be attributed to the bioclogging mechanism of the XG (i.e., XG coats the soil particles and forms viscous hydrogels that arrest the movement of water).

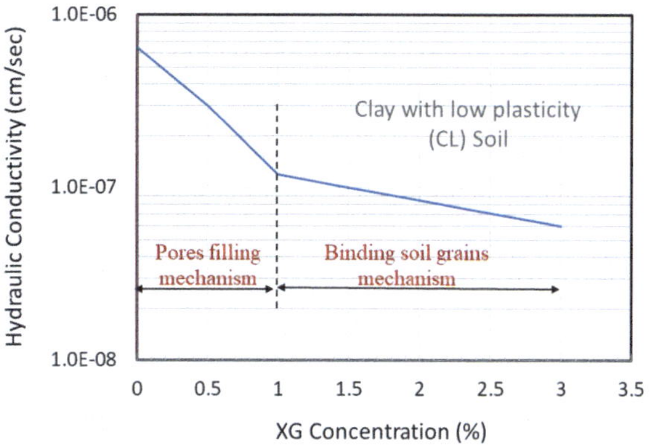

Fig. 7.26 Variations of hydraulic conductivity against XG concentration for XG-treated low plasticity clay (data from Cabalar et al. (2018))

Low Plasticity Clay

A study by Cabalar et al. (2018) investigated the effects of XG treatment on the HC of a low plasticity clay (i.e., CL, as defined by the USCS). XG contents of 0.5, 1.0, 1.5, 2.0, and 3.0% by dry weight of soil were examined based on the dry mixing approach. For HC measurements, a series of falling-head tests were performed according to ASTM D5084 (ASTM 2016a). The clay–XG mixtures were compacted into a conventional compaction mold, as per ASTM D1557 (ASTM 2013), and soaked in a water tank for saturation. Then the HCs were measured during the following 5 days. The results (presented in Fig. 7.26) indicated that higher XG contents produced lower HC values for the same hydraulic gradient. For example, the test results showed that the HC of samples decreased sharply from 6.5×10^{-7} cm/s (for the untreated clay) to 1.2×10^{-7} cm/s by adding 1.0% XG (i.e., 81.5% decrease). Cabalar et al. (2018) considered the breaking point shown in Fig. 7.26 to indicate a transition between "*filling the pores*" and "*binding the soil grains*" mechanisms, although they did not provide any additional supporting experiments, such as SEM analysis. In other words, the authors postulated that the changes in HC could be due to the filling of pores between the grains up to 1.0% XG, and binding the clay particles after this critical XG content.

The previously highlighted transition point (for an XG concentration level of 1.0%) evident in Fig. 7.26 can be evaluated using the results of Chang et al. (2015), whereby the microstructural interactions between the soil grains and XG gel were hypothesized to give insight into the clay–XG contact characteristics. This can be done by calculating the energy absorption capacity (i.e., the area under the stress–strain curve up to the peak point) of the XG-treated clay soil samples. Chang et al. (2015) also investigated low plasticity clay (CL, as defined by the USCS). Samples were compacted to achieve a density like that achieved using the modified Proctor method, as per ASTM D1557 (2013). The compacted samples were kept in plastic bags at a room temperature of 25 °C during the required curing periods of 0, 7, 28, and 56 days. The cured samples were then removed from their compaction molds, trimmed, and used for the unconfined compression testing, as per ASTM D2166

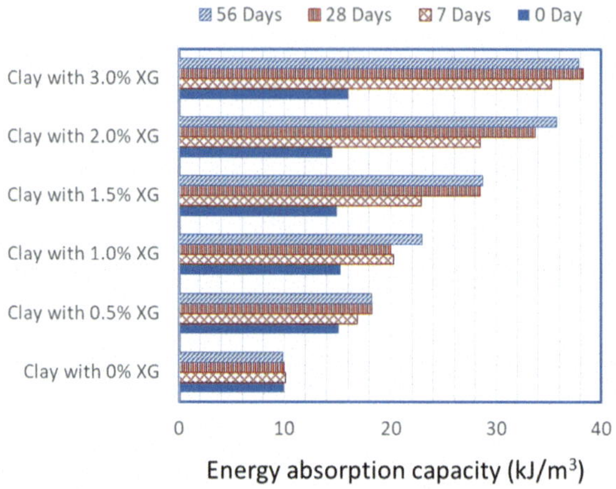

Fig. 7.27 Energy absorption capacity of XG-treated low plasticity clay at different XG concentrations and curing times (data from Chang et al. (2015))

(ASTM 2016b). The obtained energy absorption capacity results, shown in Fig. 7.27, indicated that XG treatment resulted in an increase in the energy absorption capacity due to an increase in the inelastic response to loading.

Furthermore, using the results of Fig. 7.27 and reproducing them in another form, as shown in Fig. 7.28, we can clearly establish different regions over which the clay–XG contact characteristics are changing. These regions can be characterized, depending on the toughness modulus (T), which can be defined as the ratio between the energy absorption capacity and the XG concentration. For example, for the results of 56 day curing, we can identify 3 regions: (i) Region 1 with T of about 1600 kJ/m^3 for XG concentrations ranging from 0 to 0.5 wt%; (ii) Region 2 with T of about 1133 kJ/m^3 for XG concentrations ranging from 0.5 to 2 wt%; and (iii) Region 3 with T of about 107 kJ/m^3 for XG concentrations ranging from 2 to 3 wt%. However, for zero day curing, the behavior can be characterized into 2 regions only: (i) Region 1, with T of about 1000 kJ/m^3 for XG concentrations ranging from 0 to 0.5 wt%; and (ii) Region 2 with T of 0 kJ/m^3 for XG concentrations greater than 0.5 wt% (i.e., no change of energy absorption capacity with XG concentration for the latter case). Therefore, each region would have specific associated geotechnical-engineering properties.

As an attempt to quantify the appropriate XG percent for this clay type, one can use the reported results of the free swelling tests by Cabalar et al. (2018), whereby soil samples submerged in water were allowed to change volume freely until equilibrium was reached, and then loaded and unloaded, as per ASTM D4546 (ASTM 2016c). The results showed that the calculated differential swellings were 1.125% and 2.625% for XG concentrations of 0.5% and 1.0%, respectively. Therefore, to reduce the deleterious effects of wet/drying on the aggregate stability (i.e., to limit the differential swelling), the XG concentration should not exceed 0.5% for this clay type.

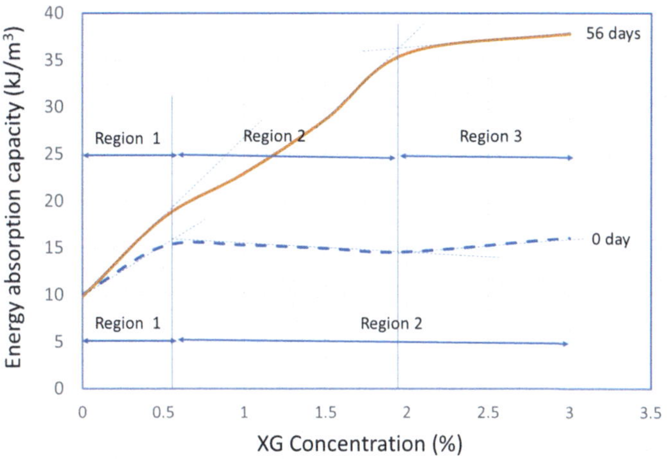

Fig. 7.28 Changes in energy absorption capacity of XG-treated low plasticity clay with XG concentration and curing time (data from Chang et al. (2015))

Kaolinite

A study by Kwon et al. (2023) investigated changes in the HC of XG-treated kaolinite clay. Specimens were prepared by mixing kaolinite, XG and deionized water (DW) according to the desired kaolinite/XG/water ratios. Kaolinite with high water content (of 80%) was used to simulate in-situ clays with high water content. Dry XG powder ($m_{XG} = 0$, 1 and 2 g) was dispersed into the target weight of DW ($m_w = 160$ g), and uniform XG solutions (with XG-to-DW mass ratios of 0, 0.625, and 1.25%) were prepared. Then, the XG solutions were hand-mixed with kaolinite ($m_s = 200$ g) to form uniform kaolinite–XG–DW mixtures (with XG-to-kaolinite mass ratios of 0, 0.5, and 1.0%). The HC of the untreated and XG-treated kaolinite samples were determined indirectly from one-dimensional consolidation tests. As evident from Fig. 7.29, the log(HC) decreases linearly in value with decreasing void ratio due to more constricted drainage paths. The HC decreased exponentially from 1.5×10^{-7} to 8×10^{-7} cm/s for the untreated kaolinite, and from 4×10^{-8} to 2×10^{-7} cm/s and 4×10^{-9} to 1×10^{-8} cm/s for 0.5% and 1% XG, respectively (Fig. 7.29). The decrease in HC was caused by the clogging of pore spaces by viscous XG hydrogels.

Sand–Bentonite Mixture

In a study by Biju and Arnepalli (2020), a sand–bentonite mixture was treated with XG, and the HC was evaluated using a modified falling-head permeability apparatus under an accelerated gravity environment. Both deionized water (DW) and synthetic leachate were used to assess the potential of the XG-treated sand–bentonite mixture

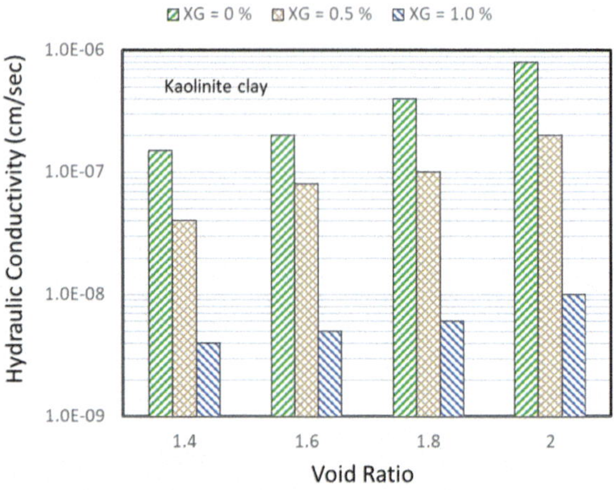

Fig. 7.29 Variations of hydraulic conductivity for XG-treated kaolinite at different void ratios (deduced from one-dimensional consolidation test data reported in Kwon et al. (2023))

as a liner material. The results indicated that the addition of XG causes aggregation in the treated soil and slightly increases the HC of the sand–bentonite mixture.

As discussed elsewhere, XG can be mixed with soil using either the wet or dry mixing. In the wet mixing approach, a solution of XG is prepared by mixing a known quantity of XG with water, and the resulting solution is then uniformly mixed with the dry disaggregated soil. Whereas, in the dry mixing method, the dry disaggregated soil and powdered XG are mixed thoroughly, and then water is added to the dry mixture. Notably, when XG is added to water in the wet mixing method, the configuration of XG changes from coiled to an extended conformation (Theng 2012). Whereas, during dry mixing, the coiled XG in the constrained void spaces cannot access as many soil particle surfaces compared to wet mixing, even after the addition of water.

Owing to its apparent advantages, Biju and Arnepalli (2020) applied different XG concentrations of 0.5, 1, 2, and 3% to their sand–bentonite soil (65% sand and 35% bentonite) using the wet mixing approach. A known quantity of XG was added to a predetermined volume of DW (i.e., pH of 6.9, ion strength of 0 mmol/l, and electrical conductivity of 5 mS/m), and mixed thoroughly in an orbital shaker for 48 h to ensure proper/uniform dispersion of XG in the suspension. The prepared suspension was then mixed with the dry sand–bentonite samples and allowed to mature by storing in airtight polyethylene bags. In addition, a strong synthetic leachate (i.e., pH of 7.1, ion strength of 83.5 mmol/l, and electrical conductivity of 950 mS/m) was also used.

Figure 7.30 shows a marginal increase in the HC with the addition of XG. The flow of permeant liquid (either DW or strong leachate) through a compacted specimen will be through inter- and intra-aggregate flow paths, with the latter contributing more to the HC development (Yong et al. 1992; Mohamed and Antia 1998; Mohamed and Paleologos 2018). Water molecules firmly attached around the clay particles inside the aggregates will create negligibly small effective flow paths, allowing flow only through the inter-aggregate paths. Thus, the size and formation of aggregates in clays considerably influence the HC.

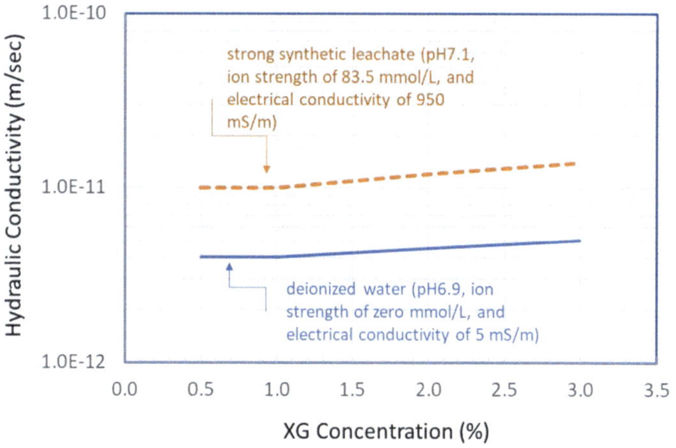

Fig. 7.30 Variations of hydraulic conductivity against XG concentration for a sand–bentonite mixture (65% sand and 35% bentonite) with deionized water and synthetic leachate as the permeant liquids (data from Biju and Arnepalli (2020))

The addition of XG causes aggregation of clay particles by binding them together, forming larger-sized aggregates (Nugent 2011). When XG is added to the soil, it forms a thin coating over the aggregates through secondary bonds. The polar components in XG may participate in the interaction with the charged surfaces of the clay particles, which in turn enhances the inter-particle adhesion. Consequently, the ionic, hydrogen and/or van der Waals bonds lead to entrapment of the soil aggregates and the formation of larger aggregates. The aggregation mechanism was supported using SEM images reported in Biju and Arnepalli (2020), whereby the aggregate sizes change with an increase in XG concentration. Smaller aggregates and pore spaces were observed in the untreated sand–bentonite, while with addition of 0.5% XG, these aggregates combine to form much larger ones (resulting in larger-sized void spaces).

Notably, the reported increase in the HC with addition of XG is well within one order of magnitude. If the sole reason for the change in HC is the change in aggregate size, the variations should have been higher than the observed values. Liu et al. (2017) described the mechanisms of soil treatment by biopolymers as a collective process of filling the void volume, enwrapping, and physicochemical interactions. The hydrogel formed by the biopolymers existing in the pore spaces will cause a reduction in the flow rate. The interactions between the sand–bentonite soil mixture and biopolymer are predominantly physical in nature. The majority of the physisorption interactions exhibit reversibility; however, the probability of desorption of the biopolymer from the soil is restrained by the "octopus" effect of the biopolymer (i.e., whereby multiple interaction points occur between the clay surface and the biopolymer) (Theng 2012). Even after the passage of numerous pore volumes of water or leachate through the soil matrix, the XGs remain anchored in place due to the described octopus effect.

Thus, a combination of widening of the flow paths, filling a portion of the voids and the octopus effect controls the HC.

Furthermore, the lower HC of XG-treated sand–bentonite for the case of DW (i.e., compared to synthetic leachate) can be attributed to the reduction in the size of the effective flow channels, which is influenced by the diffused double layer of montmorillonite mineral present in the soil (Mohamed and Paleologos 2018). The increase in cation concentration (i.e., using the synthetic leachate as a permeant liquid) results in contraction of the diffused double layer and causes broader flow paths, which in turn increases the HC. The changes in the HC ranged from 10^{-11} to 10^{-12} m/s when the permeant liquid is strong synthetic leachate instead of DW.

7.5.2.2 Soil Texture

Soil texture can also influence the effects of XG on the soil HC. XG has been found to be more effective in reducing HC in soils with finer textures (such as clay soils), compared to coarser-textured soils, likely due to the greater surface area and smaller pore sizes of fine-grained soils.

7.5.2.3 Particle-Size Distribution

The particle-size distribution of the soil being treated can also influence the effectiveness of XG in reducing the HC. XG may be more effective in moderating the HC of soils with more uniform gradation (e.g., sandy loam) than in soils with relatively wider gradation (e.g., loamy sand). This is because, in soils with wider particle-size distributions, the smaller particles may not provide sufficient surface area for the XG to bind to and form a gel.

7.5.2.4 Organic Matter Content

The presence of organic matter in the soil can affect its HC, and this behavior can be influenced by the addition of XG. The effectiveness of XG in reducing the HC is greater in soils with higher organic matter content, because XG may bind more effectively to soil particles with high organic matter, resulting in greater reductions in the HC.

7.5.2.5 Moisture Content

The moisture content of the soil being treated can influence the effectiveness of XG in reducing HC. Dry soil may be more receptive to XG treatment, as it has more available surface area for the XG molecules to bind to and form a gel, leading to greater reductions in the HC. In contrast, wet soils may already be saturated with water, making it more difficult for XG to bind to the soil particles and hence influence the HC.

7.5.2.6 Compaction

Soil compaction generally reduces the size and connectivity of the soil pores, which can in turn reduce the HC. This can lead to decreased water infiltration and increased runoff. Conversely, loose soils have higher HC due to their larger pore sizes and relatively lower bulk density.

The degree of soil compaction can also influence the effectiveness of XG in reducing the HC. In compacted soils, there may be less available surface area for the XG molecules to bind to, thereby reducing the effectiveness of XG in reducing the HC. On the other hand, in loosely compacted soils, there may be more available surface area for XG molecules to bind to and form a gel, resulting in greater reductions in the HC. Studies have shown that increasing the degree of soil compaction (from a loose to dense compaction state) decreases the HC in soils treated with XG.

A study by Ng et al. (2020) has shown that the mixture of 3% XG with a kaolinite clay decreases the gas permeability of the treated soil by up to two orders of magnitude (Fig. 7.31). Notably, the XG was mixed with the kaolinite clay using the dry mixing procedure. Moreover, the decrease in gas permeability caused by XG is more significant with increasing molding moisture content, since more moisture is available to hydrate the XG and to promote pore clogging (Joga and Varaprasad 2019).

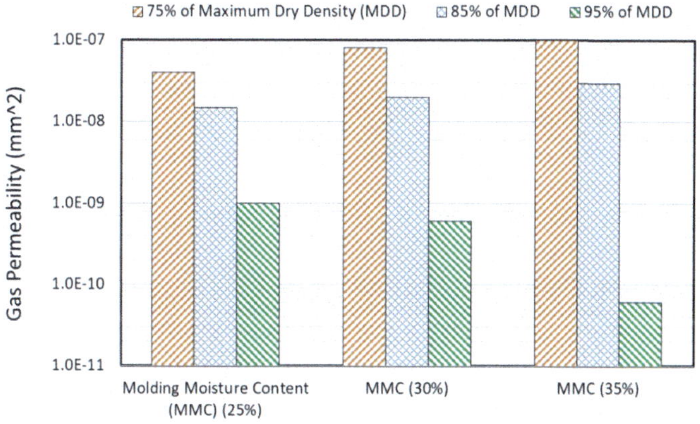

Fig. 7.31 Gas permeability variations of XG-treated kaolinite with different molding moisture contents (MMC) relative to the maximum dry density (MDD) (data from Ng et al. (2020))

Fig. 7.32 Hydraulic conductivity variations for two soils having different cation exchange capacity (CEC) and specific surface area (SSA) values treated with 0.5% XG (data from Biju and Arnepalli (2020) and Kwon et al. (2023))

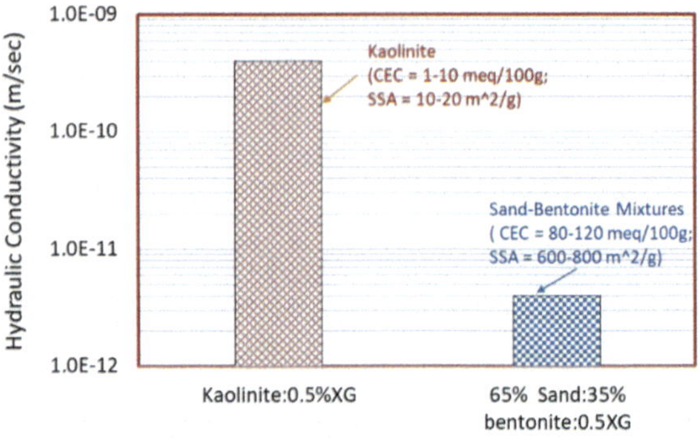

7.5.2.7 Cation Exchange Capacity

The cation exchange capacity (CEC) of soils that are treated with XG affects their HC. Soils with higher CEC tend to have lower HC due to increased soil aggregation and decreased pore size. In soils with high CEC, the XG may bind more strongly to the soil particles and lead to greater reductions in the HC. Conversely, soils with lower CEC have higher HC due to their larger pore sizes and decreased soil aggregation. In addition, Fig. 7.32 shows a comparison between two soils with different CEC values, namely kaolinite (CEC = 1–10 meq/100 g; data from Kwon et al. (2023)) and sand–bentonite mixture (CEC = 80–120 meq/100 g; data from Biju and Arnepalli (2020)), both treated with 0.5% XG. It is shown that the soil with higher CEC, i.e., the sand–bentonite, had a lower HC after XG treatment.

Notably, polymer-induced aggregation occurs through the formation of bridges between the XG biopolymer, the soil particle surfaces, and the soil electrolyte solution. Soil stabilization occurs when reactive surface sites on clay particles are saturated with XG biopolymer (Letey 1994). As such, the optimal XG concentration level hinges on the total surface area accessible (by the XG) within the colloidal suspension. In addition, sorption of anionic polymers, such as XG, onto clay particles increases with higher electrolyte concentration, since cations in solution neutralize the negative charges on the XG. XG forms a network around soil particles and does not penetrate aggregates; therefore, soil stability depends on the uninterrupted network of XG molecules around the soil aggregates (Letey 1994).

7.5.2.8 XG Conformational Structure

As discussed in Chap. 2, XG has a rigid backbone consisting of β-D-glucose, β-D-mannose, and β-D-glucuronic acid monosaccharide units linked by glycosidic bonds. The sidechains of XG are composed of a trisaccharide repeat unit containing one molecule of pyruvic acid and two molecules of acetyl groups. The sidechains extend from the backbone and confer a highly branched, anionic character to the molecule. The conformational structure of XG is determined by a balance of electrostatic repulsion, hydrogen bonding, van der Waals forces, and hydrophobic interactions. In aqueous solutions, XG can adopt a variety of conformations, ranging from highly extended (with long chains extending out from a central core), to coil-like structures, to compact globular structures. When XG is dissolved in water, the long chains can interact with each other to form a three-dimensional network, creating a gel-like structure. This network is known as the extended conformation of XG.

The extended conformation of XG is characterized by a high degree of molecular entanglement and inter-chain interactions (Fig. 7.33), which give the XG-based solution its unique rheological properties, as discussed in Chap. 3. XG solutions with an extended conformation have high viscosity and thixotropic behavior, which implies that they become more fluid when subjected to shear stress. The extended conformation of XG can create a network of polysaccharide chains that cross-link soil particles, thereby forming aggregates that reduce the soil porosity, water flow and the HC.

The extended conformational structure of XG is influenced by various factors, including pH, temperature, and concentration in solution (Fig. 7.34).

(a) *pH*: The extended conformational structure of XG is most stable at a pH level between 7 and 9. At lower pH values, the structure can collapse due to electrostatic repulsion between the negatively charged XG molecules. At higher pH values, the structure can become more compact due to the formation of additional hydrogen bonds.
(b) *Temperature*: XG typically forms an extended conformational structure at temperatures below 40 °C. At higher temperatures, the structure can become more compact due to increased molecular motion.

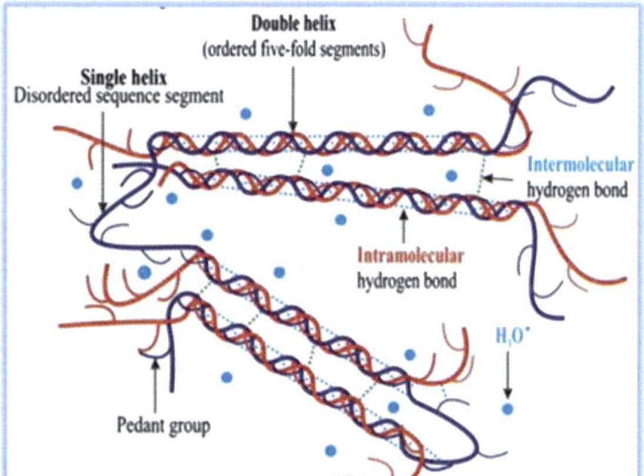

Fig. 7.33 XG extended molecular structure in water (adopted from Rwei and Nguyen (2014))

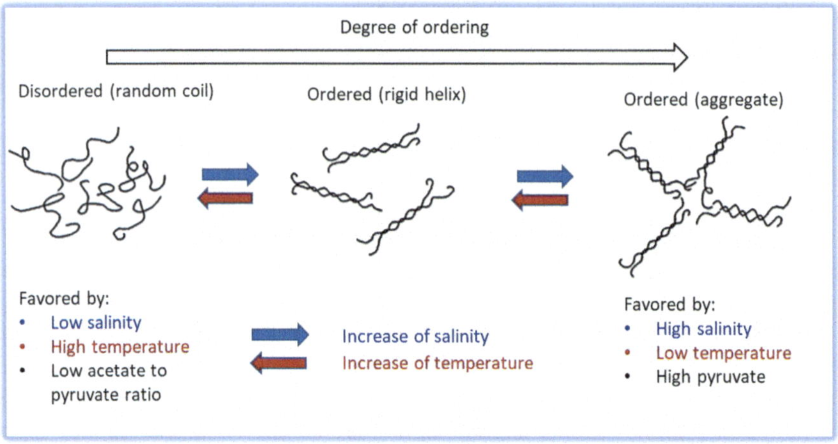

Fig. 7.34 Factors impacting the degree of XG conformational ordering in solution

(c) *Concentration*: The extended conformational structure of XG is also influenced by its concentration in solution. As discussed in Chap. 2, in the low-salt concentration regime (< 0.2 wt%), the addition of salt to a solution of XG in distilled water causes a disorder → order (random → helix) transition in the backbone conformation, which is always accompanied by a substantial decrease in molecular size and a corresponding decrease in the solution viscosity (Fig. 7.34).

In general, the preferred conditions for having an extended conformational structure of XG in solution are a pH level between pH 7 and 9, a temperature below 40 °C, and a moderate XG concentration. Even so, the specific conditions may vary depending on the intended application of the XG solution.

On the other hand, in the coil conformation, the XG chains are more randomly coiled and less tightly packed than in the extended conformation. The coil conformation of XG is typically observed at higher temperatures and lower pH levels than those required for the extended conformation. At higher temperatures, the increased molecular motion can disrupt the extended conformation, causing the chains to adopt a more random coil structure. At lower pH values, the repulsion between the negatively charged XG molecules can also lead to a more random coil structure. In the coil conformation, XG solutions have lower viscosity and less gel-like properties than in the extended conformation. Compared to the extended conformation, the coil structure is less effective in reducing both soil porosity and HC.

The coil conformation of XG leads to soil with higher HC than for the extended conformation because it does not form a three-dimensional network that can effectively reduce soil porosity and water flow. In the coil conformation, the XG chains are more randomly coiled and less tightly packed than in the extended conformation, which allows water to flow more easily through the soil. In addition, the coil conformation of XG is less effective in binding soil particles together, which can also contribute to higher HC.

In summary, the conformational structure of XG plays a critical role in its effectiveness as a soil stabilizer. XG with a more extended conformation is more effective in reducing HC, water infiltration, and promoting water redistribution in soils. This knowledge can be applied in geotechnical engineering practices to improve soil structure and water retention, which can lead to more efficient water management practices.

7.5.2.9 XG Field Application Method

The method of XG application can affect its ability to reduce the HC. Studies have found that injecting XG solution into the soil mass using a syringe or needle can be more effective than its surface application, likely due to the greater distribution and penetration of XG into the soil mass. When added directly to the soil surface, molecular weight determines the effectiveness of the polymer. For example, to maintain a high infiltration rate into the soil, a small amount of high-molecular weight polymer is required instead of a large amount of low-molecular weight polymer. Adding a source of electrolytes with polymers further reduces runoff and erosion (Graber et al. 2006). Polymers applied directly onto the soil surface must be dissolved in water and distributed by spraying. In some cases, this method is inefficient because a large volume of water is required to dissolve the polymer (Graber et al. 2006).

For example, when added to irrigation water, the polymer conformation and the quality of the surrounding water determines the effectiveness of the polymer. Synthetic polymers (e.g., polyacrylamide or PAM) are more effective than natural polysaccharides at cementing aggregates, since they are generally longer and have relatively lower capacity to absorb water (Graber et al. 2006). When applying polymers through furrow irrigation, medium- to fine-textured soils exhibit less surface sealing and increased infiltration. PAM applied through furrow irrigation results in less runoff, less soil detachment and reduced sediment transport capacity across agricultural lands (Graber et al. 2006).

7.5.2.10 XG Concentration/Application Rate

As outlined in Chap. 3, XG enhances the viscosity of polymer solutions at relatively low concentrations. XG exhibits pseudoplastic properties, whereby the viscosity of the polymer solution decreases with increasing shear rate (Nugent et al. 2009). Materials creating pseudoplastic solutions become oriented in the direction of shear and provide less resistance. When the ionic strength of a solution increases, the macromolecular interactions between the salt cations and pyruvate groups increase, inherently increasing the viscosity and pseudo-plasticity of aqueous solutions of XG. At higher XG concentrations in solutions of high ionic strength, interactions in the macromolecular dimension increase viscosity.

As previously discussed, the application of XG can result in notable changes in HC (i.e., for the same curing period, an HC reduction with an increase in XG concentration, and for the same XG concentration, an HC increase with an increase in curing period) and soil water retention across diverse soil types. However, the optimal application rate and method of XG application may vary depending on the specific soil and environmental conditions. It is also important to consider the long-term effects of XG on the soil properties, which demand further research. For example, Lee et al. (2021) used an injection technique to simulate the field condition, as follow. The dry sand, with $D_{50} = 0.47$ mm, was filled inside a cell measuring 100-mm in diameter and 200-mm high, and compacted in three layers by tamping to achieve an initial constant density with a porosity of 0.43. Then, the compacted sand was saturated by deionized water (DW) and a vertical confining pressure of 400 kPa was applied to prevent expansion of the placed soil during the injection process. Then, the XG hydrogel with concentrations of 1.25, 2.5 and 5% (by dry weight of soil) was injected into the saturated sand under a constant pressure of 300 kPa through an injection needle. During the injection process, the injection rate of the XG hydrogel was recorded. Following the injection, a DW inflow with a constant hydraulic head was applied to measure the HC of the XG-grouted sand, as per ASTM 2434–19. A total amount of 500 g of XG hydrogel was injected into a constant soil volume of 265 cm^3 in periods of 360, 620, and 1770 s for XG concentrations of 1.25, 2.5, and 5%, respectively. The XG injection rates through the XG-treated sands are illustrated in Fig. 7.35. The XG injection rates decreased as the XG concentration increased (due to the increase of XG viscosity).

Figure 7.36 shows the HC reduction of the tested sand with an increase in the concentration of the injected XG hydrogel. The HC decreased exponentially from 2.0×10^{-4} cm/s for the untreated sand to 1.03×10^{-7} cm/s for sand treated with 5% XG. This reduction is expected to be associated with the water holding capacity and pore-filling properties of the XG hydrogel. The XG that is injected into the pore spaces between the sand particles forms a viscous fluid by absorbing surrounding

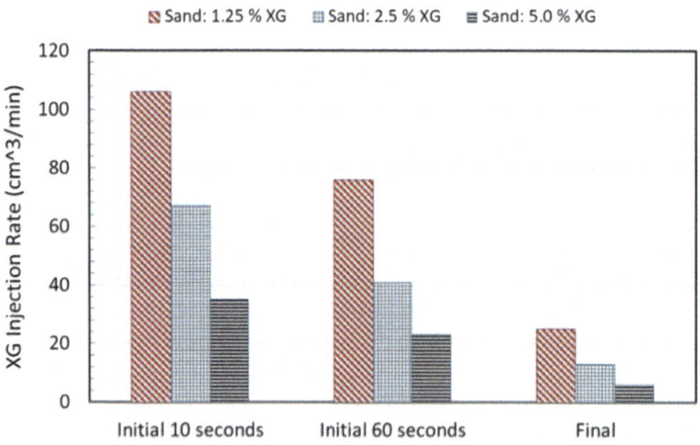

Fig. 7.35 Variations of XG injection rates for sand treated with different XG concentrations (data from Lee et al. (2021))

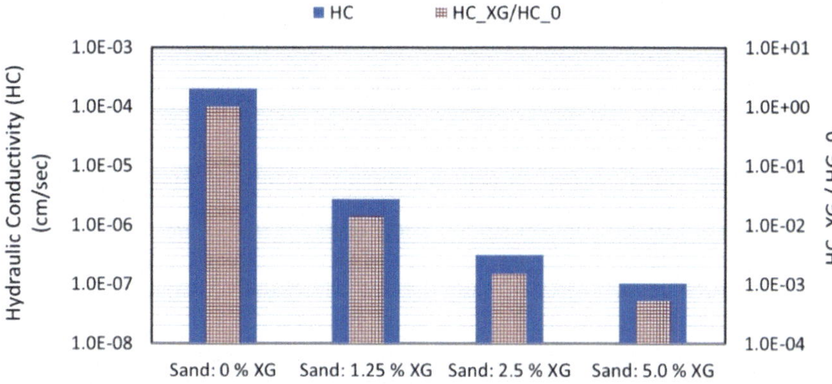

Fig. 7.36 Variations of the hydraulic conductivity for sand treated with different XG concentrations via injection (data from Lee et al. (2021)). *Note* HC_XG and HC_0 denote the hydraulic conductivities of the XG-treated and untreated sand samples, respectively

free water to create a highly viscous hydrogel. Higher concentrations of XG will reduce the available free water within the sand pores, thereby further reducing the HC of the soil.

In summary, the effects of XG concentration on the HC of soils is a complex phenomenon and depends on a variety of inter-related factors, such as soil type, curing time, and other treatment conditions. For example, in a study by Sujatha et al. (2020), a natural sandy clay soil—with geotechnical properties of $G_s = 2.28$, optimum moisture content for compaction of 10%, maximum dry unit weight of 19.33 kN/m^3, LL = 38%, PI = 18.9%, shrinkage limit (SL) = 14.8% and swelling index of 4.89%—was stabilized with XG, which was added to the test soil using the dry mixing method. The variations of HC with XG concentration and curing time are shown in Fig. 7.37. With the addition of 1% XG, the HC decreased by five and seven orders of magnitude after 7 and 90 days of curing, respectively. Such reduction rates are about 4 times higher than those discussed earlier based on the results reported by Mendonça et al. (2021).

Fig. 7.37 Variations of the hydraulic conductivity for a natural sandy clay soil treated with different XG concentrations and tested at different curing times (data from Sujatha et al. (2020))

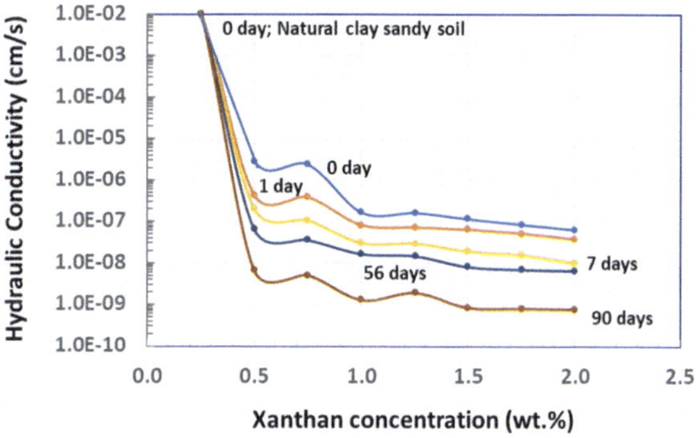

Overall, the reasons for the reduction in HC with increasing XG concentration can be explained as follows:

(a) *Soil particle-size distribution*: The effects of XG on soil HC can depend on the particle-size distribution of the soil. XG has been found to be more effective in reducing the HC of soils with finer particle sizes, such as clay soils, compared to soils with coarser gradations, such as sandy soils.
(b) *Water retention*: XG can increase the water retention of soils due to its ability to hold water molecules through hydrogen bonding. The increase in water retention can reduce the pore space available for water movement, leading to a decrease in HC.
(c) *Formation of a soil–water barrier*: XG can form a soil–water barrier that prevents the movement of water through the soil mass. The barrier can reduce the soil HC, which may affect the soil's drainage and water movement characteristics.
(d) *Reduction in soil particle mobility*: XG can bind to soil particles, reducing their mobility. The reduction in soil particle mobility can lead to the formation of a dense soil matrix, which can significantly reduce the soil HC.
(e) *Increased viscosity*: XG is known to increase the viscosity of water, and its addition to soil can lead to the formation of a gel-like substance. The gel-like substance can clog the soil pores, thereby reducing the soil HC.
(f) *Temperature dependence*: The effects of XG on soil HC can also be temperature dependent. XG solutions can become more viscous at low temperatures, leading to a decrease in HC of the treated soil. Conversely, at high temperatures, XG solutions can become less viscous, resulting in an increase in HC.
(g) *XG concentration/application rate*: The concentration of XG used for soil stabilization can clearly affect the soil HC. Studies have found that higher concentrations of XG generally lead to greater reductions in HC, although the precise relationship between XG concentration and HC can depend on the specific soil and XG properties.
(h) *Sorption*: XG can be sorbed onto the surface of soil particles, especially clays, and create a surface film that can reduce water movement through the soil pores. The sorption of XG onto soil particles can also lead to the formation of aggregates that can further reduce the HC.
(i) *Chemical interactions*: The presence of XG in soil can also alter the chemical properties of the soil, including its pH, particle surface charge and ion exchange capacity. These changes can affect soil structure and reduce the size of soil pores, leading to a decrease in HC.

7.5.2.11 XG Purity

The purity of XG used for soil stabilization can also affect its ability to reduce the soil HC. Impurities in XG, such as proteins and other organic matter, can diminish its effectiveness in reducing the HC and hence lead to variability in the expected stabilization outcomes.

7.5.2.12 XG Degradation

Over time, XG can degrade in the soil due to microbial activity and other environmental factors. As XG degrades, its ability to reduce the soil HC can diminish, leading to a gradual increase in HC over time.

7.5.2.13 Pore Fluid Salinity and Concentration

The presence of salts in the soil can also affect its HC, and this effect can be influenced by the presence of XG. In general, XG is more effective in reducing the HC of soils with higher salinity levels because of the formation of the XG extended conformational structure (Fig. 7.38).

The HC of XG-treated soils can be affected by the ionic strength of the soil solution. Ionic strength is a measure of the concentration of ions in the soil solution. When the ionic strength of the soil solution is low, there are fewer ions to interact with the XG, such that the XG can bind more tightly to soil particles and form a more cohesive network, which can improve the soil structure, thereby resulting in an increase in the soil HC.

As the ionic strength of the soil solution increases, more ions become available to interact with the XG. This can lead to the formation of ionic bridges between the XG molecules, which can reduce their ability to bind to the soil particles and form a cohesive network. This can result in a decrease in the soil HC. At high ionic strengths, XG can become more hydrated and swell, leading to an increase in its volume and a reduction in its ability to bind to soil particles; this can also reduce the HC. In addition, the increased viscosity of the fluid at higher XG concentrations can lead to the formation of a more stable and cohesive soil structure that restricts the movement of water through the soil pores.

Therefore, the influence of ionic strength on the HC of XG-treated soils is complex and depends on a variety of inter-related factors, such as the concentration and type

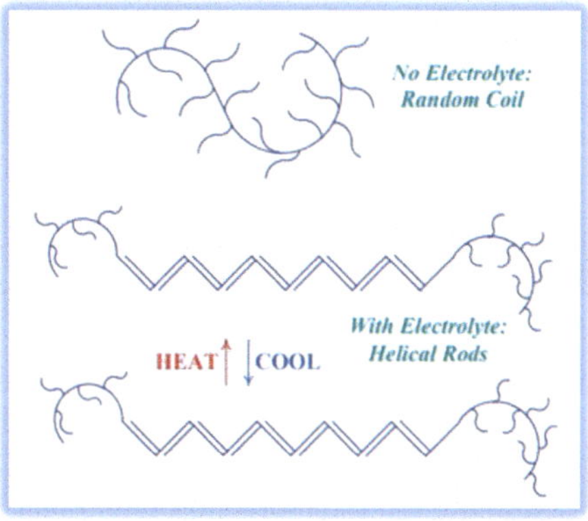

Fig. 7.38 XG structural conformation in the presence of electrolytes and heat

of ions present in the soil solution, the type and concentration of XG used, and the soil texture and structure.

Overall, the effects of ionic strength on the HC of XG-treated soils can be explained by the following reasons:

(a) *Electrostatic repulsion*: XG is an anionic polymer that can form a hydrated layer around soil particles. At low ionic strengths, the inherent repulsion between the negatively charged XG molecules allows them to bind more tightly to soil particles, creating a more cohesive network that enhances water movement through the soil.
(b) *Ion bridging*: As the ionic strength of the soil solution increases, more cations become available to interact with the XG molecules. This can lead to the formation of ionic bridges between the XG molecules favoring formation of ordered aggregates (Fig. 7.34; XG conformational ordering in solution), which can reduce their ability to bind to soil particles and form a cohesive network. The presence of ionic bridges can also cause the XG molecules to aggregate, resulting in a decrease in HC.
(c) *Swelling of XG*: At high ionic strengths, XG can become more hydrated and swell, leading to an increase in its volume and a reduction in its ability to bind to soil particles. This can also result in a decrease in HC.
(d) *Soil texture and structure*: The effects of ionic strength on the HC of XG-treated soils can also depend on the soil texture and structure. For example, in soils with high clay content, the clay particles can interact with XG molecules; thereby enhancing their ability to form a cohesive network. In addition, the soil structure can affect the distribution of water and ions in the soil mass, which can in turn influence the HC of soils.

7.5.2.14 Soil pH

Changes in HC can also be attributed to the individual and combined effects of XG and pH on several soil properties, such as soil structure, pore-size distribution, and soil–water interactions. Here are some possible reasons for the changes in HC.

(a) *Viscosity and rheology of XG solutions*: Changes in pH can influence the viscosity and rheology of XG solutions, which can in turn affect its performance as a soil stabilizer. XG is known to have high viscosity and shear-thinning behavior, which allows it to form stable gels or films when mixed with water or soil. However, the viscosity and rheology of XG solutions can be affected by pH, with, for instance, moderate acidity possibly promoting higher viscosity and greater shear-thinning behavior. This can lead to more stable XG–soil mixtures, and reduce the HC of the treated soil.
(b) *Solubility and stability of XG*: pH variations can affect the solubility and stability of XG in soil. XG is a polysaccharide that is soluble in water, but its solubility and stability can be affected by pH variations. At low pH values, XG can become less soluble and less stable, which can reduce its effectiveness as a soil stabilizer; this can lead to a lower reduction in HC.

(c) *Sorption of XG onto soil particles*: pH variations can affect the sorption of XG onto soil particles. XG can sorb onto soil particles through electrostatic and hydrophobic interaction mechanisms. The extent of sorption can be affected by the pH of the soil solution, with higher sorption occurring at lower pH values due to the increased surface charge of the clay soil particles. This can result in a greater reduction in HC at lower pH values.
(d) *Surface charge*: pH can affect soil surface charge and soil–water interactions. Soil surface charge is affected by the pH of the soil solution, with negatively charged clay soil particles becoming more positively charged at lower pH values. This can lead to increased soil–water interactions, which can decrease the HC by increasing water retention in the soil. In addition, at lower pH levels, XG can have a stronger binding effect on soil particles due to increased ionic interactions between the negatively charged XG and positively charged soil particles; this can promote greater reductions in HC.
(e) *Soil structure and reduced pore size*: XG can alter soil structure and reduce pore size, which can decrease the soil HC. When XG is added to soil, it can form a network of long-chain polymers that can bind soil particles together and reduce the size of soil pores. This can result in a reduction in the HC of soil treated with XG.
(f) *Aggregation and stability of soil particles*: pH can affect the aggregation and stability of soil particles, which can in turn influence HC. XG can aggregate soil particles and alter soil structure, which can reduce HC. However, the aggregation and stability of soil particles can be affected by pH, with lower pH values generally leading to greater particle aggregation and stability. This not only reduces HC but can also lead to a more heterogeneous soil structure and increased water retention.
(g) *Concentration and distribution of dissolved ions in soil solution*: Changes in pH can affect the concentration and distribution of dissolved ions in soil solution, which can influence the behavior of XG and its interactions with soil particles. XG can interact with soil particles and dissolved ions in soil solution through a variety of chemical and physical mechanisms, including electrostatic interactions, hydrogen bonding and hydrophobic interactions. The concentration and distribution of these ions can be influenced by pH, which can in turn affect the interactions that take place between XG and soil particles. This can lead to changes in soil structure and hence HC. Moreover, pH can affect the solubility and availability of nutrients in soil, thereby affecting plant growth. XG is often used as a soil conditioner to improve plant growth, but its effectiveness can be affected by pH variations. Lower pH values can limit plant growth. This can indirectly influence HC by altering the root distribution and hence the soil structure.
(h) *Bioavailability of XG*: The bioavailability of XG and other soil conditioners to plants and microorganisms can be affected by pH, with lower pH values generally leading to lower bioavailability. This can affect the long-term performance of XG as a soil conditioner and may lead to changes in the soil structure and hence HC over time.

(i) *Activity and functionality of soil enzymes and microorganisms*: pH can influence the activity and functionality of soil enzymes and microorganisms, which can affect soil structure and HC. Soil enzymes and microorganisms play key roles in the decomposition of organic matter, nutrient cycling, and soil structure formation. However, their activity and functionality can be impacted by pH variations, with lower pH values generally leading to lower enzyme and microbial activity. This can affect the decomposition of XG and other organic matter present in the soil mass, and thus lead to changes in soil structure and HC over time. XG can be decomposed by soil microorganisms, and the rate of decomposition can be affected by pH. At lower pH values, microbial activity can be reduced, which can lead to slower decomposition of XG. This can result in XG having a longer-lasting effect on the HC at lower pH values.

Overall, the combination of XG and pH can affect soil properties and soil–water interactions, which in turn can lead to changes in the soil HC; these factors can interact in complex, inter-related ways to affect the HC. Therefore, the effects of pH on HC can depend on the specific soil and XG properties, as well as the experimental and/or environmental conditions. These factors can contribute to the complex and multifaceted effects of pH on the HC of soils treated with XG and highlight the importance of understanding the underlying mechanisms and interactions between XG, soil, and pH variations.

7.5.2.15 Mixing Temperature

Mixing temperature affects the HC of soils treated with XG. Higher mixing temperatures tend to enhance the formation of XG random coiled structures and truncated helix due to acid hydrolysis (Fig. 7.39); hence, prompting an increase in HC due to a decrease in aggregation and an increase in pore sizes. However, lower mixing temperatures tend to decrease the soil HC due to an increase in soil aggregation and a decrease in pore sizes (Fig. 7.34).

7.5.2.16 Shear Rate

The shear rate applied during mixing of XG solution with soil can affect the XG's structure (Fig. 7.40) and its mode of interaction with the soil. In addition, higher shear rates during shearing of soils might lead to realignment of the XG molecules, thereby increasing the alignment of the gels along the shearing direction, and hence resulting in greater reductions in HC.

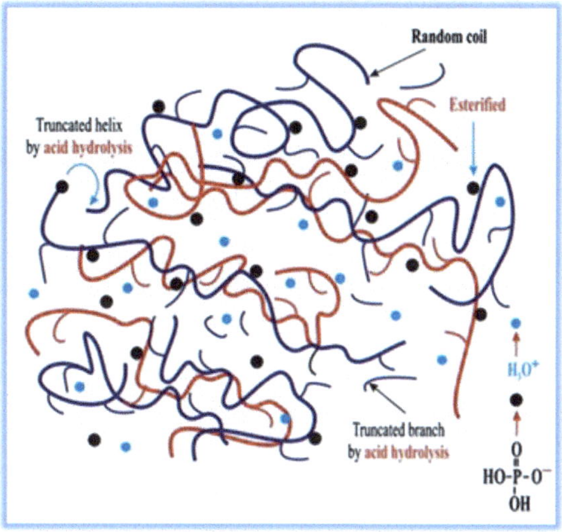

Fig. 7.39 XG structure in H_2O/H_3PO_4 at high temperature (adopted from Rwei and Nguyen (2014))

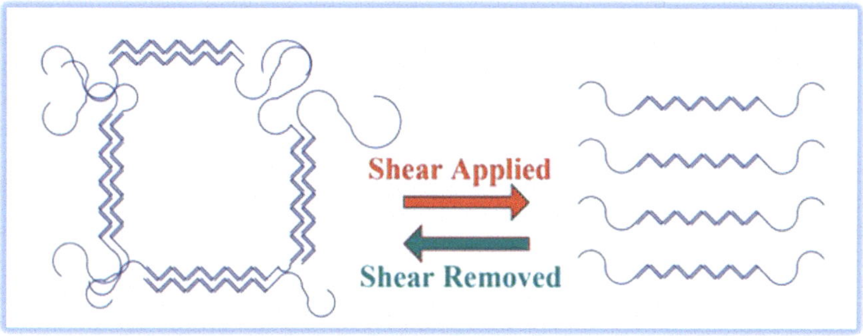

Fig. 7.40 Effect of shear on the XG biopolymer network structure

7.5.2.17 Mixing Time

The duration of mixing XG with soil can also affect the XG's ability to reduce the soil HC. Longer mixing times tend to promote more uniform distribution of XG throughout the soil mass, allowing stronger gel formation and hence greater reductions in the soil HC.

7.5.2.18 Curing Time

Mendonça et al. (2021) used 0.33% XG to stabilize a silty sand soil and evaluated the effects of curing time on HC under a hydraulic gradient of 6.9 (Fig. 7.41). The results indicated a sharp decrease in HC during the first 3 days of curing, followed by a slight

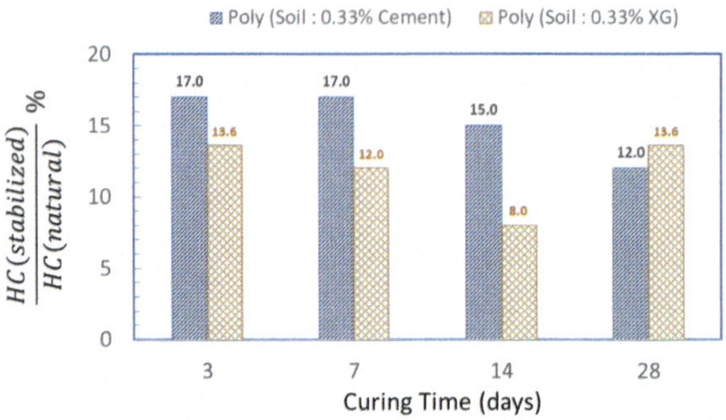

Fig. 7.41 Variations of the stabilized-to-natural hydraulic conductivity ratio (i.e., HC(stabilized)/ HC(natural)) as a function of curing time for a silty sandy soil treated with either 0.33% XG or 0.33% cement addition (data from Mendonça et al. (2021))

decrease to a minimum HC after 14 days of curing, which was attributed to the fast hydration of the XG hydrogels during the first 3 days, as well as some hydration afterward up to 14 days. Furthermore, the authors indicated that the hydration starts from the outside of the hydrogels; hence, hindering water penetration to the inner part of the hydrogels, which might indicate partial hydration of the inner zone. Comparable results were also reported by Khachatoorian et al. (2003) for a similar range of curing times.

Referring to Fig. 7.41, after 14 days of curing, a slight increase in the HC is noted, which was probably due to some biodegradation of the XG, dehydration of the hydrogels, shrinkage, and increase in pore sizes. The observed increase in HC in the long term (after 14 days) is consistent with that reported by Cabalar et al. (2017). For the same silty sand soil stabilized with 0.33% cement (see Fig. 7.41), the reduction in HC was also dependent on the curing duration, attributed to the development of the pozzolanic reactions over time. Interestingly, for a curing time of 28 days, the applications of 0.33% XG and 0.33% cement showed similar performance in terms of HC reduction; while, in the short term, the XG induced a comparatively greater reduction in the HC.

7.5.2.19 Age of XG Solution

XG solutions can change over time, which can affect their ability to reduce the soil HC. Studies have found that fresh XG can be more effective in reducing HC compared to aged XG, likely due to the degradation and loss of viscosity for the aged XG solution.

7.5.2.20 Wetting and Drying

As discussed earlier, XG has been applied to soils to improve aggregation. For instance, a soil from southeast Scotland investigated by Czarnes and Hallett (2000) experienced increased aggregation and stability due to an increase in tensile strength

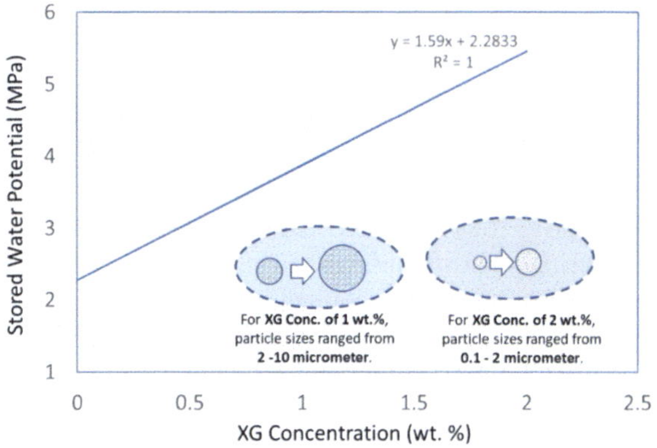

Fig. 7.42 Drying/wetting stored water potential and particle sizes of XG-treated natural silty soil (data from Zhang et al. (2022))

after XG treatment. Tensile strength describes the bond energy within a soil aggregate and how quickly energy is released when bonds break during rapid wetting. An increase in tensile strength indicates an increase in the bond energy between particles. For untreated soils, wetting releases energy that disrupts inter-particle bonds and destabilizes soil structure. In XG-treated soils, the increase in tensile strength enables soil to resist disruption and destabilization when wetted. Additionally, XG absorbs water; therefore, when an XG-treated soil is wetted, the XG expands, forming a fibrous network that stabilizes the sediment.

As discussed in Chap. 5, Zhang et al. (2022) evaluated the water potential characteristics of XG-treated natural silty soil during drying/wetting and showed that (refer to Fig. 7.42):

(a) As the XG content increases (from 0 to 2 wt%), the water retention capacity of the treated soil increases;
(b) With the addition of XG, the cumulative pore intrusion of mercury decreased by 4.95% and 6.5% for samples treated with 1% and 2% of XG, respectively, indicating a decrease in the pore-size distribution;
(c) When 1% XG was added to the natural silt (which had large pore void spaces measuring 10–80 μm in size), the pores were gradually connected by the XG hydrogels, and they reduced to medium-sized pores measuring 2–10 μm. Moreover, for 2% XG addition, those pores were further decreased to smaller pore sizes measuring 0.1–2 μm, and their numbers were also decreased; and
(d) As the XG content increased, the hysteresis effect enlarged and the stored water potential due to the hysteresis effect increased from 2.28 MPa for 0% XG, to 3.88 and 5.46 MPa for 1% and 2% of XG, respectively; meaning that the stored soil–XG water potentials have increased by about 70% and 139% for additions of 1% and 2% XG, respectively.

These changes (i.e., high water retention capacity, smaller pore sizes, high stored water potentials) would contribute to a constructive decrease in the HC of the XG-treated soil.

7.6 Summary and Concluding Remarks

XG is a natural polysaccharide produced by the bacterium *X. campestris*. It has a high molecular weight and a long, branching molecular structure that gives it unique properties, including high viscosity and excellent water-holding capacity. These properties make XG useful in a wide range of applications, including food processing, pharmaceuticals, cosmetics, petroleum, agriculture, and engineering.

In geotechnical and geo-environmental engineering, XG is used as a soil stabilizer to improve the soil physical/mechanical properties and increase its water-holding capacity. The XG can be applied to soil as a dry powder or in solution form. The addition of XG to soil can increase the soil's water-holding capacity, which can be beneficial in arid and semi-arid regions. However, excessive application of XG can lead to a reduction in the soil's HC, which can affect the soil's drainage and water movement characteristics.

HC is a measure of the ease with which water can move through a soil. It is affected by several factors, such as soil texture and gradation, porosity and structure, XG properties, pore fluid chemistry, and environmental conditions, such as pH, temperature, and humidity. The addition of XG to soil can affect these factors and thereby influence the soil HC. Increasing the XG concentration can lead to a reduction in HC due to the formation of a gel-like substance in the soil pore voids. The gel-like substance can reduce the pore sizes and increase the soil's water-holding capacity, but it can also impede water movement through the soil, leading to reductions in HC. Also, the degree of reduction in HC depends on several factors, including the concentration of XG used, soil type, and local environmental conditions.

In addition, the reduction in soil HC due to an increase in XG concentration (or application rate) is primarily due to an increase in viscosity, XG-induced soil structure alterations, changes in soil electrical properties, and the HC dependence on soil pH.

(a) *Increase in viscosity*: The addition of XG to soil can significantly increase the consistency of the soil–XG mixture due to its high molecular weight and the formation of a gel-like substance in water. The increased viscosity can cause the soil–water system to become more resistant to flow, which leads to a reduction in HC.
(b) *Alteration of soil structure*: XG can modify the soil structure by changing the arrangement of soil particles, resulting in the formation of soil aggregates. The aggregates can restrict water movement through the soil pores and hence decrease HC. XG can also reduce the soil porosity by filling of the pores and reducing the spaces available for water movement, which further reduces HC.
(c) *Changes in soil electrical properties*: XG has a negative charge and can interact with charged soil particles. This interaction can cause clay particles to flocculate and create blockages that can restrict water movement through the soil pores, hence reducing HC. Additionally, XG can change the soil's electrical properties, leading to changes in the soil's water-holding capacity, which can also affect HC.

(d) *pH dependence*: The effects of XG on soil HC can be pH dependent. At high pH, the XG–soil system tends to flocculate, potentially leading to increased HC. However, at lower pH values, XG can bind with soil particles, leading to the formation of more stable aggregates that can potentially decrease HC.

Therefore, the use of XG as a soil stabilizer can be beneficial for moderating soil hydraulic properties and increasing its water-holding capacity. However, it is essential to optimize the XG concentration to achieve the desired stabilization outcomes without compromising HC requirements. A better understanding of the factors that influence the HC of XG-treated soils can lead to the development of more effective soil management practices in agricultural, construction, geotechnical and environmental engineering.

References

ASTM D5856-15 (2015) Standard test method for measurement of hydraulic conductivity of porous material using a rigid-wall, compaction-mold permeameter. ASTM International, West Conshohocken, PA. www.astm.org

ASTM (2013) Standard test methods for laboratory compaction characteristics of soil using modified effort. ASTM D1557. ASTM, West Conshohocken, PA

ASTM (2016a) Standard test methods for measurement of hydraulic conductivity of saturated porous materials using a flexible wall permeameter. ASTM D5084. ASTM,West Conshohocken, PA

ASTM (2016b) Standard test method for unconfined compressive strength of cohesive soil. ASTM D2166. ASTM, West Conshohocken, PA

ASTM (2016c) Standard test methods for one-dimensional swell or collapse of soils. ASTM D4546. ASTM,West Conshohocken, PA

Ayeldeen KM, Abdelazim MN, Mostafa AES (2016) Evaluating the physical characteristics of biopolymer/soil mixtures. Arab J Geosci 9(5):371. https://doi.org/10.1007/s12517-016-2366-1d

Azimi M, Soltani A, Mirzababaei M, Jaksa MB, Ashwath N (2024) Biopolymer stabilization of clayey soil. J Rock Mech Geotech Eng 16(7):2801–2812. https://doi.org/10.1016/j.jrmge.2023.12.020

Biju MS, Arnepalli DN (2020) Effect of biopolymers on permeability of sand-bentonite mixtures. J Rock Mech Geotech Eng 12(5):1093–1102. https://doi.org/10.1016/j.jrmge.2020.02.004

Bouazza A, Richards S, Chan I (2001). Soil-biopolymer barriers and soil-biofilm barriers. In: Proceedings 15th international conference on soil mechanics and geotechnical engineering, Istanbul, vol 2, pp 1561–1564

Bouazza A, Gates WP, Ranjith PG (2009) Hydraulic conductivity of biopolymer treated silty sand. Géotechnique 59(1):71–72. https://doi.org/10.1680/geot.2007.00137

Cabalar AF, Wiszniewski M, Skutnik Z (2017) Effects of xanthan gum biopolymer on the permeability, odometer, unconfined compressive and triaxial shear behavior of a sand. Soil Mech Found Eng 54:356–361. https://doi.org/10.1007/s11204-017-9481-1

Cabalar AF, Awraheem MH, Khalaf MM (2018) Geotechnical properties of a low-plasticity clay with biopolymer. J Mater Civ Eng 30(8):04018170

Chang I, Im J, Prasidhi AK, Cho GC (2015) Effects of xanthan gum biopolymer on soil strengthening. Constr Build Mater 74:65–72. https://doi.org/10.1016/j.conbuildmat.2014.10.026

Chen C (1993) Clay or sand polysaccharide associations as models for the interface between microorganisms and soil: water related properties and microstructure. Geoderma 56(1–4):143–156

Chenu C, Pons CH, Robert M (1987) Interaction of kaolinite and montmorillonite with neutral polysaccharides. In: Schultz, LG, van Olphen H, Mumpton FA (eds) Proceedings of the International Clay Conference, Denver, 1985. The Clay Minerals Society, Bloomington, Indiana, pp 375–381

Clapp CE, Emerson WW (1972) Reactions between Ca-montmorillonite and polysaccharides. Soil Sci 114:210–216

Czarnes S, Hallett PD (2000) Root and microbial-derived mucilages affect soil structure and water transport. Eur J Soil Sci 51(3):435–443

Dehghan H, Tabarsa A, Latifi N, Bagheri Y (2019) Use of xanthan and guar gums in soil strengthening. Clean Technol Environ Policy 21:155–165

Dontsova K, Bigham JM (2005) Anionic polysaccharide sorption by clay minerals. Soil Sci Soc Am J 69(4):1026–1035. https://doi.org/10.2136/sssaj2004.0203

Feeney DS, Crawford JW et al (2006) Three-dimensional micro-organization of the soil-root microbe system. Microb Ecol 52(1):151–158

Finch P, Hayes MHB, Stacey M (1967) Studies on soil polysaccharides and their interactions with clay preparations. In Jacks GV (ed) Soil chemistry and fertility. Trans. II and IV Comm., Aberdeen. September 1966. International Society of Soil Science, Aberdeen, Scotland, pp 19–32

Frey SD (2005) Aggregation. In: Hillel D (ed) Encyclopedia of soils in the environment, vol 1. Elsevier Ltd., New York, pp 22–33

Gomez-Suarez C, Van Der Mei HC, Busscher HJ (2002) Adhesion, immobilization, and retention of microorganisms on solid substrata. In: Bitton G (ed) Encyclopedia of environmental microbiology, vol 1. John Wiley & Sons, Inc, New York, pp 100–112

Graber ER, Fine P et al (2006) Soil stabilization in semiarid and arid land agriculture. J Mater Civ Eng 18(2):190–205

Greenland DJ (1956) The adsorption of sugars by montmorillonite: I. X-Ray Studies. J. Soil Sci. 7:319–328

Gu B, Doner HE (1993) Dispersion and aggregation of soils as influenced by organic and inorganic polymers. Soil Sci Soc Am J 57(3):709–716

Guckert A, Valla M, Jacquin F (1975) Adsorption of humic acids and soil polysaccharides on montmorillonite. Pochvovedenie 2:41–47

Joga JR, Varaprasad BJS (2019) Sustainable improvement of expansive clays using xanthan gum as a biopolymer. Civ Eng J 5:1863–1903

Khachatoorian R, Petrisor IG, Kwan C-C, Yen TF (2003) Biopolymer plugging effect: laboratory-pressurized pumping flow studies. J Pet Sci Eng 38(1–2):13–21. https://doi.org/10.1016/S0920-4105(03)00019-6

Kumar SA, Sujatha ER (2021) An appraisal of the hydro-mechanical behaviour of polysaccharides, xanthan gum, guar gum and β-glucan amended soil. Carbohyd Polym 265:118083. https://doi.org/10.1016/j.carbpol.2021.118083

Kwon Y-M, Chang I, Cho G-C (2023) Consolidation and swelling behavior of kaolinite clay containing xanthan gum biopolymer. Acta Geotechnica. https://doi.org/10.1007/s11440-023-01794-8(0123456789(),-volV)(0123456789

Labille J, Thomas F, Bihannic I, Santaella C (2003) Destabilization of montmorillonite suspensions by Ca^{2+} and succinoglycan. Clay Miner 38:173–185

Labille J, Thomas F, Milas M, Vanhaverbeke C (2005) Flocculation of colloidal clay by bacterial polysaccharides: effect of macromolecule charge and structure. J Colloid Interface Sci 284(1):149–156. https://doi.org/10.1016/j.jcis.2004.10.001

Land L (2010) Physical and microbial responses of dredged sediment to two soil-stabilizing amendments, xanthan gum and guar gum, for use in coastal wetland restoration. MS. Thesis, Louisiana State University and Agricultural and Mechanical College

Lee M, Jooyoung Im J, Chang I, Cho G-C (2021) Evaluation of injection capabilities of a biopolymer-based grout material. Geomech Eng 25(1):31–40. https://doi.org/10.12989/gae.2021.25.1.031

Letey J (1994) Adsorption and desorption of polymers on soil. Soil Sci 158(4):244–248

Liu J, Wang Y, Lu Y, Feng Q, Zhang F, Qi C, Wei J, Kanungo DP (2017) Effect of polyvinyl acetate stabilization on the swelling-shrinkage properties of expansive soil. Int J Polym Sci. Article ID 8128020. https://doi.org/10.1155/2017/8128020
Martin GR, Yen TF, Karimi S (1996) Application of biopolymer technology in silty soil matrices to form impervious barriers. In: Proceedings 7th Australia-New Zealand conference on geomechanics, Adelaide, pp 814–819
Martin JP (1955) Soil aggregation. Adv Agron 7:1–37
Mendonça A, Morais PV, Pires AC, Chung AP, Oliveira PJV (2021) Reducing soil permeability using bacteria-produced biopolymer. Appl Sci 11:7278. https://doi.org/10.3390/app11167278
Moavad H, Guzev VS, Babyeva IP, Zvyagintsev DG (1974) Adsorption of extracellular polysaccharide of the yeast *Lipomyces lipofer* on kaolinite. Pochvovedenie 11:79–84
Mohamed AMO, Antia HE (1998) Geo-environmental engineering. Elsevier, 707p
Mohamed AMO, Paleologos EK (2018) Fundamentals of geo-environmental engineering. Elsevier
Mortensen JL (1962) Adsorption of hydrolyzed polyacrylonitrile on kaolinite. Clays Clay Miner 9:530–545
Ng CWW, So PS, Lau SY, Zhou C, Coo JL, Ni JJ (2020) Effects of biopolymers on gas permeability of unsaturated clay. Eng Geol 272(11). https://doi.org/10.1016/j.enggeo.2020.105631
Nugent RA (2011) The effect of exopolymers on the compressibility and shear strength of kaolinite. PhD Thesis, Louisiana State University, Louisiana, USA, 2011
Nugent RA, Zhang G, Gambrell RP (2009) Effect of exopolymers on the liquid limit of clays and its engineering implications. Transp Res Rec 2101:34–43. https://doi.org/10.3141/2101-05
Olness A, Clapp CE (1973) Occurrence of collapsed and expanded crystals in montmorillonite-dextran complexes. Clays Clay Miner 21:289–293
Parfitt RL (1972) Adsorption of charged sugars by montmorillonite. Soil Sci 113:417–421
Parfitt RL, Greenland DJ (1970) Adsorption of polysaccharides by montmorillonite. Soil Sci Soc Am Proc 34:862–866
Rwei S-P, Nguyen T-A (2014) Liquid crystalline phase in xanthan gum (XG)/H_2O/H_3PO_3 and XG/H_2O/H_3PO_4 tertiary systems: a thermal and rheological study. Cellulose 21(5):3231–3241. https://doi.org/10.1007/s10570-014-0358-4
Singh S, Das R (2019) Geo-engineering properties of expansive soil treated with xanthan gum biopolymer. Geomech Geoeng:1–16. https://doi.org/10.1080/17486025.2019.1632495
Soltani A, Raeesi R, Taheri A, Deng A, & Mirzababaei M (2021) Improved shear strength performance of compacted rubberized clays treated with sodium alginate biopolymer. Polymers 13(5):764. https://doi.org/10.3390/polym13050764
Song K-W, Kuk H-Y, Chang G-S (2006) Rheology of concentrated xanthan gum solutions: oscillatory shear flow behavior. Korea-Australia Rheol J 18(2):67–81
Sujatha ER, Atchaya S, Sivasaran A, Keerdthe RS (2020) Enhancing the geotechnical properties of soil using xanthan gum: an eco-friendly alternative to traditional stabilizers. Bull Eng Geol Environ. https://doi.org/10.1007/s10064-020-02010-7
Theng BKG (1976) Interactions between montmorillonite and fulvic acid. Geoderma 15:243–251
Theng BKG (1982) Clay-polymer interactions: summary and perspectives. Clays Clay Miner 30:1–10
Theng BKG (2012) Formation and properties of clay-polymer complexes, 2nd ed. Elsevier, 2012
Theng BKG, Scharpenseel HW (1975) The adsorption of ^{14}C-labelled humic acid by montmorillonite. In Bailey SW (ed) Proceedings of international clay conference, Mexico City, Mexico, 16–23 July 1975. Appl. Publ. Ltd., Wilmette, IL., p 643–653
Wallace A, Wallace GA et al (1986) Mechanisms involved in soil conditioning by polymers. Soil Sci 141(5):381–386
Warkentin BR (2008) Soil structure: a history from tilth to habitat. Adv Agron 97:239–272
Wen Y (2002) Xanthan sorption on synthetic goethite. M.S. Thesis. The Ohio State University, Columbus
Wiszniewski M, Cabalar AF (2014) New frontiers in geotechnical engineering—hydraulic conductivity of a biopolymer treated sand. American Society of Civil Engineers Geo-Shanghai

2014—Shanghai, China (May 26–28, 2014), 19–27. https://doi.org/10.1061/9780784413456.003

Yokoi H, Shiraki M, Hirose J, Hayashi S, Takasaki Y (1996) Flocculation properties of xanthan produced by Xanthomonas campestris. Biotechnol Tech 10(10):789–792

Yong RN, Mohamed AMO, Warkentin BP (1992) Principles of contaminant transport in soils. Elsevier, 327p

Zhang Z, Meng Z, Jiang T, Wang S, Zhao J, Zhao X (2022) Experimental study on the shear strength of silt treated by xanthan gum during the wetting process. Appl Sci 12(12):6053. https://doi.org/10.3390/app12126053

SPRINGER NATURE

GPSR Compliance

The European Union's (EU) General Product Safety Regulation (GPSR) is a set of rules that requires consumer products to be safe and our obligations to ensure this.

If you have any concerns about our products, you can contact us on ProductSafety@springernature.com

In case Publisher is established outside the EU, the EU authorized representative is:

Springer Nature Customer Service Center GmbH
Europaplatz 3
69115 Heidelberg, Germany

The manufacturer's authorised representative in the EU is Springer Nature Customer Service Centre GmbH, Europaplatz 3, 69115 Heidelberg, Germany. If you have any concerns regarding our products, please contact ProductSafety@springernature.com

Printed and bound by CPI Group (UK) Ltd, Croydon, CR0 4YY
01/10/2026
02225965-0006